Adaptations of Desert Organisms

Edited by J.L. Cloudsley-Thompson

Springer
Berlin
Heidelberg
New York
Barcelona
Hong Kong
London
Milan
Paris
Singapore
Tokyo

Volumes published in the series

Ecophysiology of the Camelidae
and Desert Ruminants
By R.T. Wilson (1989)

Ecophysiology of Desert Arthropods
and Reptiles
By J.L. Cloudsley-Thompson (1991)

Plant Nutrients in Desert Environments
By A. Day and K. Ludeke (1993)

Seed Germination in Desert Plants
By Y. Gutterman (1993)

Behavioural Adaptations of Desert
Animals
By G. Costa (1995)

Invertebrates in Hot and Cold Arid
Environments
By L. Sømme (1995)

Energetics of Desert Invertebrates
By H. Heatwole (1996)

Ecophysiology of Desert Birds
By G.L. Maclean (1996)

Plants of Desert Dunes
By A. Danin (1996)

Biotic Interactions in Arid Lands
By J.L. Cloudsley-Thompson (1996)

Structure-Function Relations
of Warm Desert Plants
By A.C. Gibson (1996)

Physiological Ecology of North
American Desert Plants
By S.D. Smith, R.K. Manson, and
J.E. Anderson (1997)

Ecophysiology of Small Desert Mammals
By A.A. Degen (1997)

Homeostasis in Desert Reptiles
By S.D. Bradshaw (1997)

Ecophysiology of Amphibians
Inhabiting Xeric Environments
By M.R. Warburg (1997)

Avian Desert Predators
By W.E. Cook (1997)

Carnivore Ecology in Arid Lands
By J. du P. Bothma (1998)

Ecophysiology of Economic Plants
in Arid and Semi-Arid Lands
By G.E. Wickens (1998)

Arthropods of Mediterranean-Type
Ecosystems
By G.P. Stamou (1998)

Dispersal Biology of Desert Plants
By K. van Rheede van Oudtshoorn and
M.W. van Rooyen (1999)

Spatial Ecology of Desert Rodent Communities
By G.I. Shenbrot, B.R. Krasnov and K.A.
Rogovin (1999)

Georgy I. Shenbrot Boris R. Krasnov Konstantin A. Rogovin

Spatial Ecology of Desert Rodent Communities

With 124 Figures and 38 Tables

Springer

Dr. Georgy I. Shenbrot
Dr. Boris R. Krasnov

Ramon Science Center
Jacob Blaustein Institute for Desert Research
Ben-Gurion University of the Negev
P.O. Box 194
Mizpe Ramon 80600, Isarel

Dr. Konstantin A. Rogovin
Severtzov Institute of Ecology and Evolution
Russian Academy of Science
Leninsky prospect 33
Moscow, Russian Federation

Cover illustration: Sekeetamys calurus (front cover) and Acomys russatus (front and back cover). Photographs taken by Georgy Shenbrot

ISSN 1430-9432
ISBN 978-3-642-64224-1 Springer-Verlag Berlin Heidelberg New York

Library of Congress Cataloging-in-Publication Data

Shenbrot, Georgy I., 1950- Spatial ecology of desert rodent communities / Georgy I. Shenbrot, Boris R. Krasnov, Konstantin A. Rogovin. p. cm. -- (Adaptations of desert organisms, ISSN 1430-9432) Includes bibliographical references (p.) and index.
ISBN-13:978-3-642-64224-1 e-ISBN-13:978-3-642-60023-4
DOI: 10.1007/978-3-642-60023-4

1. Rodents--Ecology. 2. Rodents--Geographical distribution.
3. Desert animals--Ecology. 4. Desert animals--Geographical distribution. I. Krasnov, Boris R., 1955- . II. Rogovin, Konstantin A., 1955- , III. Series. QL737.R6S46 1999 599.35 1756--dc21

Softcover reprint of the hardcover 1st edition 1999

Cover design: *design & production* GmbH, Heidelberg
Camera ready by: Emanuel Rachl

SPIN 10573607 31/3137 - 5 4 3 2 1 0 - Printed on acid-free paper

In memory of our teacher, the late Professor Nikolai Naumov

Preface

The authors of this volume were fortunate in being introduced to zoological studies by their teachers at the Department of Zoology and Comparative Anatomy of Terrestrial Vertebrates, Moscow State University. This department has a long tradition of field animal research in various life zones. In 1954, Professor Nikolai Naumov was appointed head of the department. Under his leadership, the department began intensive ecological studies and has become the alma mater of several generations of Russian animal ecologists. Another stimulus for rodent ecology research in the department was created by the problem of the epidemiology of plague and leishmaniasis in arid zones, which necessitated detailed studies of rodents, the main reservoirs of these diseases. Our teachers include the noted Russian zoologists, namely N.P. Naumov, V.G. Heptner, I.A. Shilov, V.S. Lobachev, N.N. Vorontsov, L.N. Eliseev, V.E. Flint, I.M. Gromov, V.V. Kucheruk, V.E. Sokolov and S.A. Shilova. Almost all of them were connected in one way or another with the department and we have learned much from them.

Georgy Shenbrot began his scientific career studying the taxonomy and ecology of jerboas. Decades of experience with this desert taxon stimulated his interest in the general rules governing the organization and functioning of communities of desert rodents. In his research, he has spent much time in the different arid regions of the world, from the Gobi to Chihuahuan deserts. Boris Krasnov studied the behavioral mechanisms that influence the spatial structure of rodent populations in different landscapes, from the tundra and arctic shore of the Chukchii Peninsula to the rainforests of south Vietnam. His interest in the spatial organization of rodent populations led him to study the spatial organization of their communities. Konstantin Rogovin began his acquaintance with desert animals by studying the ethology of jerboas and later broadened his scope to include the behavioral adaptations of desert rodents in general. Thus, the individual experiences of these authors are complementary. This interaction has been strengthened by the fact that they have been friends since their student days (Krasnov and Rogovin even earlier). Shenbrot and Rogovin collaborated for 7 years at the Institute of Animal Ecology and Evolutionary Morphology, at the Russian Academy of Science in Moscow, Russia (now the Institute of Ecology and Evolution Problems). For the last 7 years, Shenbrot and Krasnov have worked together at the Ramon Science Center. This small scientific unit is located in the heart of the Negev Desert of Israel and is affiliated with the Jacob Blaustein Institute for Desert Research, Ben-Gurion University of the Negev.

The ideas elucidated in this book have been discussed over the years with our colleagues. These include (in chronological and geographical order) L.Y. Borkin, I.Y. Pavlinov, D.V. Semenov, A.V. Surov, I.M. Fokin, A.M. Gilyarov, Y.A. Puzachenko, V.V. Ivanitsky, E.N. Panov, E.G. Potapova, V.S. Gromov, A.V. Chabovsky, S.V. Popov (Russia), V.M. Mazin (Kazakhstan), O.V. Mitropolsky (Uzbekistan), J.H. Brown, E.J. Heske, D.A. Kelt, J.S. Brown (USA), D.W. Morris (Canada), B.J. Fox, C.R. Dickman (Australia), P. Marquet (Chile), D. Ward, Z. Abramsky, U. Safriel, A.A. Degen, M. Kam, B.P. Kotler, I.S. Khokhlova, T. Dayan, Y. Yom-Tov (Israel), M. and A. Stubbe (Germany), I. Prakash, M. Idris (India), G. Diaz (Argentina), J. Nel (South Africa). Technical assistance in data collection was provided by (in chronological order) G. Kulikova, N. Bodyak, L. Bar-Sagi, and M. Laniado.

We thank John Cloudsley-Thompson for inviting us to write this book and for taking on the gargantuan task of editing it. His editing was most helpful and we owe him much for clarifying our statements. We also thank David Ward, the Director of the Ramon Science Center. Dave encouraged us to write this book. He made invaluable comments on the manuscript and worked hard on improving our English prose.

Financial support during the writing of this book was provided by the Ministry of Science of Israel, the Ministry of New Immigrant Absorption of Israel and the Local Council of Mizpe Ramon (for Georgy Shenbrot and Boris Krasnov) and by the Russian Academy of Science and the Russian Foundation of Basic Researches (for Konstantin Rogovin).

Ramon Science Center
Jacob Blaustein Institute for Desert Research
Ben-Gurion University of the Negev

Dr. Georgy Shenbrot
Dr. Boris Krasnov

January 1999

Contents

CHAPTER 1

Introduction

Rodents are conspicuous and important components of the desert biome. Their populations are frequently great and many of them inhabit large areas. They also provide the diet for numerous carnivorous and scavenging species. Furthermore, they provide habitats for many vertebrate and invertebrate animals as a consequence of their burrowing activity. Additionally, they regulate the aeration and water regime of the soil, and provide a suitable environment for the estivation and early development of the seeds of numerous plants.

Many general concepts in modern community and behavioral ecology are based on desert rodents as a main model. The study of spatial structure is one of the most common approaches to the community ecology of terrestrial vertebrates. Comparative descriptive studies of vertebrate, particularly rodent, communities are essential because they are free of many of the limitations inherent in experimental studies (Schoener 1983, 1986; Brown et al. 1986; Diamond 1986; Price 1986; Dueser et al. 1989; Gurevitch et al. 1992). They are also partly free of the difficulties that arise in the construction of null models designed to test specific hypotheses based on field observations (Bowers and Brown 1982; Harvey et al. 1983; Simberloff 1983; Gilpin and Diamond 1984; Connor and Simberloff 1984; Strong 1986; Wiens 1989). Comparisons of community spatial organization in different regions with similar environments, such as the different deserts of the world, can aid in understanding which patterns are universal and which are determined by the unique taxonomic composition and/or evolutionary history of a community. Comparisons of communities occurring in more or less similar environments, but possessing different histories and taxa, can provide information about general directions in the evolution of community organization, as well as identifying the specific features of each community.

Conservation of biodiversity is impossible without understanding the rules of the natural patterns of species distribution within the landscape. Understanding the response of species assemblages to environmental factors has long been a central aim of community ecology (e.g. Arrhenius 1921; Elton and Miller 1954). Few would deny that animal species live in a characteristic range of habitats; within this range they are most abundant around their environmental optima. Species abundance, as well as the composition of biotic communities, changes along environmental gradients as a result of changes in the environmental optima of the various species. These gradients do not necessarily possess physical authenticity as continua in either space or time, but rather are an abstraction for explaining the dispersion and distribution of organisms in space and time (Austin 1985; ter Braak and Prentice 1988). Hence, replacements of species occur as a function of changes in the environmental optimum of each species (Whittaker 1972; Austin 1985). In spite of the considerable success of such efforts, attempts to explain changes in the distributions of species along environmental gradients have frequently been confounded either by: (1) evidence of discontinuities in the gradients caused by the meeting of biogeographic zones (e.g. Ward and Olsvig-Whittaker 1993) or, in some cases, (2) dissimilar patterns along similar gradients in different regions or in the same region among different taxa (see e.g. Ricklefs 1987, reviewed in Ricklefs and Schluter 1994a). In the past, such inconsistencies have been attributed to "noisy" data inherent in variable ecosystems. However, following an influential review of the topic by Ricklefs (1987), many community ecologists have adjusted the emphasis of their studies to examine the role of historical (or taxonomic) factors in conjunction with geographical/environmental ones. Consideration of these two sets of factors in tandem has enhanced the resolution of our knowledge of the structure of communities by considering discontinuities and anomalies in species-environmental gradients as possible effects of biogeographic and geological factors that changed in times past, and persist into the present (Ricklefs and Schluter 1994b).

All the above-mentioned considerations suggest that regular compilations of original data and generalizations should be made. The present volume is an attempt to compile and generalize such data on the spatial structure of desert rodent communities, taking into account both global (biogeographical) and local (ecological) patterns. The book is based on 25 years of experience in studies of rodents in different deserts of the Northern Hemisphere (Karakum, Kyzylkum, Bet-Pak-Dala, North Caspian deserts, Gobi, Thar, Chihuahua, Negev) as well as on analysis of the literature.

The borders of the definitions of what is "a desert" and, consequently, what species should be called "desert animals" are somewhat obscure (Degen 1996). So, we need to confine our consideration to particular areas and to particular species. We have chosen to restrict ourselves to those areas that are defined as "arid" and "hyper-arid" (see below). Thus, semi-arid areas will remain outside the boundaries of our consideration. Furthermore, we will not consider cold deserts at high altitudes (more than 3000 m a.s.l.) on account of the specificity of the adaptations of their occupants.

Given the difficulty of the definition of "desert", the attribution of a particular species to either a "desert" or "non-desert" category is no less difficult. This assignment is absolutely clear in the simplest cases. For example, nobody doubts that the collared lemming *Dicrostonyx groenlandicus* is a non-desert species, while the bush-tailed jird *Sekeetamys calurus* which occurs only in areas with under 100 mm of annual precipitation, is a desert species. However, there is no agreement about how to deal with species that inhabit both desert and steppe zones or with species distributed in several different zones including deserts. A lot of species exhibit such "biome opportunism" (e.g. *Spermophilus pygmaeus* in Asia, *Acomys cahirinus* in North Africa, *Peromyscus maniculatus* in North America, *Pedetes capensis* in South Africa, *Calomys callosus* in South America, *Notomys mitchelli* in Australia). The lists of desert rodent species that have been published are therefore profoundly different. For example, Mares (1993) lists 18 desert rodent genera in the arid regions of North America, whereas Kelt et al. (1996) consider there to be 12. We present the inventory of rodent species that we regard as being "desert species" in the Appendix. The main criterion for including a particular species in this list was that its geographic range lies entirely (e.g. *Acomys russatus*) or mostly (i.e. > 50%) (e.g. *Allactaga sibirica*) within a desert zone. In doubtful cases, species that have specific adaptation(s) to arid conditions (e.g. water independence in *Calomys musculinus*, Mares 1977a) were included in the list. Contrarily, the criterion for excluding a particular species from the list was its occurrence only in intrazonal habitats such as river valleys and lake shores with simultaneous absence from "true" desert habitats (e.g. *Brachiones przewalskii* in the river valleys of the southern Gobi desert). Some species (e.g. *Mus musculus*, *Mus spretus*, *Eliomys quercinus*) do not conform to the above-mentioned criteria. However, they were included in our consideration on account of their important role in desert rodent communities. Inclusion or exclusion of species into the list necessitates a very high degree of familiarity with a particular fauna. Our species inventories for most Asian deserts and the Chihuahuan desert of North America (the deserts of authors' field studies) can be regarded as being almost definitive. In contrast, the inventories for other deserts are based on information from the literature, and therefore may contain some errors.

We will begin our presentation with a description of the main features of the deserts of the world. Then, we will attempt to classify them according to the composition of their rodent faunas so that further comparisons of uniform rodent faunal units can be made. We will compare these biogeographic units in order to understand the global patterns of rodent distribution in the different deserts of the world and the causes that underlie these patterns. The following chapter will be dedicated to analyses of the distribution of rodent species of different sizes, dietary habits and locomotory modes both within and among desert faunas. The delineation of global patterns will permit a better understanding of local patterns of the spatial distribution of desert rodents. The spatial distribution of a particular species on a local scale is determined by the manner and detail with which it perceives its environment, and how it responds to continuous changes in the expression of particular environmental factors. Thus, the set of environmental gradients that has the strongest effect on the distribution and abundance of each species must be identified. In other words, we need to identify the microhabitat preferences of the species. At the same time, a simultaneous analysis of the relationships between all the species of a community and all the environmental gradients there permits us to understand the response of a community as a whole unit to its environment. Furthermore, the relationships of co-occurring species placed in the space of environmental variables represent the spatial structure of a community and reveal the principles of resource partitioning among coexisting species. However, data collected over a short period of time can only represent a time slice of a community. The processes of density-dependent habitat selection (Rosenzweig 1987a, 1991) may drastically change some aspects of community organization over time. We will therefore describe the temporal dynamics of a spatial structure of a rodent community to clarify how temporal dynamics of separate species influence changes in general community parameters over time. Finally, based on the above, we will attempt to define the main gaps in our knowledge of spatial organization of desert rodent communities and to outline the horizons for the future studies.

CHAPTER 2

Deserts of the World

2.1 Common Features

The diverse character of deserts has led to a plethora of definitions (see Thomas 1989). The most general definition is that deserts are areas of aridity, and they are usually defined by some measure of water shortage. This is commonly based on the relationship between water gained from precipitation and water lost by evaporation or transpiration (see Cooke et al. 1993). Five major climatic causes of aridity are recognized (Cooke et al. 1993). These are continentality (distance from marine or other water sources), dynamic anticyclonic subsidence, orographic influences, coastal upwelling of cold water associated with cold equatorward-flowing offshore currents and, finally, high reflectivity (albedo) of the desert surfaces themselves. In addition, the enormous variability of all desert characteristics (climate, substrate, vegetation, lithology etc.) makes it difficult to define the boundaries of arid areas and to divide them into smaller typological units. Consequently, the boundaries of arid areas and their typology are usually based on a relatively small number of environmental parameters. As a result, a number of different maps and classifications of deserts are available, some of which are based on climatic data (e.g. Meigs 1953), while others are based on pedological parameters (e.g. Dregne 1976), vegetation patterns (e.g. Shantz 1956), relief (e.g. Thomas 1989) and so on. One of the most widely used schemes (Meigs 1953) is based on Thornthwaite's (1948) indices of moisture availability (*Im*):

$$Im = (100S - 60D)/PE\,,$$

where *PE* is potential evapotranspiration, calculated from meteorological data, and *S* and *D* are, respectively, the moisture surplus and moisture deficit, aggregated on an annual basis from monthly data and taking stored soil moisture into account.

Meigs (1953) identified three levels of aridity and, thus, three types of arid environment: semi-arid ($-40 \leq Im < -20$); arid ($-56 \leq Im < -40$); and hyper- (or extremely) arid ($Im < -156$). Arid areas are variable in relation to

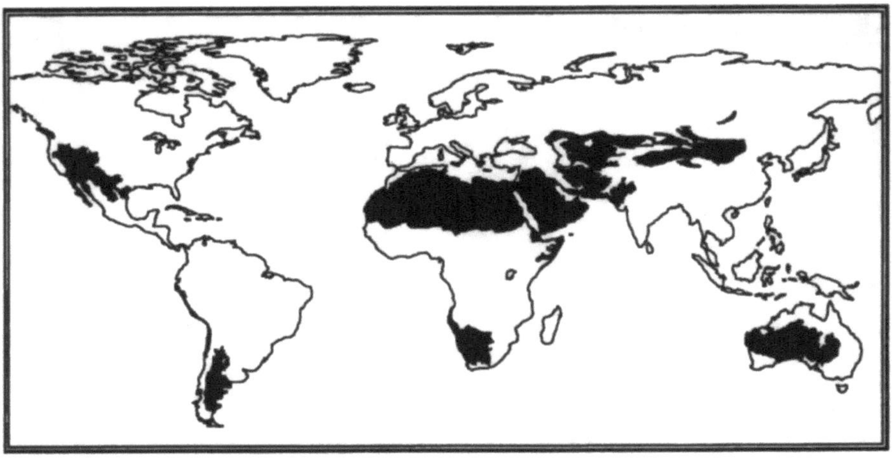

Fig. 2.1. The world's arid and hyper-arid regions

their precipitation regime. There are areas with precipitation in winter only or in summer only, areas with two periods of precipitation and also areas without a distinct rainy period. Hyper-arid areas have no seasonal precipitation regime and occur where 12 consecutive months without precipitation have been recorded (Meigs 1953). As mentioned above, we will consider only arid and hyper-arid areas in this volume and will refer to them as deserts (Fig. 2.1).

2.2 Regional Features

Deserts occur on all continents except for Antarctica, and are located in three belts, namely temperate, subtropical and tropical. Deserts occupy different proportions of different continents; for example, 49% of Australia is desert, whereas deserts comprise only 10% of North America (Petrov 1976). The fundamental features of deserts (location, elevation, predominant landscape, geological and geomorphological structure, climate etc.) as well as biogeographic characteristics (vegetation) are highly variable, resulting in many attempts to classify the deserts of the world (see Meigs 1953; McGinnies et al. 1968; Petrov 1976; Walter 1979; Babaev et al. 1986 and references therein). In this book, we do not follow any particular classification but rather consider deserts according to their geographic locations. Detailed geographic descriptions of deserts can be found in a number of monographs, for example Petrov (1976), Evenari (1985), Babaev et al. (1986), Cooke et al. (1993). We outline below some general information on physical geography and geomor-

Table 2.1. Climatic parameters and main plant associations of different desert regions

Region	Mean January temperature (°C)	Mean July temperature (°C)	Composition of main plant associations
Central Asian deserts	-17 to -16	18 to 27	*Haloxylon ammodendron, Tamarix* spp., *Calligonum mongolicum, Timouria villosa, Artemisia ardossica, Karelinia caspica, Piptanthus mongolicus.* Stony and sandy-gravel plains – *Stipa glareosa, Stipa gobica, Allium polyrrhyzum, Artemisia frigida, Anabasis brevifolia, Reamuria soongorica, Salsola passerina, Nanophyton erinaceum, Ephedra przewalskii, Nitraria sphaerocarpa, Haloxylon ammodendron, Iljinia regelii, Sympegma regelii, Zygophyllum xanthoxylon.*
Southern Turan deserts	-13 to +5	25 to 32	*Haloxylon persicum, Calligonum* spp., *Ammodendron conolli, Eremosparton flaccidum, Astragalus* spp., *Aristida* spp. (sandy deserts); *Haloxylon aphyllum, Anabasis salsa, Salsola gemmascens, Artemisia herba-alba* (clay deserts); *Carex pachystylis* and *Poa bulbosa* with numerous ephemerals (loess deserts).
North-eastern Turan deserts	-15 to -12	24 to 26	*Artemisia terrae-alba, Anabasis aphyllum, Nanophyton erinaceum, Salsola arbuscula, Atriplex cana.*
North-western Turan deserts	-10 to -7	24 to 26	*Artemisia arenaria, Elymus giganteus, Callygonum aphyllum* with *Artemisia terrae-alba, Kohia fruticosa, Suaeda microphylla, Halocnemum strobilaceum, Aleurops littoralis* on salty-clay patches.
Deserts of the Indian Peninsula	7 to 9	40 to 41	*Leptadenia pyrotechnica, Cenchrus setigerus, Tephrosia purpurea, Crotalaria burhia, Eleusine compressa, Aristida* spp., *Acacia tortilis, Prosopis cineraria, Capparis decidua* (sand plains); *Euphorbia caducifolia, Ziziphus nummularia, Commiphora wightii* (rocky patches); *Euphorbia caducifolia, Fagonia cretica, Calotropis procera, Capparis decidua* (gravel plains).
Deserts of the Iranian Highlands	-8 to +1	28 to 33	*Calligonum leucocladum, Haloxylon persicum, Stipagrostis pennata* (sands); *Petrosimonia glauca, Halostachys belangeriana, Salsola arbuscula,* (solonchaks); *Salsola orientalis, Zygophyllum eurypterum, Poa bulbosa* (clay-gravel plains).
Northern Arabian deserts	7 to 14	27 to 34	*Anabasis articulata, Zilla spinosa, Zygophyllum dumosum* (hammadas); *Acacia* spp., *Tamarix* spp., *Retama raetam, Moricandia nitens; Hammada salicornica* (dry riverbeds); *Calligonum comosum, Artemisia monosperma, Haloxylon persicum* and *Pancratium sickenbergeri* (sands); *Haloxylon persicum, Hammada salicornica, Atriplex halimus, Anabasis articulata, Salsola tetrandra, Seidlitzia rosmarinus* (clay-salty deserts)

Table 2.1. continued

Region	Mean January temperature (°C)	Mean July temperature (°C)	Composition of main plant associations
Southern Arabian deserts	18 to 24	30 to 34	Similar to that of northern Arabia.
North African deserts	10 to 23	28 to 37	*Anabasis aretioides, Limonastrum feei* (hammadas); *Acacia* spp., *Ziziphus lotus* (serirs); *Calligonum azel, Calligonum comosum, Retama raetam, Ephedra alata* (sands); *Suaeda* spp., *Salicornica herbacea, Halocnemum strobilaceum* (solonchaks). Atlantic Sahara has unique vegetation due to the high humidity (*Euphorbia echinus, Euphorbia balsamifera, Rhus oxyacantha* and *Lycium intricatum*).
African Horn deserts	27 to 30	27 to 30	*Atrocnemum* spp., *Halopeplis* spp., *Suaeda* spp., *Salsola* spp. (salty soils); *Sporobolus* spp. (gypsum soils).
Great Basin deserts	-0.3	20.8	*Artemisia tridentata, Artemisia nova, Artemisia rigida, Eurotia lanata, Atriplex canescens, Atriplex confertifolia*. Stony habitats of the southern part of the Great Basin deserts are covered by *Coleogyne ramosissima*, while *Chrysothamnus nauseosus* predominates on light sandy soils of the northern and central part.
Mojave Desert	7 to 11	31 to 39	*Yucca brevifolia, Yucca schidigera, Prosopis cubescens, Franseria dumosa* (stony-gravel patches); *Larrea tridentata, Franseria dumosa, Hilaria rigida* (sands); *Atriplex* spp., *Dondia* spp. (salty-clay soils).
Sonoran Desert	7 to 11	31 to 39	*Carnegia gigantea, Opuntia* spp., *Ferocactus wislizenii, Larrea tridentata, Cercidium microphyllum, Fouquieria splendens* (bajadas); *Prosopis velutina, Acacia greggi, Celtis pallida, Clematis drummondi* (arojos); *Franseria dumosa, Larrea tridentata, Ephedra trifurca, Hilaria rigida* (sand dunes).
Chihuahuan Desert	7 to 11	26 to 34	*Larrea tridentata, Fouquieria splendens, Yucca elata, Agave* spp., *Opuntia* spp. (bajadas); *Prosopis glandulosa* (dry riverbeds); *Yucca elata, Jatropha spathula* (sands); *Hilaria mutica, Suaeda nigrescens* (clay patches on playas).
Kalahari Desert	26 to 28	14 to 15	*Aristida* spp., *Eragrostis superba, Anthephora pubescens, Acacia* spp., *Acanthosicyos horrida* (sands); *Pentzia* spp., *Salsola* spp., *Suaeda monoica, Frankenia hirsuta* (stony patches); *Atriplex* spp., *Suaeda* spp., *Salicornia* spp. (pans).
Namib Desert	16 to 22	12 to 18	*Mesembranthemum salicornioides, Hydrodea bossiana, Drosanthemum paxianum, Aizoon dinteri, Stipagrostis* spp., *Zygophyllum simplex* (coastal sands); *Aristida* spp., *Trichocaulon dinteri, Hoodia currori, Euphorbia brachiata, Aloe* spp., *Welwitschia mirabilis* (gravel and stony habitats of northern Namib).

Table 2.1. continued

Region	Mean January temperature (°C)	Mean July temperature (°C)	Composition of main plant associations
Australian deserts	28 to 30	12 to 14	*Triodia basedovi, Eucalyptus* spp., *Atriplex vesicaria*, (rocky slopes); *Zygochloa paradoxa, Triodia* spp., *Eucalyptus eudesmeoides, Casuarina decaisneana, Callitris* spp (sands); *Atriplex* spp., *Kochia aphylla, Bassia* spp., *Rhagodia* spp. (gravel plains).
South American Pacific deserts	18 to 26	12 to 22	*Pennisetum* spp., *Cenchrus* spp., *Anthephora* spp., *Chloris* spp. (near-shore zone); *Cereus* spp., *Echinocactus cinereus, Eophorbia lactiflora, Helianthocereus pasacana* (stone and gravel habitats).
Monte Desert	25 to 26	7 to 8	*Larrea cuneifolia, Larrea divaricata, Trichocereus terschekii* (stony-gravel plains); *Prosopis* spp. (dry valleys); *Atriplex* spp., *Suaeda* spp., *Salicornica* spp. (salinas).

phology of different desert regions. Data on mean temperatures and main vegetation types are listed in Table 2.1.

2.2.1 The Saharo-Gobian Desert Belt

The Saharo-Gobian desert belt stretches from 18° W to 114° E and occupies the area between 36° and 11° N in the west and 52° and 36° N in the east. It encompasses all the deserts of the Northern Hemisphere of the Old World, including the deserts of the Horn of Africa. We can divide this desert belt into several geographic units as follows.

2.2.1.1 The Central Asian Deserts

Central Asia is characterized by the presence of elevated plains and uplands surrounded by mountain ridges. Large and deep depressions began to appear in the Palaeozoic, and were completed in the Cenozoic era. Sandy deserts predominate in the south and west, whereas stony and sandy-gravel deserts occur in the north-east. The main landscapes of the southwestern part of Central Asia are shaped by the sediments of large rivers (the Tarim, Huang Ho and Edzin Ghol), which formed sand massifs after aeolian reworking. One of the largest sandy deserts of Central Asia is the Takla Makan, located in the Tarim basin and surrounded by the high mountain ridges of Tien Shan, Pamir and Altun Tagh. The sands of Takla Makan are up to 300 m in depth. Sand dunes are the main element of the landscape, while sandy-gravel ramps

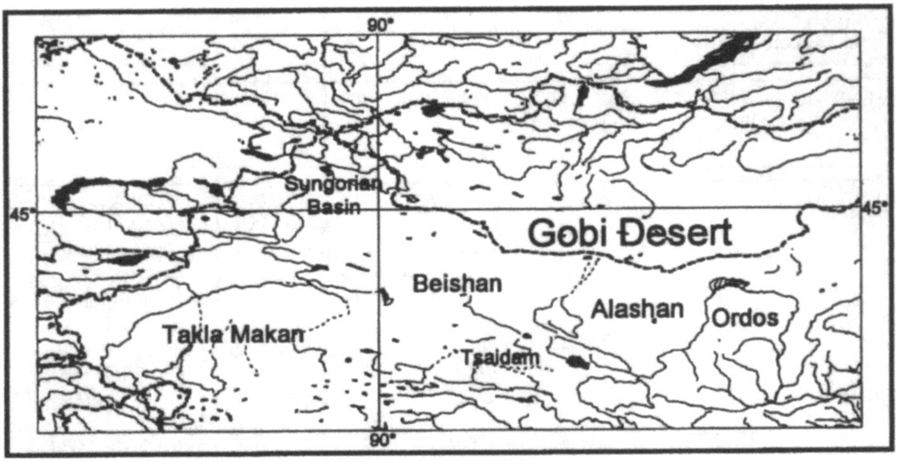

Fig. 2.2. The deserts of Central Asia

form the margins. The Ordos, Alashan and Sungorian Basin constitute the other sandy deserts of the region (Fig. 2.2). In addition to sandy deserts, there are two large non-sandy deserts in the southwest of Central Asia, namely the Tsaidam and Beishan. The Tsaidam desert is a giant saline depression (*solonchak*), surrounded by the high mountains of Tibet. The Beishan desert is the result of an ancient upland massif that has been eroded to stony desert.

The northeastern parts of the Central Asian deserts (the Gobi Desert) are represented by chains of intermontane depressions. Sandy-gravel and stony ramps are the dominant landscape types, whereas sand occurs in small patches, usually in lake basins.

The climate of the Central Asian deserts is determined by the relatively high elevation together with the isolation by high mountains on the south and the west and the influence of Pacific monsoons. As a result, the climate is temperate-cold. Maximum precipitation occurs in summer (June-August), except for the westernmost regions (Sungoria, western Kashgharia). The latter are under the influence of Atlantic cyclones and receive precipitation evenly all year-round.

2.2.1.2 The Turan Deserts

The Turan desert region can be subdivided into three main parts. The southern part includes the Turanian plain with four large deserts, Karakum, Kyzylkum, Moinkum and the desert of the Ily basin (Sary-Ishik-Otrau), and two intermontane arid basins, namely Ferghana Valley and the Southern Tadjik Depression. The deserts of the northeastern part of Turan are the Kazakh

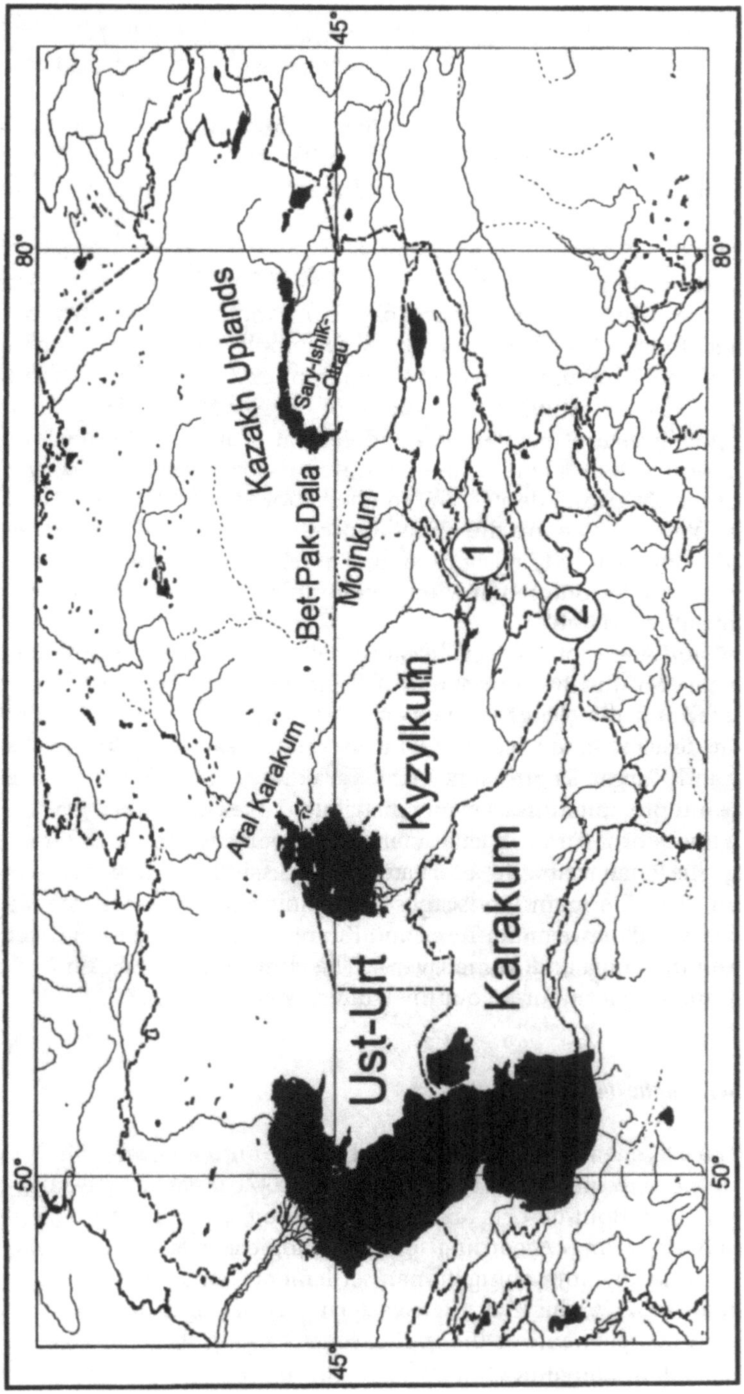

Fig. 2.3. The deserts of Turan and Kazakhstan. 1 – the Ferghana Valley, 2 – the southern Tadjik Depression

Uplands, Bet-Pak-Dala Desert, Northern Aral (Aral Karakum) Desert and the Ust-Urt Plateau (Fig. 2.3). The northwestern part includes arid lands surrounding the northern Caspian shore.

The landscapes of the southern Turan region have been formed on the base of low ancient mountains (Balkhans, Ak Tau, Buten Tau, Kara Tau) and the Neogene plateau. In the Quaternary, these ancient landscapes were transformed by the wandering of the large rivers (Amu Darya, Syr Darya, Zerafshan, Chu, Ily, Murgab, Tedjen), which flow from the high mountains of Tien Shan, Pamir and Paropamyz. The sediments of former river deltas shaped by aeolian processes created large complexes of sand dunes alternating with dry bottoms of temporal lakes (*takyrs*). These complexes are the dominant landscapes of the Karakum, Kyzylkum, Moinkum and Sary-Ishik-Otrau Deserts. Massive areas consisting of the remains of the Neogene plateau can be found in the northwestern (Kaplankyr) and central (Zaunguz Karakum) parts of the Karakum Desert and in the central and southwestern (Gazly plateau) parts of the Kyzylkum desert. The southern border of the region is composed of loess plains, which are the main landscape elements of the Ferghana Valley and Southern Tadjik Depression. The climate of the southern part of Turan is strongly continental with maximum precipitation occurring in winter (November-March).

The landscapes of the northeastern part of the Turan region are formed on the bases of the low mountains of Palaeogene age (Kazakh Uplands, Bet-Pak-Dala) and the Neogene plateaux (Ust-Urt) and represent mainly stony and clay deserts. Sands occupy relatively small areas and occur in the Sarysu Valley and Turgai Depression (Aral Karakum) only. The climate is continental, and precipitation is evenly distributed over the whole year.

Northwestern Turan is characterized by very young landscapes created during the Quaternary by alternate transgressions and regressions of the Caspian Sea. The main landscapes are sandy hills (between the Volga and Ural rivers and between the Terek and Kuma rivers) and salty-clay complexes (between the Volga and Kuma rivers). The climate of the region is similar to that of northeastern Turan, but the winters are not so cold.

2.2.1.3
The Deserts of the Indian Peninsula

The Thar Desert and the adjacent Thal Desert on the north (Fig. 2.4) are located at the extreme east of the subtropical-tropical part of the Saharo-Gobian desert belt. Both deserts are situated to the east of the Indus River. They are relatively young. According to archaeological data, these recent deserts were occupied by flourishing human settlements about 5000-10000 years ago (Krishnan 1952; Wadia 1960). The eastern part of the Thar Desert is situated on the eroded elements of the Indian Shield and is characterized by a complex of aeolian sand dunes intermixed with rocky outcrops. The western Thar

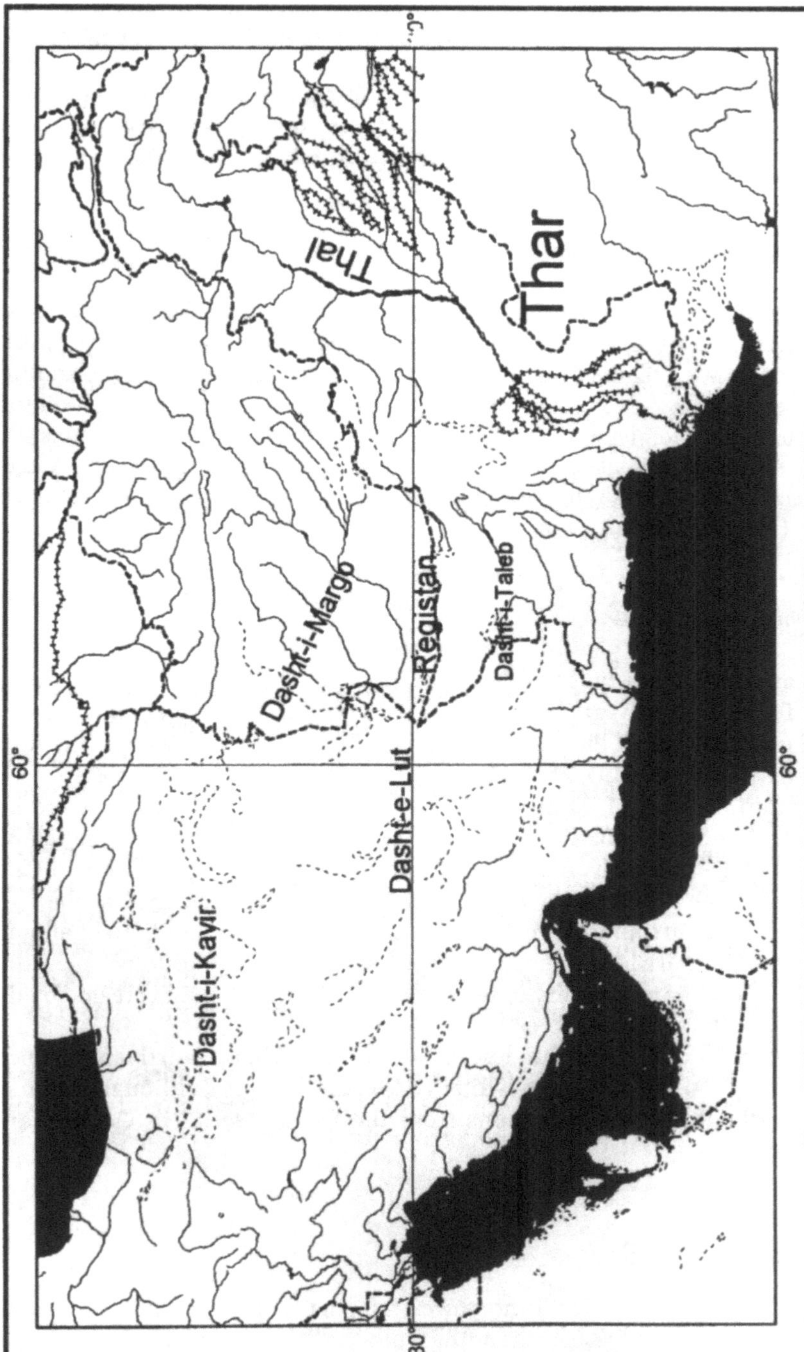

Fig. 2.4. The deserts of the Iranian Highlands and the Indian Peninsula

Desert and the entire Thal Desert are located in the ancient Indus Valley. The main landscape consists of sand fields created by aeolian reworking of alluvial sediments.

The climate of the region is typical of tropical deserts with warm winters and hot summers. Rainfall varies from 80 to 425 mm/year. Ninety percent of the rain falls during the monsoon season, from July to September.

2.2.1.4 The Deserts of the Iranian Highlands

The main deserts of the Iranian Highlands are those of Dasht-i-Kavir, Dasht-e-Lut, Dasht-i-Margo, Registan and Dasht-I-Taleb (Fig. 2.4). They represent a system of uplifted plains in enclosed depressions separated from one another by low mountain ridges. The principal landscape types are aeolian-alluvial sand plains (the Registan and Dasht-e-Lut Deserts), solonchaks (the Dasht-i-Kavir and Dasht-i-Taleb Deserts) and clay-gravel plains (the Dasht-i-Margo Desert). The climate is relatively cold for subtropical deserts and rain falls in winter (December-March) only.

2.2.1.5 The Arabian Deserts

There are three main types of desert on the Arabian Peninsula and adjacent areas. These are stony-gravel deserts (the Syrian Desert, Negev-Sinai desert, Al-Hijara Desert, Njad Plateau, Hadhramaut Plateau), sandy deserts (the Rub-al-Khali, Al Nafud and Dekhna Deserts) and clay-salty desert (the Mesopotamian plain) (Fig. 2.5). Stony-gravel deserts developed on ancient eroded mountains and plateaux. The major landforms in these deserts are flat plains (*hammadas*) and dry riverbeds (*wadis*). In contrast to the Turanian and Central Asian deserts, the sandy deserts of Arabia are the result of aeolian reworking of the surrounding rocky massifs and plateaux rather than of alluvial sediments. They are located in large depressions between mountain plateaux. The Mesopotamian clay-salty deserts are typical alluvial plains created by the Tigris and Euphrates rivers.

The climate of the Arabian deserts is subtropical in the north and tropical in the south. Rainfall occurs in winter (November-March), although rain may also be brought by the monsoons from the Indian Ocean in October and April-May.

2.2.1.6 The North African Deserts

All North Africa, except for its northwestern part (High Atlas), is occupied by the gigantic Saharan Desert extending from the Atlantic coast to the Nile Val-

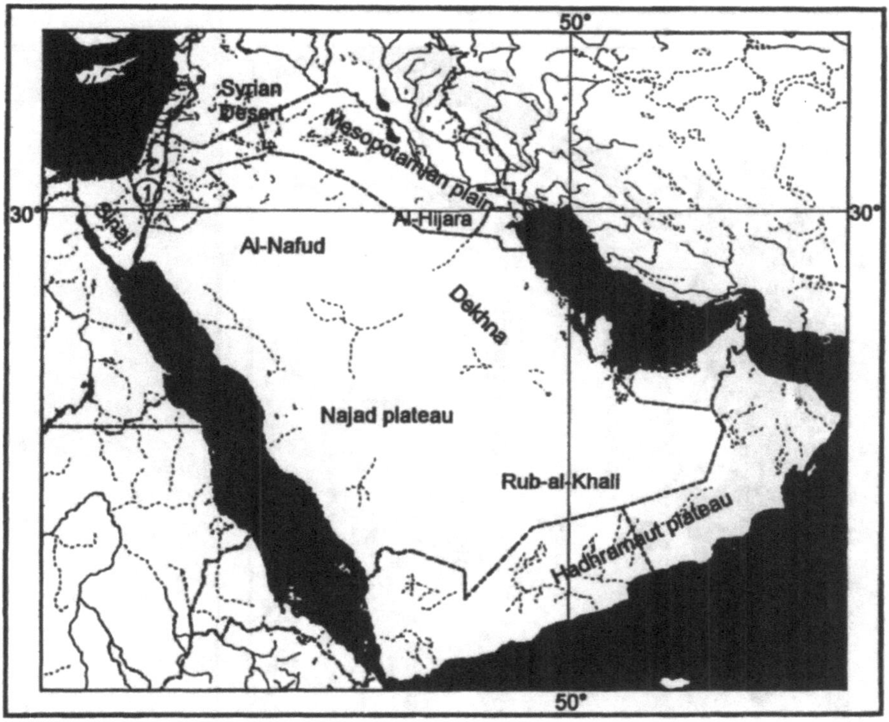

Fig. 2.5. The deserts of Arabia. 1 – the Negev Desert

ley (Fig. 2.6). Arid lands continue to the east, where two relatively small deserts are situated between the Nile Valley and Red Sea [the Arabian (Eastern) Desert and Nubian Desert]. In its geostructural aspect, the Sahara represents ancient dry land occupying the northern part of the African Plate.

Eighty percent of the Sahara area consists of stony (*serirs*), gravel (*hammadas*) and sandy (*ergs*) plains with small islands of ancient mountains and uplands (Ahaggar Massif, Tibesti, Adrar Plateau, Air Plateau etc.). Stony and gravel plains are located on the uplifted plateaux, while ergs occur in the large depressions between the mountains and plateaux. The plains are crossed by a system of dry riverbeds (*wadis*). There are young (Quaternary) closed depressions and shore lowlands in the north of the Sahara occupied by solonchaks.

The Arabian (Eastern) and Nubian Deserts are stony-gravel plains situated on elevated plateaux. Sand massifs occur only in the extreme west of the Nubian Desert.

The climate of North Africa is characterized by warm winters and hot summers. There are four regimes of precipitation. Rains in the northern Sahara are determined by Atlantic cyclones and occur from autumn (October) to spring (April). Rainfall in the southern Sahara is connected with the

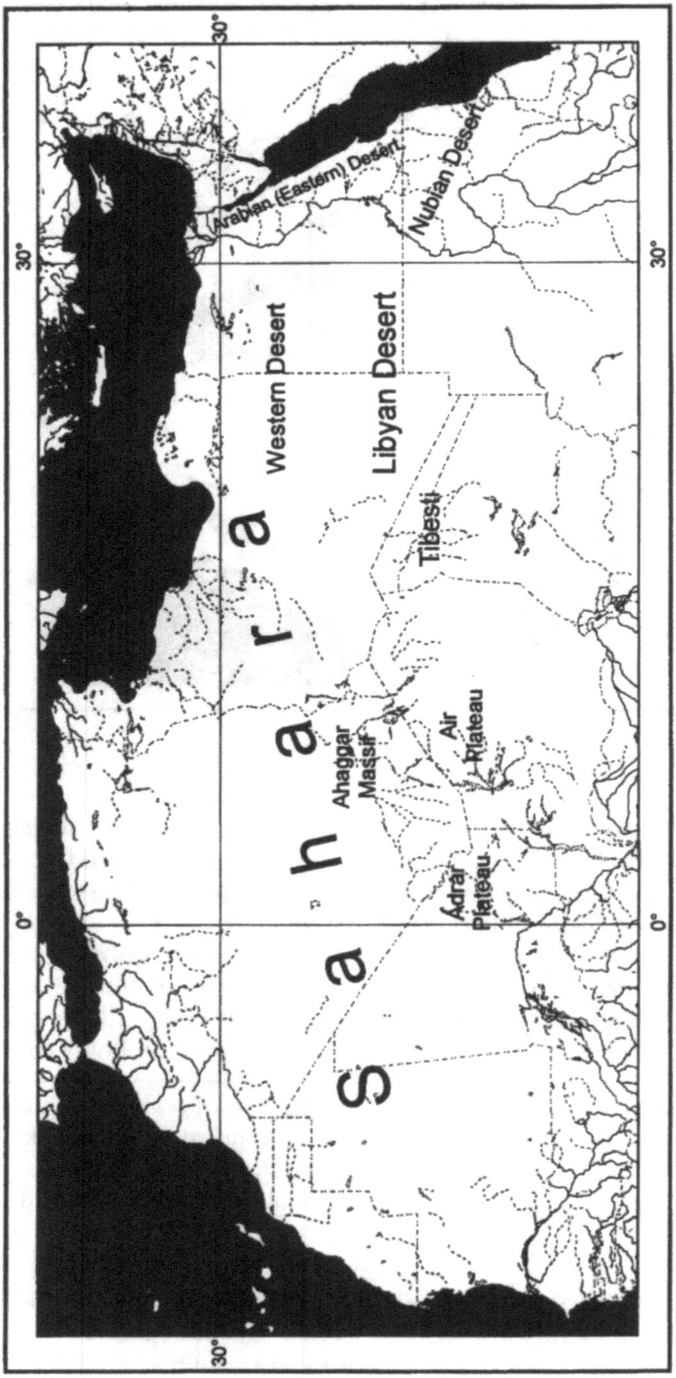

Fig. 2.6. The deserts of North Africa

monsoons of the Bay of Guinea and have a summer maximum (June-August). The central Sahara receives irregular and extremely low precipitation. The western (Atlantic) Sahara receives both winter-spring Mediterranean rains and summer tropical rainstorms.

2.2.1.7
The Deserts of the Horn of Africa

Deserts of the Horn of Africa are situated on the large escarpment which forms the northern edge of the eastern portion of the Ethiopian Highlands. The Northern Mountains are ragged and rise high above the hyper-arid north coast to capture moisture and create small pockets of land with subhumid climates and perennial water supplies. The Central and Southern Plateaux are situated south of the mountain zone and present a long monotonous upland thorn-scrub desert.

The soils of the region are largely skeletal with some moderate development in the reddish soils of the Kenya-Somalia border region. Deep soils of clay occur on the upland plains in the vicinity of the Guiba and Scebeli Rivers. The alluvial soils on the flood plains of these two rivers are calcareous, medium to fine in structure and sometimes saline. Sand dunes are restricted mainly to coastal areas, especially along the Indian Ocean, while stony plains are located in the central regions. Excess salt accumulation occurs in the large Danakil salt pan west of the Danakil Mountains and in the coastal, gypsiferous, and irrigated soils.

The climate of the region is equatorial, with temperatures almost constant all year round. The annual regime of precipitation is characterized by two peaks (March-May and October-November).

2.2.2
The North American Deserts

The North American deserts (Fig. 2.7) are located within the temperate and subtropic zones between 22 and 44° N. They do not occupy a large area but are distributed as small massifs in wide intermontane depressions of the southern portions of the Rocky Mountains and Plateau of Mexico, which are crossed by folded ridges, and in the valley and delta of the Colorado River. There are a number of different geographic subunits of these deserts as follows.

2.2.2.1
The Great Basin Deserts

The Great Basin deserts encompass a large area between the Pacific mountain chains (Cascade Range, Sierre Nevada) in the west, the Colorado Plateau in the east and south, and the Columbian Plateau in the north. The Great Basin

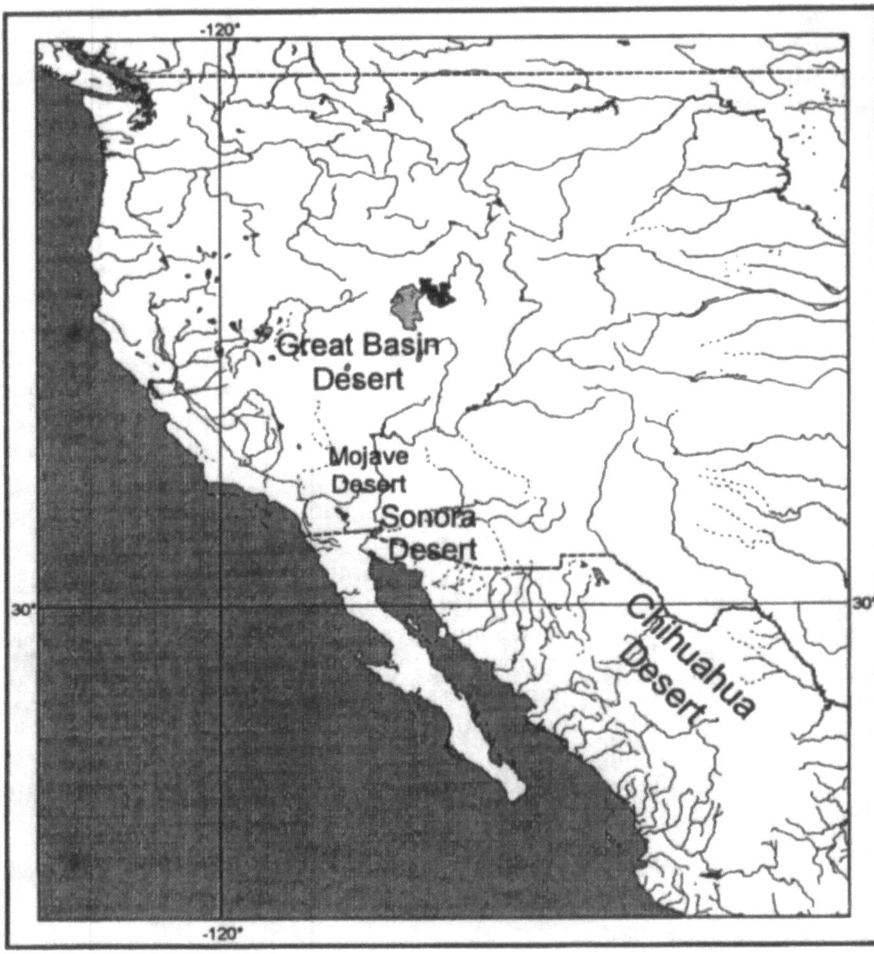

Fig. 2.7. The North American deserts

represents a plateau with an average elevation of about 1200 m with a large number of parallel mountain ridges running in a longitudinal direction which have been created by crustal fractures. Intermontane depressions (*bolsons*) contain relatively small, closed salty basins (*playas*) developed on the sites of former lakes. Thus, soils are salty and have developed on substrates varying in their mechanical composition from clay to stone. The arid regions of the Great Basin do not exhibit any flow because of their complicated orography.

The climate of the Great Basin deserts is characterized by cold winters and relatively cool summers. The relative isolation of the region and its remoteness from sources of water results in scarce and irregular precipitation.

2.2.2.2
The Mojave Desert

The Mojave desert includes the most arid regions of North America. This desert is an uplifted plain at an elevation of 600-1000 m. The main landscape type consists of alternate low mountain ridges and intermontane depressions filled with alluvial deposits. These deposits are partially transformed by wind into elements of aeolian relief.

The climate of the Mojave is subtropical with warm winters and hot summers. Precipitation is extremely low and occurs irregularly.

2.2.2.3
The Sonoran Desert

The typical landscape of the Sonoran desert consists of wide intermontane valleys with remnant flat-top hills (*mesas*). A large proportion of the area of these valleys (75%) consists of foothills (*bajadas*), which descend into flat alluvial plains. The valleys are cross-cut by channels of temporary (*arojos*) and perennial rivers.

There are two periods of precipitation, rains in winter (December-March) and rainstorms in summer (July-September). Rains are brought to the region by moist air masses from the Gulf of Mexico and the Gulf of California.

2.2.2.4
The Chihuahuan Desert

The relief of the Chihuahuan Desert is similar to that of the Great Basin deserts. The main landscape types are bolsons with playas on the bottom surrounded by small patches of sand. Beside these, there are sand massifs (White Sands) that are not connected with recent bolsons but originated from gypsum and other sulphates deposited as sediment on the shores of ancient warm lakes.

The temperature regime of the Chihuahuan Desert is similar to that of the Mojave and Sonoran deserts. However, the summers are cooler than those in the Sonoran desert. Precipitation occurs mainly in summer (June-September).

2.2.3
The South African Deserts

2.2.3.1
The Kalahari Desert

The Kalahari Desert is located to the east of the Great Western Escarpment in a large closed depression at an elevation of 800-1000 m and surrounded by plateaux and mountains (Fig. 2.8). The depression is filled by fluvial sedi-

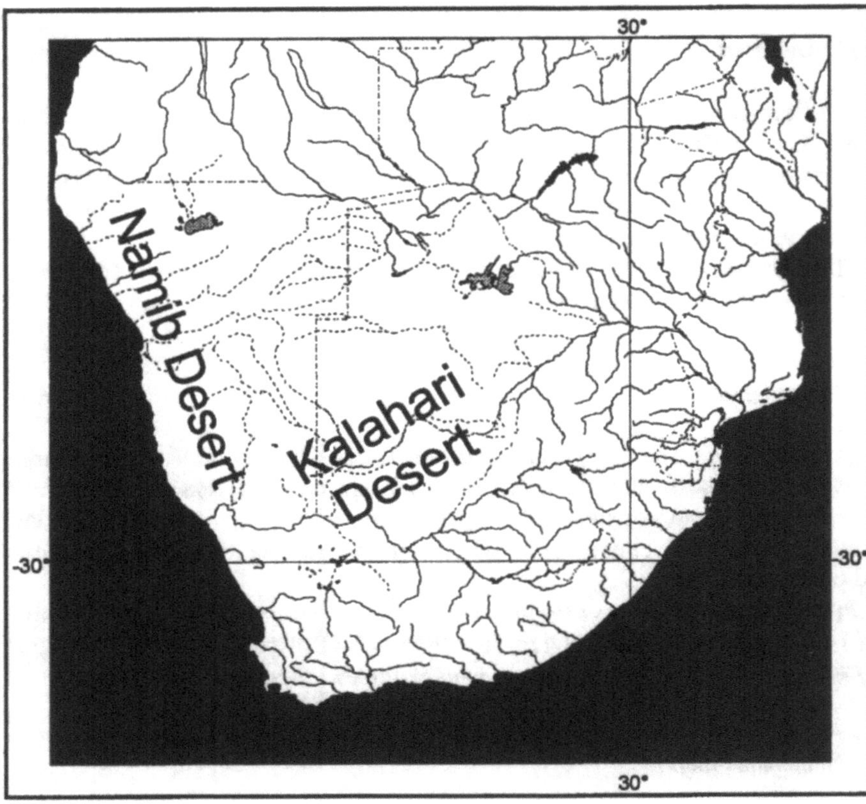

Fig. 2.8. The South African deserts

ments of the Molopo and Nossob Rivers. Fluvial sediments have been reworked by wind and almost the entire area of the Kalahari Desert is now covered by a mantel of red and gray sands of different thickness. The main landscapes are represented by sand dunes of different shapes with rocky outcrops and small depressions with clay bottoms (*pans*).

The climate of the region is continental subtropical with warm winters and relatively mild summers, although summer temperature is high in the inland Kalahari. Rainfalls occur mainly in summer (December-March).

2.2.3.2
The Namib Desert

The coastal Namib Desert is situated on the coast of Namibia (southwestern Africa) between the Orange and Kunene Rivers (Fig. 2.8). The desert occupies a sloping foothill plain composed of marine, proluvial and alluvial deposits.

Gravel surfaces predominate in the northern portion of the Namib Desert, whereas sandy ridges represent the main landscape element in the southern part.

Proximity to the Atlantic ocean with its cold Benguela Current determines the specifics of the Namib climate, which can be identified as oceanic hyper-arid. Summers here are relatively cool (although there is a steep gradient of increasing temperature as one moves inland), while the winters are relatively warm. Precipitation is extremely low (15-43 mm per annum). The heaviest rains occur in May-September in the far south of the Namib, while most rains are in December-March. However, the air humidity is high due to proximity to the ocean. The result of the high air humidity is that there are many foggy mornings (27 days per month), resulting in condensation on the soil surface, thereby moistening the top layer of the soil.

2.2.4
The Australian Deserts

Deserts occupy approximately 49% of area of the continent of Australia. Those of western and central Australia are located on the Australian Shield, while the eastern portion of the Australian deserts occupies the Sedimentary Basins (Fig. 2.9). The main types of deserts in Australia are rocky deserts (the Pilbara and Hammersley Range, Kimberley Plateau, McDonnel and Musgrave Ranges), sandy deserts (the Simpson, Great Sandy, Tanami and Great Victoria Deserts) and stony-gravel deserts (the Gibson and Sturt Deserts and the Nullarbor Plain). In addition, there are clay deserts in the valleys of the Darling, Thompson, Diamantina and Georgina Rivers.

Two types of climate can be distinguished in the Australian deserts. Most of these are characterized by a tropical continental climate, while the southernmost arid areas (the southern portion of the Great Victoria Desert and the Nullarbor Plain) have a subtropical continental climate. The precipitation regime has a summer maximum in the northern deserts (November-March) and a winter maximum in the southern deserts (April-July). However, seasonal peaks in rainfall are weakly expressed and highly variable between years and locations.

2.2.5
The South American Deserts

South American deserts are located in two separate regions, namely the Pacific deserts (the Atacama, Sechura and Pampa del Tanarugal Deserts; Fig. 2.10) on the Pacific shore of the continent, and the Argentinian deserts on the eastern slopes of the Andes (Fig. 2.11). There is also a high-altitude cold desert of Puna, which is not considered in this book.

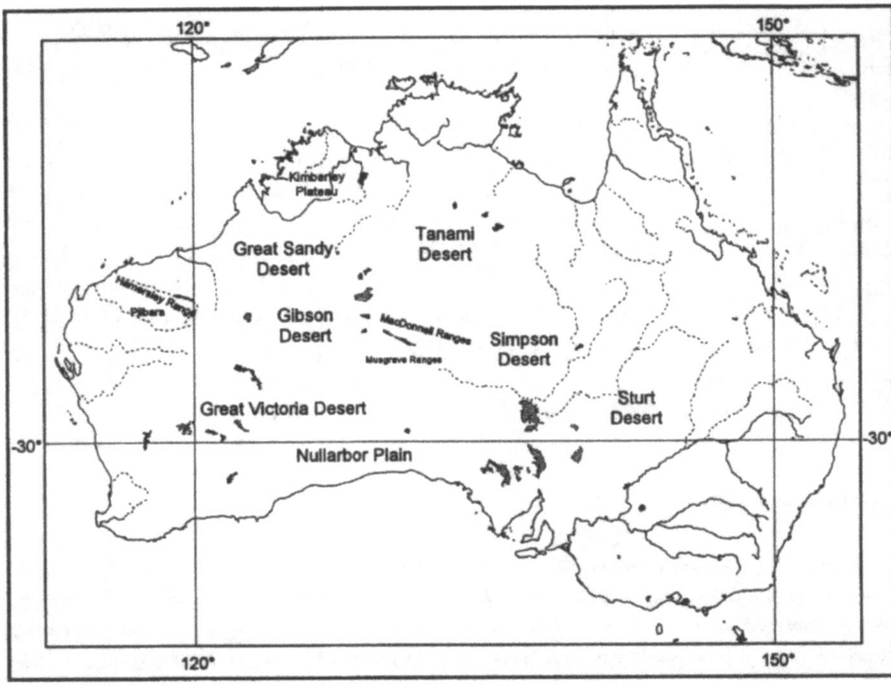

Fig. 2.9. The Australian deserts

2.2.5.1
The Pacific Deserts

The Pacific deserts occupy the narrow strip between the Pacific coast and the western slopes of the Andes from Peru in the north to central Chile in the south (Fig. 2.10). The cause of arid conditions in the region is similar to that in the Namib and western Australian deserts, namely a cold west coast current, here called the Peruvian Current. Stony desert and solonchaks (*salinas*) occupy the main portion of the region. Patches of gravel plains and large massifs of moving sands can also be found.

The climate of the Pacific deserts is marine hyper-arid. It is similar to that of the Namib Desert with warm winters and relatively cool summers. Rainfall is extremely rare; the main sources of water are dense fogs and dew.

2.2.5.2
The Argentinian Deserts

The principal desert of this region is the Monte Desert (Fig. 2.11). This is situated on the northern portion of the Patagonian Shield at an elevation of

Fig. 2.10. The Pacific deserts of South America. 1 – the Pampa del Tanarugal Desert

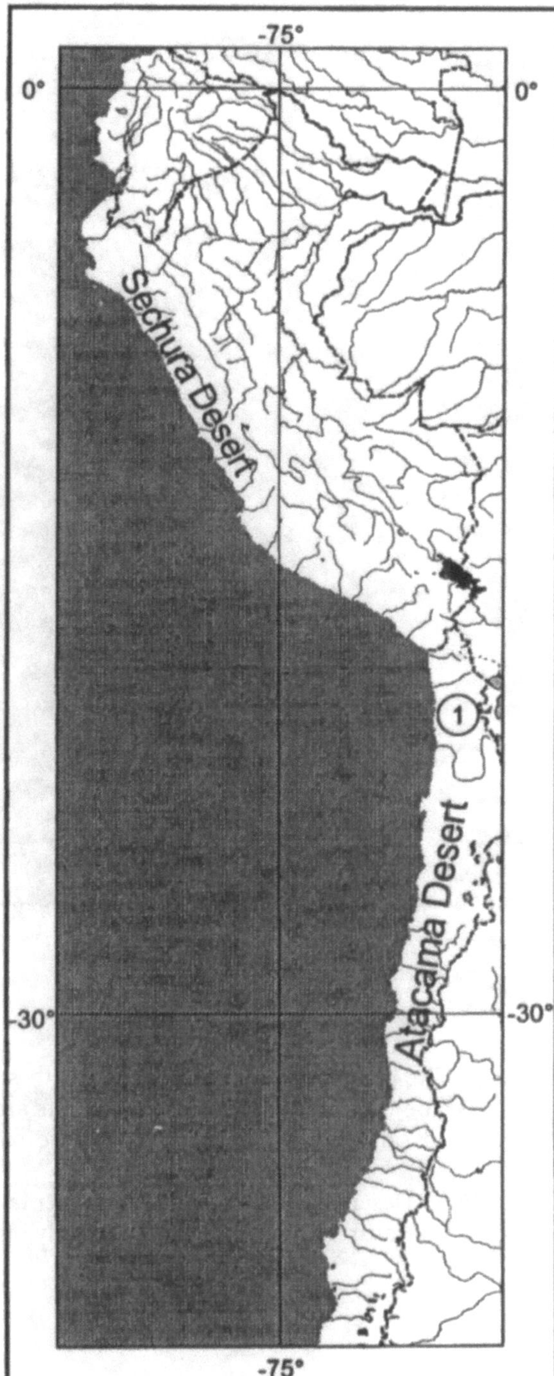

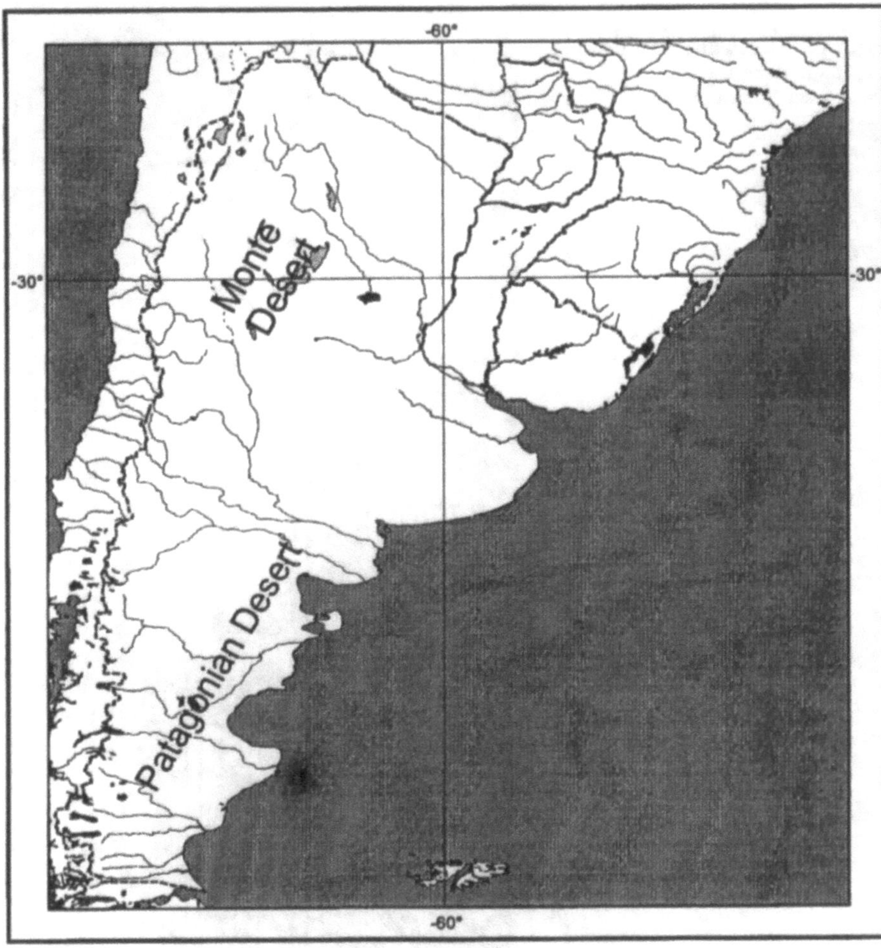

Fig. 2.11. The Argentinian deserts of South America

600-800 m. The plateau is an uplifted plain with stony-gravel surfaces as the main landscape type. Other landscape elements are solonchaks in closed depressions and sand dunes which delineate the southeastern margin of the Monte Desert.

The Monte Desert has a subtropical continental climate with weak fluctuations of temperature. Rains are brought mainly by western air masses.

CHAPTER 3

Composition of the Rodent Faunas of Different Desert Regions

3.1 Biogeographic Classification Using Rodents as a Key Taxon

Patterns of species distribution are key factors in all biogeographic analyses, including biogeographic classification and regionalization (Myers and Giller 1988). Indeed, the concept of biogeographic regions, one of the oldest concepts in biogeography (Nelson and Platnick 1981), is based mainly on patterns of species distribution. This concept now incorporates other concepts including the consideration of a province as a response to both ecological or historical processes (Rosen 1988). Nevertheless, different aspects of biogeographic regionalization have been severely criticized (see review in Hengeveld 1992). One of the main points of criticism is the problem of choice of taxa for regionalization, because regionalization results may differ depending on the taxonomic group selected. Comparison of schemes of regionalization based on different taxa is beyond the scope of this book. Our sole aim has been to denote the regions with relatively homogeneous rodent faunas (provinces) for subsequent comparisons between them.

Two possible directions of biogeographic classification can be considered. The first is a classification of geographic regions based on their faunal composition (regionalization), while the second is a classification of species according to their geographic distribution (areographic analysis). In mathematical terms this is equivalent to R/Q analyses of a presence/absence matrix (Pielou 1984).

We wish to emphasize that our goal has been to analyze the arid-adapted component of deserts and not the whole species assemblages found within them. Consequently, the resulting classifications (for both regions and species) may differ from those obtained using broader sets of either regions or species.

The principles of species selection for such a classification were elaborated by Kucheruk (1959) in his analysis of the mammalian fauna of the Palaearctic steppes. Because the paper cited is written in Russian and has become a bibliographic rarity, we have adopted and describe these principles here. There are two main points, namely selection of species for the whole set analyzed, and an assignment of each species to a particular type of geographic range.

First, we selected species which we consider to belong to desert faunas. As mentioned above, the geographic ranges of these species are either entirely or mostly located in the desert and they occupy zonal desert habitats rather than intrazonal ones (mountains, river valleys and lake shores). For example, *Microtus limnophilus* is found only in the Gobi desert, but it inhabits exclusively densely-vegetated lake shores, and does not penetrate into true desert habitats; so, it cannot be considered to be a desert species and was omitted from analysis.

The next step was to make a classification of regions. There have been many attempts to classify different regions based on distribution of their mammalian faunas. For example, Hagmeier and Stults (1964) and Hagmeier (1966) divided the entire continent of North America into different regions using the entire mammalian fauna. Another approach was demonstrated in the series of papers published by Neronov and colleagues devoted to the analyses of separate desert regions of the Saharo-Gobian belt (Neronov 1976; Neronov and Arsenieva 1980; Neronov and Lushchekina 1980; Bobrov and Neronov 1988; Neronov and Bobrov 1990; Varshavsky et al. 1995, 1997). We used the same technique as in all the papers cited above. This technique consists of the three steps. First, we divide a map of a region into quadrates 2° by 2°, and compile a list of rodent species for each quadrate. Second, we combine adjacent quadrates that have the same species set and calculate values of Jaccard index for each pair of the resulting units. Finally, we perform cluster-analysis of these units based on the matrix of Jaccard indices. However, our results differ from those previously described in the relative ranking of the regions of the final hierarchical scheme. The reasons are as follows: (1) in contrast to Hagmeier and Stults, we examined only one taxonomic-ecological group of mammals. Consequently, the resulting ranks of the regions are significantly lower than those of Hagmeier and Stults (1964) and Hagmeier (1966); (2) contrary to Neronov and his colleagues (cited above), we did not consider each separate desert, but rather the whole desert zone within each biogeographic region. It does not seem useful to analyze two or more regions together if there are no shared species. Moreover, we did not consider the entire spectrum of rodent species, but only those that appear to be desert-specific. As a result, Neronov's schemes and ours are not comparable, even though their objectives are similar.

The final stage of the analysis consisted of the classification of species by their geographic ranges. The classification of geographical ranges is equivocal because the ranges of different species are not related in a linear hierarchy. Two main approaches to areographic classification can be distinguished. One is based on the concept of the centers of origin (e.g. De Lattin 1957 in Hengeveld 1992; Mekaev 1987). This approach can only be applied to well-known groups because the results would otherwise be doubtful on account of speculative and unprovable assumptions (see Shenbrot 1989). Another approach is purely phenomenological. It is based on the contemporary similarity of the ranges (e.g. Kucheruk 1959; Kryzhanovsky 1965; Kulik 1980). We used a combination of both approaches. Thus, we consider a type of geographic range as a group of species' geographic ranges similar in their position, shape and origin. Actually, each type of geographic range is specific for one of the zoogeographic provinces determined in the previous analysis (regionalization). Consequently, species with a particular type of geographic range are those endemic or subendemic to a given province, or else they are widely distributed species, if there is paleontological or ecological evidence of their origin in that province. While it is clear what the endemic species and the paleontological evidence of origins are, two other terms require explanation. A species was considered to be subendemic to a province if its range includes this province in its entirety and only parts of an adjacent province or provinces. For example, *Cardiocranius paradoxus* was considered to be a subendemic to the Gobian province because its distribution includes the whole Gobian province and parts of the Sungorian and Balkhash provinces. We incorporated patterns of species' habitat distribution as ecological evidence of its origin. We suggest that a species occupies a wide spectrum of habitats in the center of its origin whereas in the regions of recent penetration it becomes more stenotopic. For example, *Dipus sagitta* has a broad distribution throughout the whole Turano-Gobian subrealm. There is no clear paleontological evidence as to the place of its origin. However, this species is strictly psammophillous in most of its range but in the Gobi desert alone it occupies practically all desert habitats (Shenbrot et al. 1995). So, we assigned it to the Gobian range type.

Data on the geographic distribution of rodents were obtained from the following regional monographs: North Africa – Ranck (1968), Osborn and Helmy (1980), Le Berre (1990), Kowalski and Rzebik-Kowalska (1991); Southwest Asia – Lay (1967), Hassinger (1973), Roberts (1977), Harrison and Bates (1991), Corbet and Hill (1992), Shalmon (1993), Qumsiyeh (1996); Central Asia (former USSR) – Bobrinsky et al. (1965), Sludsky et al. (1969, 1977, 1978), Yanushevich et al. (1972), Gromov and Erbajeva (1995); Mongolia and China – Bannikov (1954), Sokolov and Orlov (1980), Ma et al. (1987); South Africa – De Graaff (1981), Skinner and Smithers (1990); Australia – Strahan (1983); North America – Hall (1981) and South America – Redford and Eisenberg (1992). Some information was also obtained from the taxonomic monographs of Wilson and Reeder (1992), Pavlinov et al. (1990), Genoways and Brown (1993) and Shenbrot et al. (1995).

The resultant zoogeographic subdivision based on desert rodent distribution is presented in Figs. 3.1, 3.2 and 3.3. The hierarchical scheme of faunistically-homogenous areas (provinces) is as follows.

I.	**Palaearctic region**
IA.	Saharo-Gobian realm
IA1.	Saharo-Arabian subrealm
IA1a.	North Saharan superprovince
IA1a1.	Sub-Maghrebian province
IA1a2.	Libyan province
IA1b.	Central Saharan superprovince
IA1b1.	Central Saharan province
IA1b2.	Nubian province
IA1c.	South Saharan superprovince
IA1c1.	South Saharan province
IA1d.	Arabian superprovince
IA1d1.	North Arabian province
IA1d2.	South Arabian province
IA2.	Irano-Sindian subrealm
IA2a.	Syrian superprovince
IA2a1.	Syrian province
IA2b.	Irano-Mesopotamian superprovince
IA2b1.	Iranian province
IA2b2.	Mesopotamian province
IA2c.	Sindian superprovince
IA2c1.	South Pakistan province
IA2c2.	Sindian province
IA3.	Turano-Gobian subrealm
IA3a.	Turanian superprovince
IA3a1.	Turanian province
IA3b.	Kazakhstan superprovince
IA3b1.	Central Kazakhstan province
IA3b2.	Balkhash province
IA3c.	Gobian superprovince
IA3c1.	Sungorian province
IA3c2.	Gobian province
IA3c3.	Kashgharian province
IA3c4.	Tzaidam province
IB.	African Horn realm
IB1.	Somalian desert province

II.	**Palaeotropic region**
IIA.	South African desert realm
IIA1.	Namibian province
IIA2.	Kalahari province
III.	**Australian region**
IIIA.	Ausralian desert realm
IIIA1.	Ausralian desert province
IV.	**Neoarctic region**
IVA.	North American desert realm
IVA1.	Great Basin province
IVA2.	Sonoran province
IVA3.	Chihuahuan province
V.	**Neotropic region**
VA.	Atacama realm
VA1.	Atacama province
VB.	Monte-Patagonian realm
VB1.	Monte province
VB2.	Patagonian province

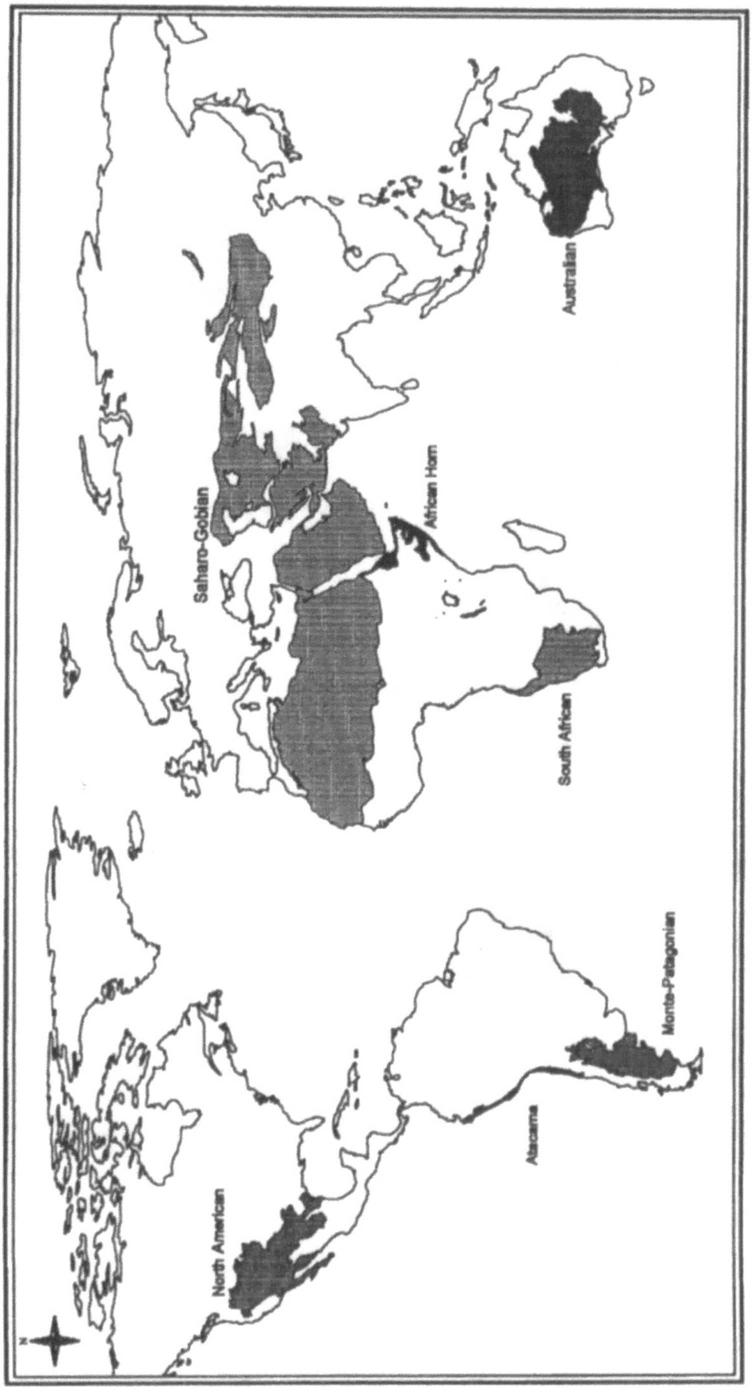

Fig. 3.1. High rank zoogeographic units of the world's deserts based on rodent distribution

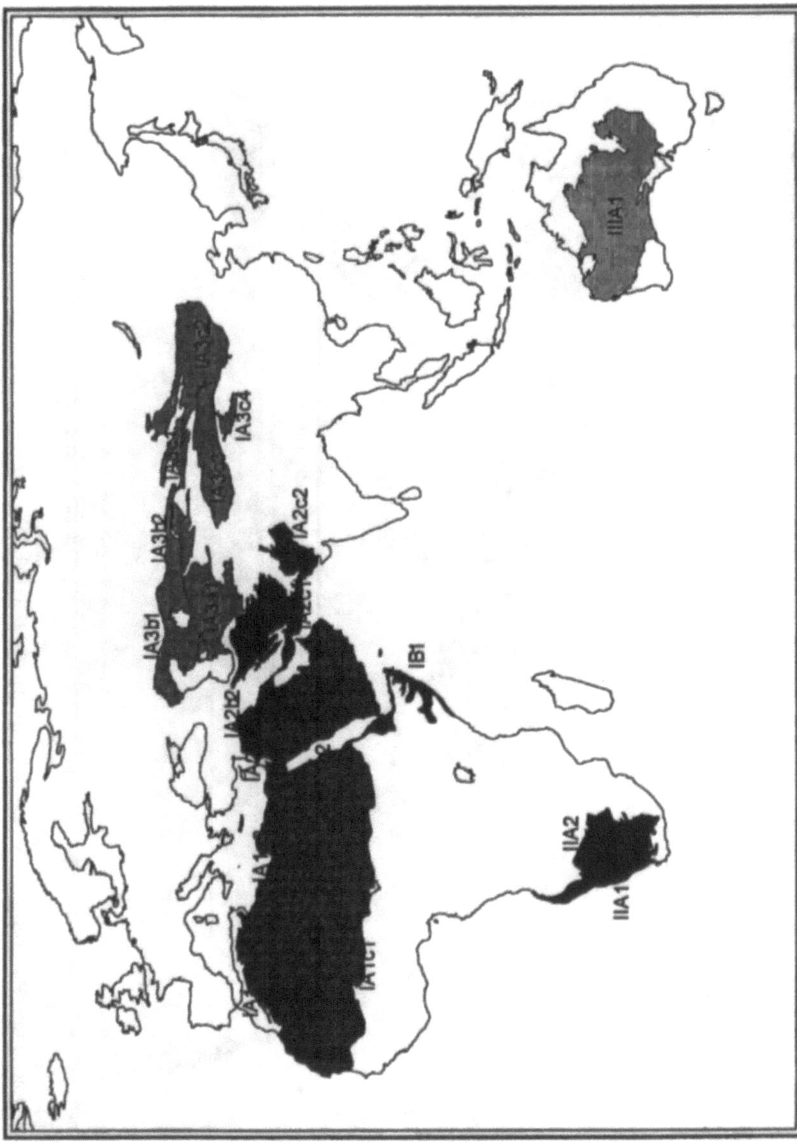

Fig. 3.2. Zoogeographic subdivision of the Old World deserts based on rodent distribution. See text for designations of the provinces

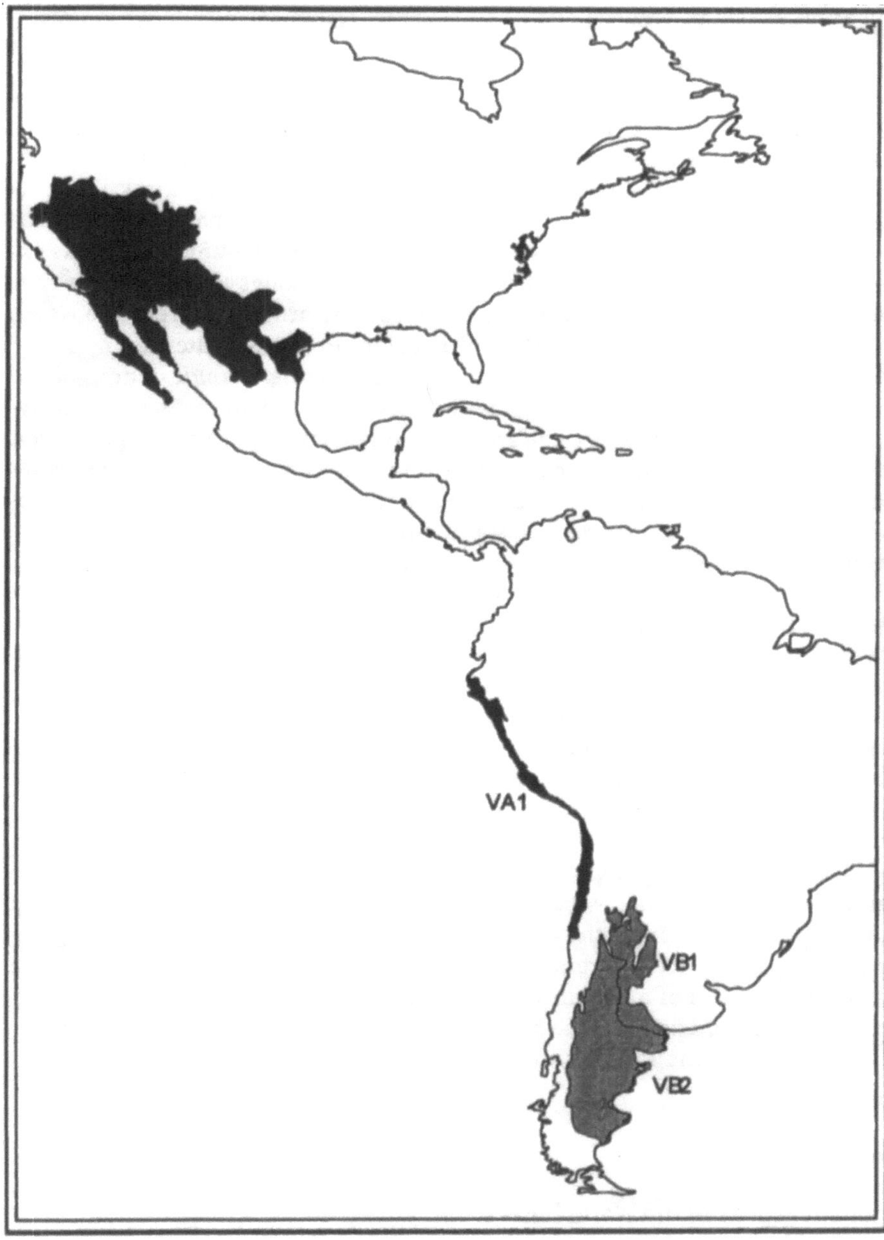

Fig. 3.3. Zoogeographic subdivision of the New World deserts based on rodent distribution. See text for designations of the provinces

As mentioned above, each zoogeographic province is inhabited mainly by species possessing the type of geographic range that is specific for this province. However, there can also be species with other types of geographic ranges. The latter have usually penetrated from adjacent provinces. Sometimes, albeit rarely, the presence of widespread species may mask this pattern. In such cases, a species recorded in one province may have a type of geographic range whose center is located far away in another. Such a case is exemplified by *Meriones erythrourus*. This species belongs to the Iranian type of geographic range, but is distributed in several provinces, from the North Arabian province in the west to the Sungorian province in the east.

In four cases, a respective type of geographic range is not distinguished because there are no endemic species in a province. In two cases, such a province can be considered as a transition zone between adjacent provinces. These are the Syrian province (a transition between the Iranian, Mesopotamian and South Arabian provinces) and the South Pakistan province (a transition between the Iranian, Mesopotamian and Sindian provinces). The remaining two cases are marginal provinces, those of Kashgharia and Tzaidam. Both include the depauperate Gobian fauna. However, the species reductions point in different directions. The Kashgharian depression is a hyper-arid sandy desert. Consequently, its fauna has only retained the psammophillous elements of the Gobian species assemblage. In contrast, the Tzaidam depression is a transitional zone between the central Asian lowland deserts and the high-mountain deserts of the Tibetan Plateau. The main landscapes of the Tzaidam are solonchaks and piedmont plains, so only the non-psammophillous fraction of the Gobian species complex can invade the region.

Each type of geographic range has a distinct geographic core. This coincides with the province with the largest number of species belonging to its own type. In addition, each type of geographic range has a geographical periphery expressed to different extents. This periphery can either extend far from the borders of a province (for example, the Gobian type) or be practically absent (for example, the Mesopotamian type). In the latter case, a given type of geographic range is only characteristic of the endemic species of a particular province.

3.2 Rodent Faunas of the World Deserts

Descriptions of provincial faunas of the deserts throughout the world are presented below. The results of our analyses of the composition of rodent faunas according to the geographic origin of the species can be found in Tables 3.1-3.29. Types of geographic range are presented in Figs. 3.4-3.10. There is no figure for the Nubian type because it is represented by only a

Table 3.1. Rodent species with different types of geographic range in the fauna of the Sub-Maghrebian province

Type of geographic range	Species
Endemic	*Atlantoxerus getulus, Meriones grandis, Ctenodactylus gundi, C. vali*
Subendemic	*Pachyuromys duprasi*
Libyan	*Gerbillus simoni, Meriones shawi, Jaculus orientalis*
Central Saharan	*Gerbillus campestris, G. nanus, G. henleyi, G. pyramidum, G. gerbillus, Meriones crassus, M. libycus*
Western Mediterranean	*Mus spretus, Eliomys quercinus*
Sindian	*Mus musculus*
Sahelian	*Hystrix cristata*

single (endemic) species (*Gerbillus mackilligini*) nor for the Australian type which cannot be considered to be definite because of the recent extinction of many species.

3.2.1 The Saharo-Gobian Realm

3.2.1.1 The Sub-Maghrebian Province

Most of the 19 species that comprise the fauna of the Sub-Maghrebian province belong to the subfamily Gerbillinae (58%). Others are in the Xerinae, Murinae, Myoxidae, Dipodinae, Hystricidae, or Ctenodactylidae (one to two species in each taxon). Four endemic species in the province are connected with the southern desert slopes or internal dry valleys of the Atlas Mountains.

The fauna of the province is composed of species belonging to six types of geographic range (Table 3.1). Representatives of the Central Saharan type are the most abundant, species with local geographic ranges occupy the second position (four endemics and one subendemic), while species of the Libyan type of geographic range occupy the third position. There are also four species which have invaded the region from afar. Two species (*Mus spretus* and *Eliomys quercinus*) have a Western Mediterranean origin (Marshall and Sage 1981, Darviche and Orsini 1982, Macholan 1990 and Filippucci et al. 1988, respectively), whereas *Mus musculus* originated from northwest India (Din et al. 1996). *Mus musculus* occurs across all deserts of the world. However, we prefer not to take this species into account for the biogeographic analyses of deserts that lie outside the borders of the Palaearctic region because we assume that *M. musculus* superspecies (*M. m. musculus* and *M. m. domesticus*) was introduced there during the last 3-4 centuries. The fourth species, *Hystrix cristata*, is distributed mainly in the western Mediterranean

Table 3.2. Rodent species with different types of geographic range in the fauna of the Libyan province

Type of geographic range	Species
Endemic	*Gerbillus amoenus, G. latastei, Allactaga tetradactyla*
Subendemic	*Gerbillus andersoni, G. simoni, Meriones shawi, Jaculus orientalis*
Sub-Maghrebian	*Pachyuromys duprasi*
Central Saharan	*Gerbillus campestris, G. gerbillus, G. henleyi, G. pyramidum, Meriones crassus, M. libycus, Psammomys obesus, Jaculus jaculus*
Western Mediterranean	*Mus spretus, Eliomys quercinus*
Sindian	*Mus musculus*
Sahelian	*Hystrix cristata*

and in the northern part of tropical Africa south of the Saharan desert. This pattern of distribution enables us to suppose a Sahelian origin for this species.

3.2.1.2
The Libyan Province

The taxonomic composition of the Libyan fauna is characterized by the absolute dominance of Gerbillinae (13 out of 20 species), whereas Murinae, Dipodinae, Myoxidae and Hystricidae each are represented by one to two species. Three of the Libyan species are endemic (Table 3.2); they all occur in mesic habitats on the Mediterranean coast. Species with a local type of geographic range and those of the Central Saharan origin are represented almost equally in the fauna. The remainder is composed of a single species from the Sub-Maghrebian province and four intruders from afar (*Mus musculus, Mus spretus, Eliomys quercinus* and *Hystrix cristata*).

3.2.1.3
The Central Saharan Province

This province is the biogeographic core of the Saharo-Arabian subrealm. The species that originated in the Central Sahara penetrate all other provinces of this subrealm and even filter into the neighboring Irano-Sindian subrealm. More than half the fauna is represented by gerbils and jirds (62%). There are also two murine, one dipodine and two ctenodactylid species.

Almost all species (excluding the Sindian *Mus musculus*) have a local origin and their geographic ranges are of the Central Saharan type. However, the geographic distribution of most of these species is wide (Fig. 3.4c) and only two, both ctenodactylids, are strictly endemic. All other species of local origin are either subendemic or widely distributed (Table 3.3).

Table 3.3. Rodent species with different types of geographic range in the fauna of the Central Saharan province

Type of geographic range	Species
Endemic	*Felovia vae, Massoutiera mzabi*
Subendemic	*Gerbillus campestris, G. pyramidum, G. gerbillus, Meriones libycus*
Widely distributed species of a local origin	*Gerbillus nanus, G. henleyi, Meriones crassus, Psammomys obesus, Acomys cahirinus, Jaculus jaculus*
Sindian	*Mus musculus*

Table 3.4. Rodent species with different types of geographic range in the fauna of the Nubian province

Type of geographic range	Species
Endemic	*Gerbillus mackgilligini*
Central Saharan	*Gerbillus nanus, G. pyramidum, G. gerbillus, Meriones crassus, Psammomys obesus, Acomys cahirinus, Jaculus jaculus*
Sindian	*Mus musculus*

3.2.1.4
The Nubian Province

The fauna of the Nubian province can be considered as a depauperated Central Saharan fauna (without ctenodactylids and some gerbils and jirds). The only important characteristic that determines the provincial rank of the region is the substitution of *Gerbillus campestris* by the Nubian endemic *Gerbillus mackgilligini*. Other species occurring in the province are of either Central Saharan or Sindian origin (Table 3.4).

3.2.1.5
The South Saharan Province

As in all North African desert regions, gerbils dominate in the South Saharan fauna (seven species out of 12). The rest of the fauna is represented by one ground squirrel (Xerinae), two spiny mice (Murinae), one three-toed jerboa (Dipodinae) and one porcupine (Hystricidae).

Species of the South Saharan type of geographic range compose 33% of the whole species assemblage. They are all endemics and do not venture beyond the borders of the province. Furthermore, species with a Central Saharan type of geographic range penetrate the province from the north, while rodents from the Sahel invade the South Sahara from south (Table 3.5).

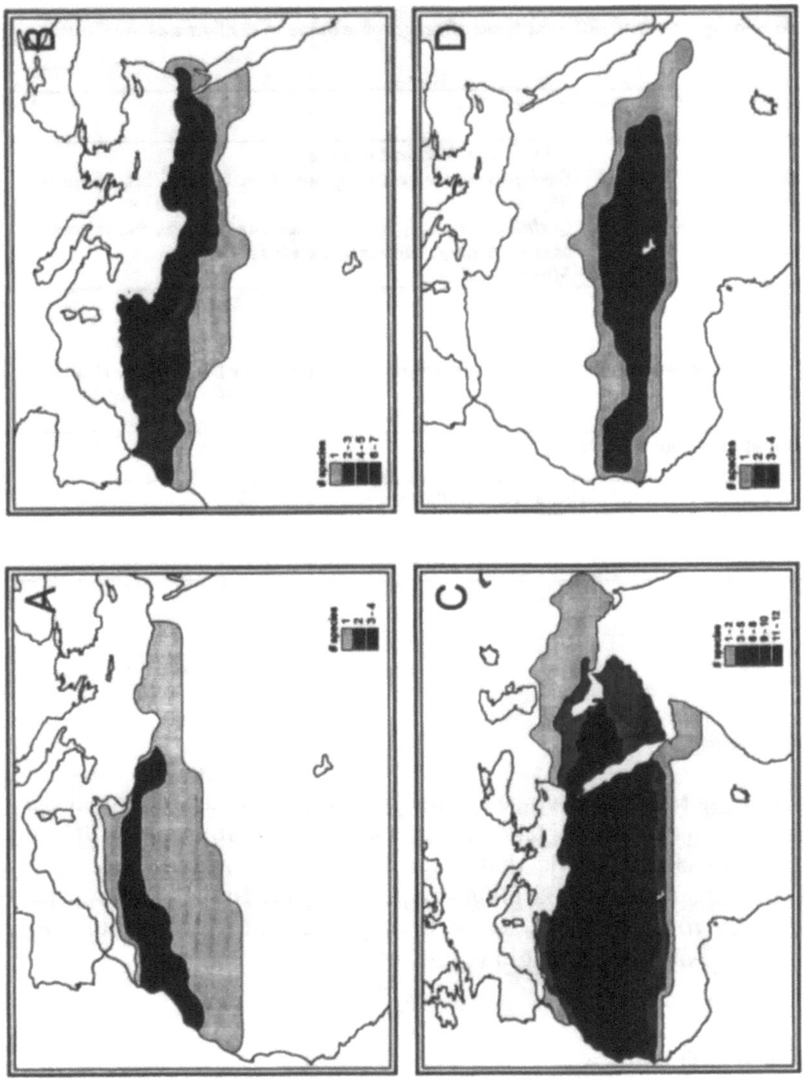

Fig. 3.4. Geographic distribution of the number of rodent species for Sub-Maghrebian (A), Libyan (B), Central (C) and South Saharan (D) types of geographic range

Table 3.5. Rodent species with different types of geographic range in the fauna of the South Saharan province

Type of geographic range	Species
Endemic	*Gerbillus agag, Monodia mauritania, Desmodilliscus braueri, Acomys cinereacus*
Central Saharan	*Gerbillus nanus, G. henleyi, G. pyramidum, G. gerbillus, Acomys cahirinus, Jaculus jaculus*
Sindian	*Mus musculus*
Sahelian	*Xerus erythropus, Hystrix cristata*

The South Sahara is the only Palaearctic province from which the ubiquitous *Mus musculus* is absent. The explanation may be that *Mus musculus* sometimes does not survive in most desert habitats, and finds refugia in relatively mesic intrazonal habitats. At the border of the Palaearctic and Palaeotropic regions, this species is replaced by tropical African *Mus* species (*Mus haussa, Mus triton, Mus tenellus, Mus mattheyi* etc.) which do not inhabit desert habitats, but nevertheless do not permit *Mus musculus* to occupy intrazonal habitats.

3.2.1.6
The North Arabian Province

The taxonomic composition of the fauna of North Arabia is characterized by the absolute dominance of gerbils and jirds (12 of 19 species). There are also three murine species (two *Acomys* and one *Mus*), two species of Dipodinae and one species each of Myoxidae and Hystricidae.

The provincial rodent fauna consists of one strict endemic and four subendemic species (Table 3.6). Other rodent species either have Central Saharan, South Arabian, or Libyan types of geographic range. In addition, there are three invaders from the distant provinces, namely the Iranian *Meriones erythrourus* and the Sindian *Mus musculus* and *Hystrix indica.*

3.2.1.7
The South Arabian Province

The fauna of the South Arabian province is similar to that of the North Arabian province in that gerbillines enjoy absolute dominance (eight out of 14 species). The dipodine, myoxid and hystricid species in South Arabia are almost the same as in North Arabia, except for the fact that the greater Egyptian jerboa *Jaculus orientalis* is absent.

One third of the species is endemic or subendemic. The rest of the species assemblage consists of widely-distributed species of Central Saharan, North Arabian and Sindian (*Mus musculus, Hystrix indica*) origin (Table 3.7).

Table 3.6. Rodent species with different types of geographic range in the fauna of the North Arabian province

Type of geographic range	Species
Endemic	*Meriones sacramenti*
Subendemic	*Gerbillus dasyurus, Sekeetamys calurus, Acomys russatus, Eliomys melanurus*
Libyan	*Gerbillus andersoni*
Central Saharan	*Gerbillus nanus, G. henleyi, G. gerbillus, G. pyramidum, Meriones crassus, Psammomys obesus, Acomys cahirinus, Jaculus jaculus*
South Arabian	*Gerbillus cheesmani, Jaculus orientalis*
Iranian	*Meriones erythrourus*
Sindian	*Mus musculus, Hystrix indica*

Table 3.7. Rodent species with different types of geographic range in the fauna of the South Arabian province

Type of geographic range	Species
Endemic	*Gerbillus famulus, G. poecilops, Merionex rex*
Subendemic	*Gerbillus cheesmani*
Central Saharan	*Gerbillus nanus, G. henleyi, Acomys cahirinus, Jaculus jaculus*
North Arabian	*Gerbillus dasyurus, Sekeetamys calurus, Acomys russatus, Eliomys melanurus*
Sindian	*Mus musculus, Hystrix indica*

3.2.1.8
The Syrian Province

The faunal composition of the Syrian province is determined by its intermediate position between the Iranian, Mesopotamian and South Arabian desert provinces on one hand and its peripheral location at the border between desert and mountain-steppe zones on the other. This province is not strictly desert but rather a mosaic of desert and patches of dry steppe. Consequently, a relatively large number of non-desert species occur here (for example, *Meriones tristrami, Meriones vinogradovi, Mus macedonicus, Spalax ehrenbergii*). These are not considered in this book. As a result, the desert fauna of the province appears diverse both taxonomically and by the geographic ranges of the various species. There are two species of Gerbillinae, two species of Cricetinae, and one species each of Murinae, Dipodinae, Allactaginae, Myoxidae and Hystricidae. From the areographic point of view, the fauna consists of species with Central Saharan, North Arabian, Iranian, Mesopotamian, Sindian or Balkhash origins (Table 3.8).

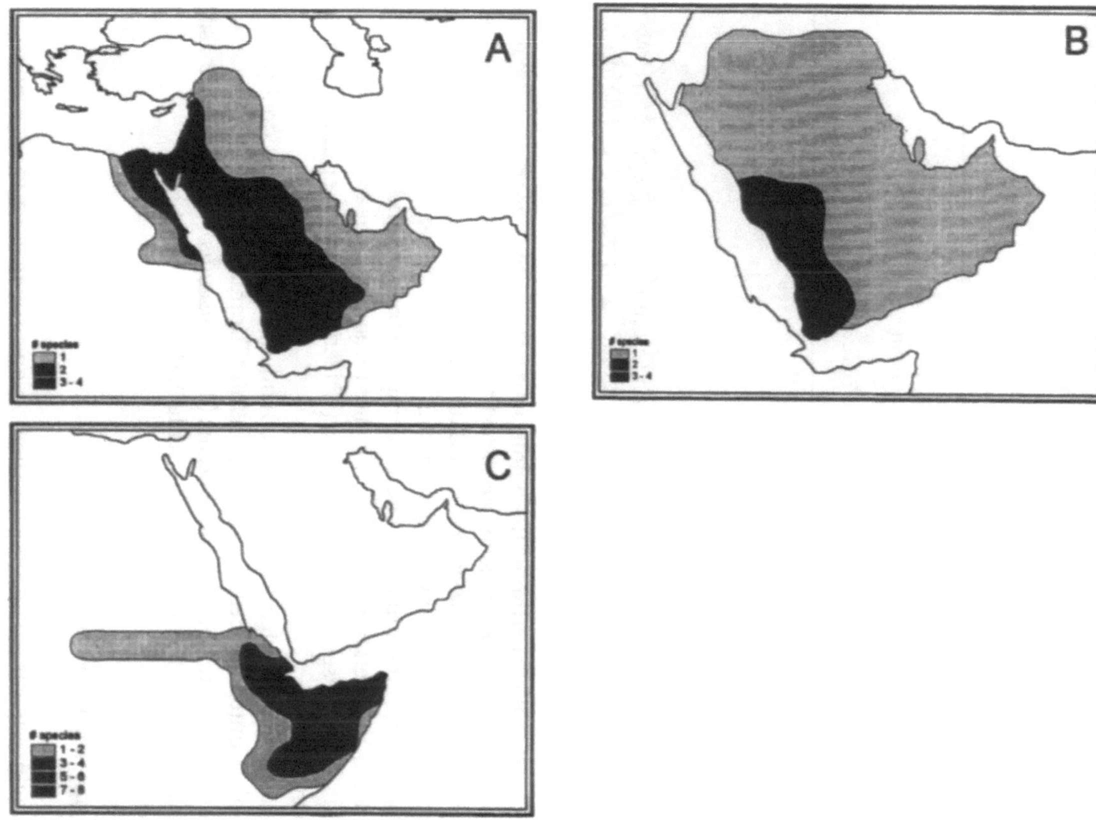

Fig. 3.5. Geographic distribution of the number of rodent species for North Arabian (A), South Arabian (B) and Somalian (C) types of geographic range

Table 3.8. Rodent species with different types of geographic range in the fauna of the Syrian province

Type of geographic range	Species
Central Saharan	*Jaculus jaculus*
North Arabian	*Gerbillus dasyurus, Eliomys melanurus*
Iranian	*Meriones erythrourus, Calomyscus bailwardi*
Mesopotamian	*Allactaga euphratica*
Sindian	*Mus musculus, Hystrix indica*
Balkhash	*Cricetulus migratorius*

3.2.1.9 The Iranian Province

Gerbillinae predominate in the Iranian fauna (eight out of 17 species). However, in contrast to the North African and Arabian provinces, jirds (*Meriones, Rhombomys, Tatera*) prevail over gerbils (*Gerbillus*; six versus two species). Furthermore, jerboas (Cardiocraniinae, Dipodinae, Allactaginae) also represent a considerable fraction of the whole assemblage (four species). This is a reflection of the general Palaearctic desert pattern of an increase in species richness of jerboas from west to east and the opposite trend in the gerbillines (see below). As in other desert provinces of the southwestern Palaearctic, the house mouse and the Indian porcupine comprise the remainder of the fauna. In addition, the neighborhood of the northeastern Palaearctic deserts is reflected by the presence of two hamsters (Cricetinae) and one ground squirrel (Marmotinae).

There are three endemic and three subendemic species. Additionally, one widely-distributed species of local origin, four Turanian species, two widely-distributed Central Saharan, one Sindian (*Mus musculus*) and one Balkhash species inhabit the Iranian province (Table 3.9).

3.2.1.10 The Mesopotamian Province

This province is similar to other Afro-Arabian provinces in that gerbillines are dominant (seven out of 13 species). Other species are presented by Murinae (two species), Cricetinae, Dipodinae, Allactaginae, Myoxidae and Hystricidae (one species in each taxon).

Only one endemic (*Gerbillus mesopotamiae*) and one subendemic (*Allactaga euphratica*) species are characteristic of the province. Other species have Central Saharan, Sindian, Iranian, South Arabian, North Arabian or Balkhash types of geographic ranges (Table 3.10).

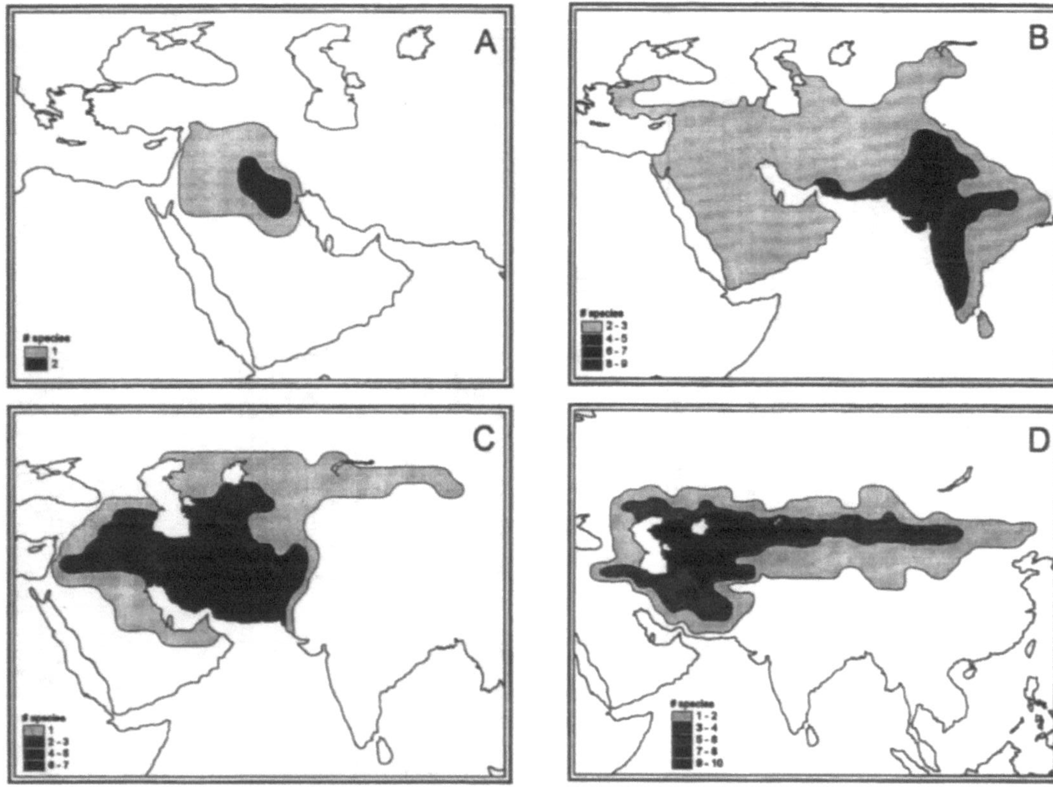

Fig. 3.6. Geographic distribution of the number of rodent species for Mesopotamian (A), Sindian (B), Iranian (C) and Turanian (D) types of geographic range. Sindian type of geographic range is shown without *Mus musculus* outside of the Arabo-Sindian geographic region

Table 3.9. Rodent species with different types of geographic range in the fauna of the Iranian province

Type of geographic range	Species
Endemic	*Gerbillus aquillus, Salpingotus michaelis, Allactaga hotsoni*
Subendemic	*Meriones persicus, Calomyscus bailwardi, Jaculus blanfordi*
Widely distributed species of a local origin	*Meriones erythrourus*
Central Saharan	*Gerbillus nanus, Meriones crassus*
Sindian	*Mus musculus*
Turanian	*Spermophilus fulvus, Meriones meridianus, Rhombomys opimus, Allactaga elater*
Balkhash	*Cricetulus migratorius*

Table 3.10. Rodent species with different types of geographic range in the fauna of the Mesopotamian province

Type of geographic range	Species
Endemic	*Gerbillus mesopotamiae*
Subendemic	*Allactaga euphratica*
Central Saharan	*Gerbillus nanus, Meriones crassus, Acomys cahirinus, Jaculus jaculus*
North Arabian	*Eliomys melanurus*
South Arabian	*Gerbillus cheesmani*
Iranian	*Meriones persicus, M. erythrourus*
Sindian	*Tatera indica, Mus musculus, Hystrix indica*
Balkhash	*Cricetulus migratorius*

3.2.1.11 The South Pakistan Province

Taxonomically, the faunal composition of this province is almost equally divided between four gerbilline and five murine species. The only representative of other taxa is *Hystrix indica*. The absence of jerboas is a noteworthy indication of the relatedness of this province to the Sind. The transitional position of the province is reflected by the absence of endemic species.

Most of this fauna's members are of Sindian origin. Two species invaded from the Central Sahara and one from Iran (Table 3.11).

Table 3.11. Rodent species with different types of geographic range in the fauna of the South Pakistan province

Type of geographic range	Species
Central Saharan	*Gerbillus nanus, Acomys cahirinus*
Iranian	*Meriones erythrourus*
Sindian	*Tatera indica, Meriones hurrianae, Mus musculus, M. booduga, M. saxicola, Millardia gleadowi, Hystrix indica*

Table 3.12. Rodent species with different types of geographic range in the fauna of the Sindian province

Type of geographic range	Species
Endemic	*Gerbillus gleadowi, Cremnomys cutchicus*
Subendemic	*Meriones hurrianae, Mus booduga, M. saxicola, Millardia gleadowi*
Widely distributed species of a local origin	*Tatera indica, Mus musculus, Hystrix indica*
Central Saharan	*Gerbillus nanus*

3.2.1.12
The Sindian Province

The taxonomic composition of the Sindian fauna, at the family/subfamily level, is the same as that of the South Pakistan province. The differences between these two regions are that two species, *Meriones erythrourus* and *Acomys cahirinus*, do not penetrate the Sind while the two other species, *Gerbillus gleadowi* and *Cremnomys cutchicus*, do not occur outside the Sind.

Almost all species are of a local origin and are either endemic or subendemic, or widely distributed (Table 3.12). The only species of non-Sindian origin is the Central Saharan *Gerbillus nanus*. However, it is possible that the local form, *G. n. indus*, may have specific rather than subspecific status.

3.2.1.13
The Turanian Province

The taxonomic composition of the Turanian fauna of desert rodents differs drastically from that of the Afro-Arabian assemblages. Half the species (ten out of 20) are jerboas belonging to four subfamilies (Cardiocraniinae, Dipodinae, Paradipodinae and Allactaginae), while the number of gerbilline species is relatively small. There are only three species of jirds (*Meriones meridianus, M. erythrourus* and *Rhombomys opimus*), while gerbils are completely absent. Other species belong to Xerinae, Marmotinae, Murinae, Cricetinae

Table 3.13. Rodent species with different types of geographic range in the fauna of the Turanian province

Type of geographic range	Species
Endemic	*Salpingotus heptneri, Paradipus ctenodactylus, Allactaga bobrinskii*
Subendemic	*Spermophilopsis leptodactylus, Spermophilus fulvus, Eremodipus lichtensteinii, Allactaga severtzovi, A. vinogradovi*
Widely distributed species of a local origin	*Meriones meridianus, Rhombomys opimus*
Iranian	*Meriones erythrourus, Jaculus blanfordi*
Sindian	*Mus musculus, Hystrix indica*
Central Kazakhstan	*Ellobius talpinus*
Balkhash	*Cricetulus migratorius, Ellobius tancrei, Pygeretmus pumilio*
Gobian	*Dipus sagitta*

(one species in each subfamily), Arvicolinae (two species) and Hystricidae (one species).

Endemic species are found among jerboas only, whereas the subendemic category is divided between ground squirrels and jerboas. Some species of local origin have a wide distribution, penetrating adjacent regions to the south and northwest and distant areas to the east of the region. Two species invaded Turan from Iran, whereas the Balkhash species are represented by three and Central Kazakhstan and Gobian species by one species each (Table 3.13). In addition, as in most other Palaearctic deserts, there are two widely-distributed Sindian rodents (*Mus musculus* and *Hystrix indica*).

3.2.1.14
The Central Kazakhstan Province

Eleven of the province's fauna of 22 species are jerboas belonging to three subfamilies (Cardiocraniinae, Dipodinae, Allactaginae): five-toed jerboas are the most diverse (seven species). Other representative taxonomic groups are Xerinae (one species), Marmotinae (two species), Gerbillinae (four species), Murinae (one species), Cricetinae (two species) and Arvicolinae (one species). The only endemic species is *Salpingotus pallidus*, but there are five subendemic species. The largest section of the fauna consists of species with the Turanian and Balkhash types of geographic ranges (Table 3.14). There are also one Iranian, two Gobian and one Sindian species.

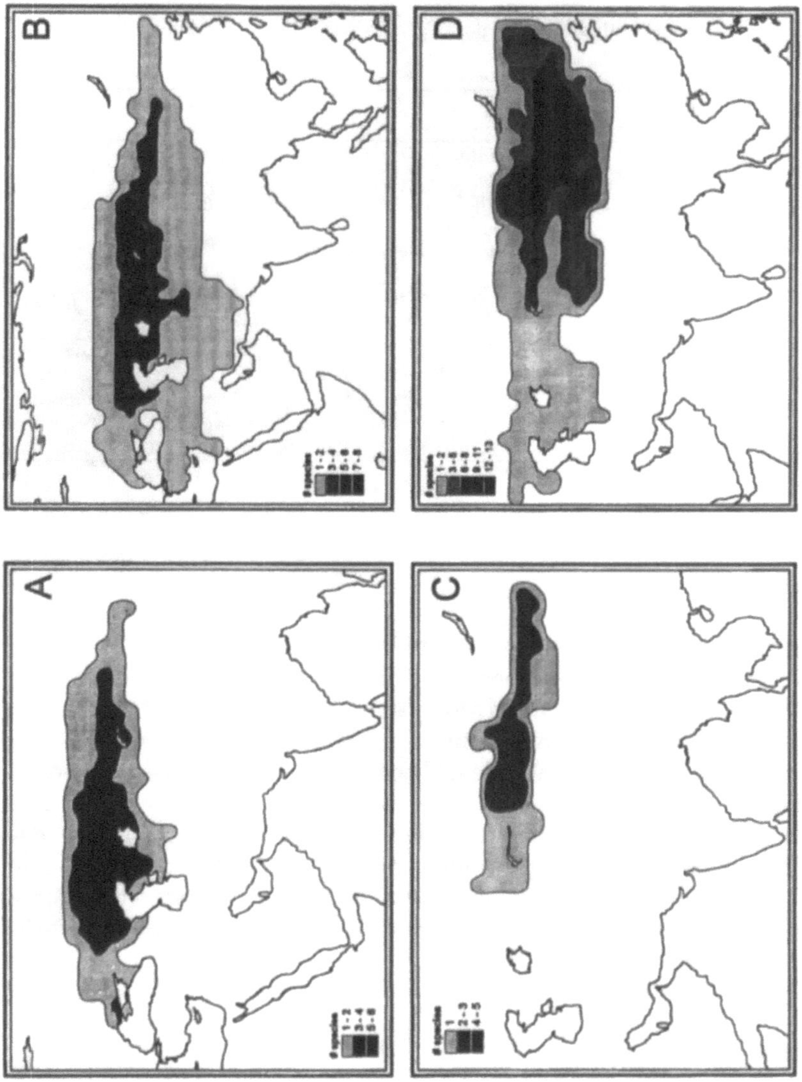

Fig. 3.7. Geographic distribution of the number of rodent species for Central Kazakhstan (A), Balkhash (B), Sungorian (C) and Gobian (D) types of geographic range

Table 3.14. Rodent species with different types of geographic range in the fauna of the Central Kazakhstan province

Type of geographic range	Species
Endemic	*Salpingotus pallidus*
Subendemic	*Spermophilus pygmaeus, Meriones tamariscinus, Ellobius talpinus, Stylodipus telum, Pygeretmus platyurus*
Iranian	*Meriones erythrourus*
Sindian	*Mus musculus*
Turanian	*Spermophilopsis leptodactylus, Spermophilus fulvus, Meriones meridianus, Rhombomys opimus, Eremodipus lichtensteinii, Allactaga severtzovi, A. vinogradovi, A. elater*
Balkhash	*Cricetulus migratorius, Allocricetulus eversmanni, Allactaga major, Pygeretmus pumilio*
Gobian	*Dipus sagitta, Allactaga sibirica*

3.2.1.15
The Balkhash Province

Taxonomically, more than half of the fauna of the Balkhash province (nine out of 17 species) is composed of jerboas (the same three subfamilies as those of the Central Kazakhstan). Other rodents consist of one jird (Gerbillinae), one ground squirrel (Marmotinae), one mouse (Murinae), two hamsters (Cricetinae), one mole-vole and one vole (Arvicolinae) and one desert dormouse (Seleviniinae).

Three species are strictly endemic. There are no subendemics. Other species with a Balkhash type of geographic range are widely distributed. The Gobian and Central Kazakhstan geographic ranges are represented by three species each. There are also some species of Turanian, Sungorian and Sindian origin (Table 3.15).

3.2.1.16
The Sungorian Province

The taxonomic composition of the Sungorian fauna can be characterized by its intermediate position between the Kazakhstan superprovince with an absolute dominance of dipodids and the Gobian province which is also dominated by dipodids with a subdominance of hamsters. Moreover, the neighborhood of Turan with its dominance by jerboas and subdominance by jirds has an influence. Seven of 16 rodents of the Sungorian fauna are jerboas (Cardiocraniinae, Dipodinae, Allactaginae), four species are jirds (Gerbillinae) and three species are hamsters (Cricetinae). The rest of the fauna consists of one

Table 3.15. Rodent species with different types of geographic range in the fauna of the Balkhash province

Type of geographic range	Species
Endemic	*Alticola streltzowi, Selevinia betpakdalensis, Pygeretmus zhitkovi*
Widely distributed species of a local origin	*Allocricetulus eversmanni, Cricetulus migratorius, Ellobius tankrei, Allactaga major, Pygeretmus pumilio*
Sindian	*Mus musculus*
Turanian	*Allactaga elater*
Central Kazakhstan	*Meriones tamariscinus, Stylodipus telum, Pygeretmus platyurus*
Sungorian	*Spermophilus erythrogenis*
Gobian	*Cardiocranius paradoxus, Dipus sagitta, Allactaga sibirica*

Table 3.16. Rodent species with different types of geographic range in the fauna of the Sungorian province

Type of geographic range	Species
Endemic	*Eolagurus luteus, Stylodipus sungorus*
Subendemic	*Spermophilus erythrogenis, Salpingotus crassicauda*
Iranian	*Meriones erythrourus*
Sindian	*Mus musculus*
Turanian	*Meriones meridianus, Rhombomys opimus, Allactaga elater*
Central Kazakhstan	*Meriones tamariscinus*
Balkhash	*Cricetulus migratorius, Ellobius tancrei, Pygeretmus pumilio*
Gobian	*Allocricetulus curtatus, Phodopus roborovskii, Cardiocranius paradoxus, Dipus sagitta, Allactaga sibirica*

species of Marmotinae, one species of Murinae and two species of Arvicolinae.

The only strict endemics of the Sungorian fauna are *Eolagurus luteus* and *Stylodipus sungorus*, although the former has become endemic in the Sungoria only in the last century. In the 1940s it was widely distributed to the west (as far as Central Kazakhstan) whereas, in the nineteenth century, its geographic range continued to the Volga River (Formozov 1938; Dmitriev 1981; Kiseleva 1982). At the present time this species has become extinct over a large part of its former geographic range.

The Sungorian subendemics are represented by *Spermophilus erythrogenis* and *Salpingotus crassicauda*. The largest proportion of the province's fauna is represented by species of Gobian origin. Species with Balkhash, Central Kazakhstan, Turanian, Iranian and Sindian types of geographic ranges also occur in the Sungorian fauna (Table 3.16).

Table 3.17. Rodent species with different types of geographic range in the fauna of the Gobian province

Type of geographic range	Species
Endemic	*Spermophilus alashanicus, Alticola barakschin, Stilodypus andrewsii, Allactaga balicunica, A. bullata*
Subendemic	*Cricetulus barabensis, C. longicaudatus, C. curtatus, Phodopus cambelli, Eolagurus przewalskii, Euchoreutes naso, Salpingotus kozlovi*
Widely distributed species of a local origin	*Phodopus roborovskii, Cardiocranius paradoxus, Dipus sagitta, Allactaga sibirica*
Sindian	*Mus musculus*
Turanian	*Meriones meridianus, Rhombomys opimus*
Balkhash	*Cricetulus migratorius, Ellobius tancrei, Pygeretmus pumilio*
Sungorian	*Spermophilus erythrogenis, Salpingotus crassicauda*

3.2.1.17
The Gobian Province

The Gobian fauna is one of the richest among those of the Palaearctic deserts (24 species). Two main Gobian groups are jerboas (one species of Euchoreutinae, three of Cardiocraniinae, two of Dipodinae and four of Allactaginae) and hamsters (six species of Cricetinae). Other taxa are represented by Marmotinae, Gerbillinae (two species of each subfamily), Arvicolinae (three species) and Murinae (one species).

There are five endemic and seven subendemic species in the Gobian province. Some species of Gobian origin are widely distributed. Others invaded the province from Sungoria, the Balkhash province and from Turan (Table 3.17). The Sindian *Mus musculus* also penetrated the Gobian province as in the case of all other Palaearctic provinces.

3.2.1.18
The Kashgharian Province

The Kashgarian province is a marginal portion of the Turano-Gobian subrealm. It is surrounded by the high mountains of Tien-Shan and Tibet and is connected with the Gobian province by a relatively narrow corridor consisting of the low Beishan mountains. Consequently, the Kashgarian fauna is derived from the Gobian fauna and contains no endemic species. The fauna is only half as rich as the Gobian fauna (nine species). There are three species of jerboas (one species of each of Euchoreutinae, Cardiocraniinae and Dipodinae), three of hamsters (Cricetinae), one jird (Gerbillinae), one mouse (Murinae) and one species of vole (Arvicolinae).

Table 3.18. Rodent species with different types of geographic range in the fauna of the Kashgharian province

Type of geographic range	Species
Sindian	*Mus musculus*
Turanian	*Meriones meridianus*
Balkhash	*Cricetulus migratorius*
Gobian	*Cricetulus longicaudatus, Phodopus roborovskii, Eolagurus przewalskii, Euchoreutes naso, Salpingotus kozlovi, Dipus sagitta*

Table 3.19. Rodent species with different types of geographic range in the fauna of the Tzaidam province

Type of geographic range	Species
Sindian	*Mus musculus*
Turanian	*Meriones meridianus, Rhombomys opimus*
Balkhash	*Cricetulus migratorius*
Gobian	*Cricetulus longicaudatus, Phodopus roborovskii, Eolagurus przewalskii, Euchoreutes naso, Cardiocranius paradoxus, Dipus sagitta, Allactaga sibirica*

Most species have a Gobian type of geographic range. The Turanian, Balkhash and Sindian types of geographic range are represented by one species each (Table 3.18).

3.2.1.19
The Tzaidam Province

This province is very similar to the Kashgharian province in its geographic position. It is likewise an intermountain depression (with an area much less than that of Kashgharia) and is connected with the Gobian province by a narrow passageway through piedmonts of the Nan-Shan and Altyn-Tag ridges. The species assemblage can be regarded as a depauperated Gobian fauna with four species of jerboas (Euchoreutinae, Cardiocraniinae, Dipodinae, Allactaginae), three of hamsters (Cricetinae), one jird (Gerbillinae), one mouse (Murinae) and one species of vole (Arvicolinae).

There are no endemic or subendemic species in the province. The largest part of the fauna has a Gobian origin. Two species have a Turanian origin while the Balkhash and the Sindian types of geographic range are represented by the same species as in Kashgharia (Table 3.19).

Table 3.20. Rodent species with different types of geographic range in the fauna of the Somalian desert province

Type of geographic range	Species
Endemic	*Ammodillus imbellis, Gerbillus watersi, G. pusillus, Microdillus peeli, Acomys louisae, Pectinator speeki*
Subendemic	*Gerbillus rosalinda*
Central Saharan	*Gerbillus nanus*
Sahelian	*Xerus erythropus, Hystrix cristata*

3.2.2 The African Horn Realm

3.2.2.1 The Somalian Desert Province

Traditionally, the fauna of the Horn of Africa has been considered to be a part of the Palaeotropic species complex (e.g. Darlington 1957). However, our analysis demonstrates that the desert rodent fauna of the Horn of Africa has strong Palaearctic connections and should be classified as a realm within the Palaearctic region.

The fauna is similar to those of the Saharan provinces in the absolute dominance of Gerbillinae (six out of ten species) and in the presence of a ctenodactylid component (one species), but differs in the absence of jerboas. The rest of the fauna consists of one species each of Xerinae, Murinae and Hystricidae.

The almost complete isolation of the Somalian desert province from the rest of the Palaearctic (excluding an extremely narrow strip along the Red Sea coast) leads to its faunal specificity. The majority of species are endemic or subendemic (Table 3.20). The Central Saharan *Gerbillus nanus* has invaded the province from the north and two Sahelian species, *Xerus erythropus* and *Hystrix cristata*, from the south.

3.2.3 The South African Desert Realm

The South African desert realm is isolated from all other desert realms, although it is closer to the Palaearctic than to any other region. There are three rodent genera (*Xerus*, *Mus*, *Hystrix*) as well as five subfamilies and families (Xerinae, Gerbillinae, Murinae, Myoxidae and Hystricidae) in common with the Palaearctic region. This realm has specific taxa that do not occur in any other desert (Pedetidae, Bathyergidae, Petromuridae, Petromyscinae, Otomyinae, Dendromurinae). *Tatera* also seems to be a genus that is common to the Palaearctic and the Palaeotropical regions. However, the most recent

Table 3.21. Rodent species with different types of geographic range in the fauna of the Namibian desert province

Type of geographic range	Species
Endemic	*Xerus princeps, Gerbillurus tytonis, G. setzeri, Petromyscus collinus, Petromus typicus, Bathyiergus janetta*
Subendemic	*Gerbillurus vallinus, Parotomys brantsii, P. littledalei, Graphiurus ocularis*
Kalahari	*Xerus inaurus, Gerbillurus paeba, Tatera brantsii, Desmodillus auricularis, Malacothrix typica, Aethomys chrysophylus, A. namaquensis, Rhabdomys pumilio, Pedetes capensis, Hystrix africaeaustralis, Cryptomys hottentotus*

revision of the taxon indicates that the Asian and the African species of *Tatera*, sensu lato, belong to different genera (Asian *Tatera*, sensu stricto, and African *Gerbilliscus*, Pavlinov et al. 1990).

The South African desert realm is weakly differentiated. Only two biogeographic provinces are usually distinguished (Rautenbach 1978); the results of our analysis based on rodent distribution support this classification.

3.2.3.1
The Namibian Province

The taxonomic composition of the province is very diverse. There are 21 representatives of 11 families and subfamilies in the Namibian fauna. The richest taxon is Gerbillinae (with six species), whereas other taxa are represented by one to three species (Xerinae, Dendromurinae, Pertomyscinae, Otomyinae, Myoxidae, Pedetidae, Hystricidae, Petromuridae, Bathyergidae). Some of these taxa consist mainly of tropical forest or savanna species with only a few representatives penetrating arid regions. This is especially true of the Dendromurinae and Otomyinae.

It is possible to distinguish three groups of species by their types of geographic range. The largest group is represented by species with a Kalahari type of geographic range that enter the Namibian province. Six species are Namibian endemics and four are Namibian subendemics that also occur in the Kalahari province (Table 3.21).

3.2.3.2
The Kalahari Province

The taxonomic composition of the Kalahari province is almost as diverse as that of the Namibian province. The difference between the two provinces lies: (1) in the absence of petromurid and petromurine species, and (2) in an increase in murines in the Kalahari (five out of 17 species). The absolute

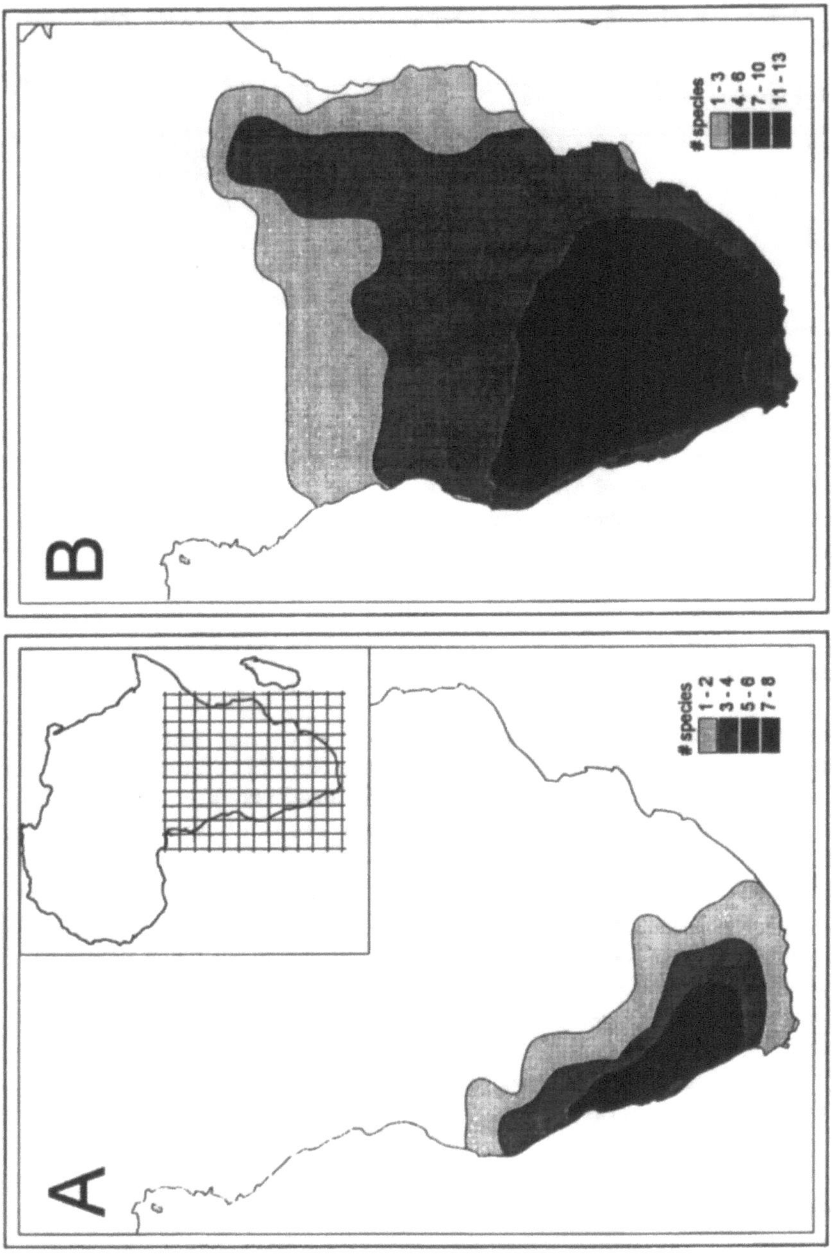

Fig. 3.8. Geographic distribution of the number of rodent species for Namibian (A) and Kalahari (B) types of geographic range

Table 3.22. Rodent species with different types of geographic range in the fauna of the Kalahari province

Type of geographic range	Species
Endemic	*Zelotomys woosnami, Mus minutoides*
Subendemic	*Xerus inaurus, Gerbillurus paeba, Tatera brantsii, Desmodillus auricularis, Malacothrix typica, Aethomys chrysophylus, A. namaquensis, Rhabdomys pumilio, Pedetes capensis, Hystrix africaeaustralis, Cryptomys hottentotus*
Namibian	*Gerbillurus vallinus, Parotomys brantsii, P. littledalei, Graphiurus ocularis*

number of gerbilline species in the Kalaharian province is lower than that in the Namib (four versus six species, respectively) although the fraction of this taxon is almost the same in both faunas (23% versus 28%, respectively). In addition, the Kalahari fauna consists of two species of Otomyinae and one species each of Xerinae, Dendromurinae, Myoxidae, Hystricidae, Pedetidae and Bathyergidae.

Thirteen of 17 species of the fauna have the Kalahari type of geographic range. Of these, 11 species can be characterized as subendemics while only two are endemic (Table 3.22). Some of the Kalahari subendemics have a broad distribution across the dry savannas of southern and eastern Africa (up to Uganda and Tanzania). This seems to reflect the ancient connections between the faunas of the southern and northern African deserts. In addition, four subendemic species of the Namibian fauna also occur in the Kalahari province (Table 3.22).

3.2.4 The Australian Desert Realm

3.2.4.1 The Australian Desert Province

The rodent fauna of the Australian deserts differs sharply from those of all other desert regions. There is only a single species (*Mus musculus*) and a single genus (*Rattus*) in common with other parts of the Old World. *Mus musculus* was introduced to Australia relatively recently, and should not be considered in biogeographic analyses. However, in contrast to its status in the South African and American deserts, it has become one of the most widespread desert species in Australia, occupies an important place in desert rodent communities and should be taken into account when local patterns of distribution are examined (see Morton and Baynes 1985). In addition, *Mus musculus* apparently replaced several native species which have become extinct since settlement by Europeans two centuries ago. Some other desert

Table 3.23. Rodent species with different types of geographic range in the fauna of the Australian desert province

Type of geographic range	Species
"Old" endemic	*Zyzomys argurus, Z. pedunculatus, Pseudomys australus, P. desertor, P. hermannsburgensis, P. chapmani, Leggadina forresti, Notomys alexis, N. fuscus, N. cervinus*
"New" endemic	*Rattus velosissimus*
Sindian	*Mus musculus*

species are not yet extinct but have declined drastically in their range (Morton and Bayens 1985; Morton 1993).

Other Australian desert rodents have also originated elsewhere. There have been at least two waves of prehistoric migration of rodents into Australia (Strahan 1983). Approximately 5-10 million years ago some murine rodents managed to enter Australia from the region of New Guinea and launched a further radiation forming two recent tribes, Hydromyini and Conilurini. This radiation resulted in a set of species that are called "old endemics". Later, no more than a million years ago, some *Rattus* species entered Australia and evolved into species called "new endemics".

Thus, the taxonomic composition of the rodent fauna of the Australian deserts is uniform. All species belong to Murinae, although most of them (excluding *Rattus velosissimus* and *Mus musculus*) represent the tribe Conilurini. The fauna is also uniform from the areographic point of view. Except for the new immigrant, *Mus musculus*, all the others are either "old" or "new" Australian endemics (Table 3.23). The internal subdivision of Australian endemics by type of geographic range is now impossible because the previous distribution of most species is weakly documented and the recent extinctions have resulted in the non-equilibrium state of the modern fauna.

3.2.5
The North American Desert Realm

The desert fauna of North America (the southwest of USA and northern Mexico) is well studied (see review in Hall 1981). After Hagmeier and Stults (1964) and Hagmeier (1966), there is a common opinion that each of the North American deserts has a high biogeographic rank ("superprovince"). According to our analysis, the biogeographic rank of the strictly desert faunas (without the species assemblages of surrounding mountains) is lower than this. These faunas are relatively similar and attain no more than provincial rank.

Table 3.24. Rodent species with different types of geographic range in the fauna of the Great Basin province

Type of geographic range	Species
Endemic	*Spermophilus townsendi, Microdipodops megacephalus*
Subendemic	*Perognathus parvus, Dipodomys ordii, D. microps, Reithrodontomys megalotis, Peromyscus crinitus, Onychomys leucogaster*
Sonoran	*Ammospermophilus leucurus, Thomomys umbrinus, Perognathus longimembris, P. formosus, Microdipodops pallidus, Dipodomys merriami, D. deserti, Neotoma lepida*
Chihuahuan	*Spermophilus variegatus, Peromyscus boylii, Onychomys torridus, Neotoma mexicana*

3.2.5.1
The Great Basin Province

Four taxa compose the fauna of the Great Basin province. The majority of species belong to Heteromyidae and Sigmodontinae (nine and seven out of 20 species, respectively). Two other taxa are Marmotinae (three species) and Geomyidae (one species).

Species with Great Basin and Sonoran geographic ranges are equally represented in the fauna. Those of the Great Basin type are mostly subendemics; there are only two Great Basin endemics. Eight Sonoran species have entered the Great Basin province. The rest of the fauna has a Chihuahuan origin (Table 3.24).

3.2.5.2
The Sonoran Province

The Sonoran fauna is the richest of all the North American desert provinces, although proportional taxonomic composition is the same as that in the Great Basin province. Heteromyidae and Sigmodontinae compose 76% of the fauna (15 and 11 of 34 species, respectively). Marmotinae take third place (seven species), while Geomyidae are represented by a single species only.

The geographic position of the Sonoran province between the Great Basin and Chihuahuan provinces determines the mixed composition of its fauna according to the type of geographic range. Only six species are endemics. Other species with the Sonoran type of geographic range also occur either in the Great Basin desert (five species) or in the Chihuahuan desert (two species). Two Sonoran species, *Thomomys umbrinus* and *Dipodomys merriami*, are widely distributed and inhabit all the deserts of North America. In addition, there are six Great Basin subendemics (see above) and also rodents with a Chihuahuan type of geographic range (Table 3.25).

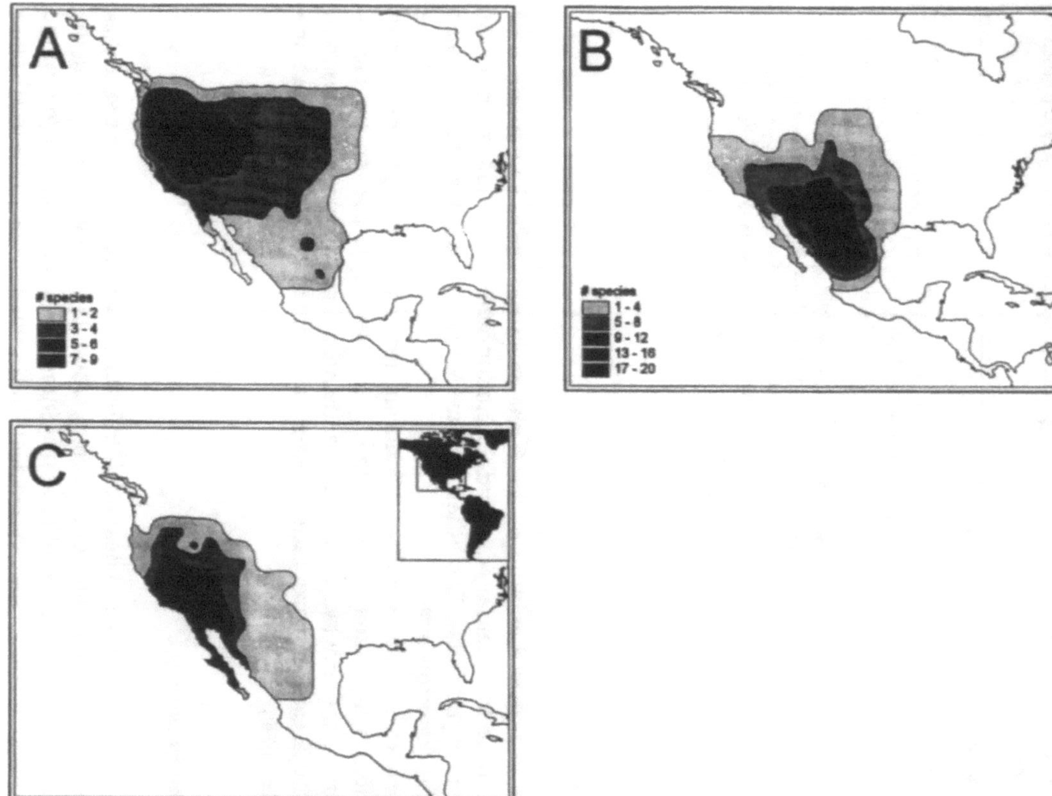

Fig. 3.9. Geographic distribution of the number of rodent species for Great Basin (A), Chihuahuan (B) and Sonoran (C) types of geographic range

Table 3.25. Rodent species with different types of geographic range in the fauna of the Sonoran province

Type of geographic range	Species
Endemic	*Ammospermophilus nelsoni, Spermophilus mohavensis, S. tereticaudus, Perognathus amplus, Chaetodipus fallax, Peromyscus merriami*
Subendemic	*Ammospermophilus leucurus, A. harrisii, Thomomys umbrinus, Perognathus longimembris, P. formosus, Chaetodipus baileyi, Microdipodops pallidus, Dipodomys merriami, D. deserti, Neotoma lepida*
Widely distributed species of a local origin	*Thomomys umbrinus, Dipodomys merriami*
Great Basin	*Perognathus parvus, Dipodomys ordii, D. microps, Reithrodontomys megalotis, Peromyscus crinitus, Onychomys leucogaster*
Chihuahuan	*Spermophilus spilosoma, S. variegatus, Perognathus flavus, Chaetodipus penicillatus, C. intermedius, Dipodomys spectabilis, Reithrodontomys fulvescens, Peromyscus eremicus, P. boylii, Onychomys torridus, Neotoma albigula, N. mexicana*

3.2.5.3
The Chihuahuan Province

This fauna is only slightly less rich than that of the Sonora (29 species) and is composed of the same four taxa as are other North American desert faunas with 75% Heteromyidae and Sigmodontinae (10 and 12 species, respectively). Five marmotine and two geomyid species supplement the species list.

A major part of the fauna is represented by species with a Chihuahuan type of geographic range, seven of which are Chihuahuan endemics. The Chihuahuan subendemics also occur either in the Sonoran (see above) or across all the North American desert provinces. The Great Basin and Sonoran species (see above) comprise the remainder of the Chihuahuan fauna (Table 3.26).

3.2.6
The Neotropic Desert Realms

The rodent fauna of the South American deserts is relatively poorly known. However, the general situation regarding the Neotropic rodents has begun to change drastically during last 10-15 years due to intensive studies by mammalogists from both North and South America. For example, a new genus, *Salinomys* with a new species *S. delicatus*, and a new species (*Andalgalomys roigi*) of phyllotine rodents from the Monte desert of Argentina were described recently (Braun and Mares 1995; Mares and Braun 1996). Nevertheless, our knowledge of the taxonomy (especially of the genera *Phyllotis* and

Table 3.26. Rodent species with different types of geographic range in the fauna of the Chihuahuan province

Type of geographic range	Species
Endemic	*Ammospermophilus interpres, Chaetodipus nelsoni, Dipodomys nelsoni, Peromyscus pectoralis, Sigmodon fulviventer, S. ochrognathus, Neotoma goldmani*
Subendemic	*Spermophilus spilosoma, Perognathus flavus, Chaetodipus penicillatus, C. intermedius, Dipodomys spectabilis, Reithrodontomys fulvescens, Peromyscus eremicus, Neotoma albigula*
Widely distributed species of a local origin	*Spermophilus variegatus, Peromyscus boylii, Onychomys torridus, Neotoma mexicana*
Great Basin	*Dipodomys ordii, Reithrodontomys megalotis, Onychomys leucogaster*
Sonoran	*Ammospermophilus harrisii, Chaetodipus baileyi, Dipodomys merriami, Thomomys umbrinus*

Eligmodontia, e.g. Hillyard et al. 1997) and distribution of the Neotropic rodents is far from perfect, so the results of our analysis of this fauna should be regarded as only preliminary.

The faunal composition of the Pacific and Argentinian lowland deserts is quite different. Our analysis has demonstrated that the Neotropical desert faunas belong to two different biogeographic units of relatively high rank ("realms"). The Atacama realm includes only a single province, while the Monte-Patagonian realm consists of two provinces, namely the Monte and Patagonian.

3.2.6.1 The Atacama Province

The desert rodent fauna of the Atacama province is one of the poorest of all desert faunas. It consists only of eight species, half of which belong to Sigmodontinae. There is also one chinchillid, two octodontid and one abrocomid species. The northern regions of the province are even poorer. The central and northern parts of the Peruvian Pacific coast are inhabited only by a single rodent species each (*Phyllotis amicus* and *Phyllotis gerbillus*, respectively).

Three quarters of the species of the provincial fauna have the Atacama type of geographical range, most of them being strictly endemic (Table 3.27). Only one species with this type of range (*Phyllotis darwini*) is subendemic. It is also found in the Monte province. In addition, two Patagonian species also occur in the Atacama fauna.

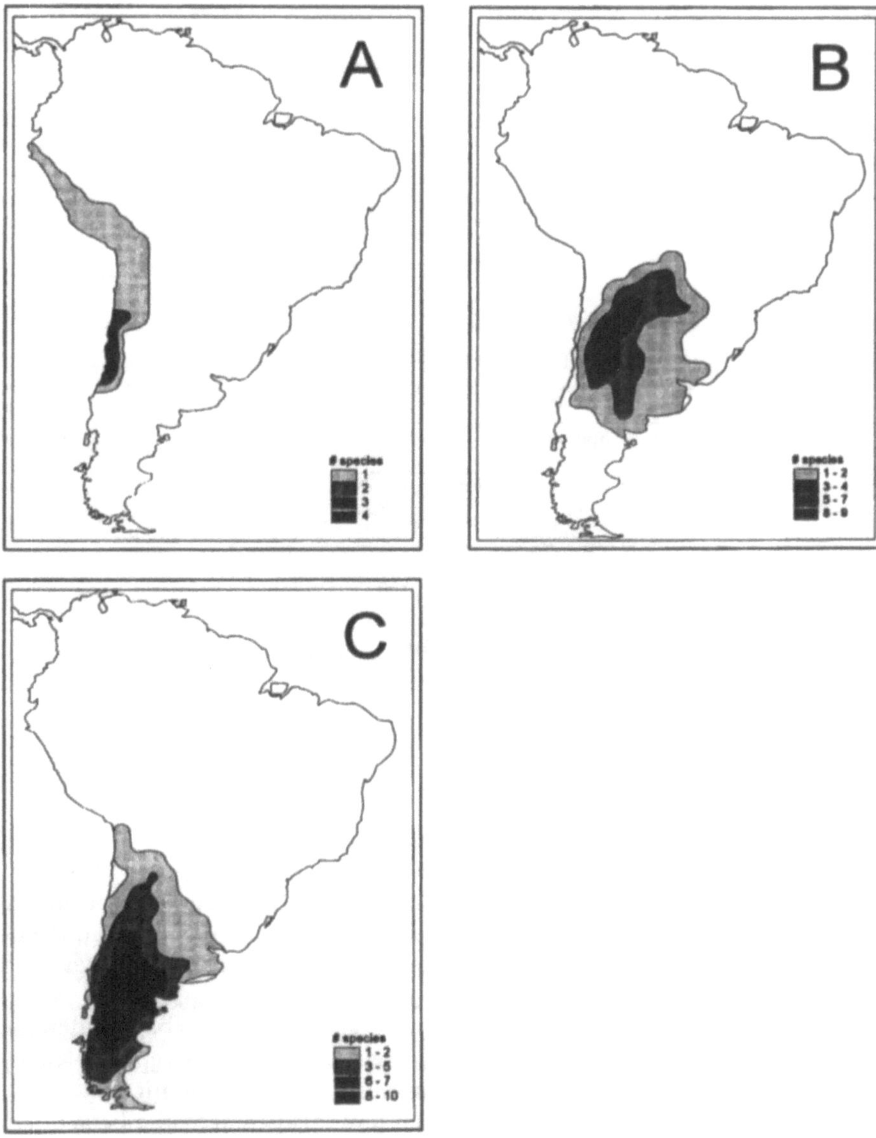

Fig. 3.10. Geographic distribution of the number of rodent species for Atacama (A), Monte (B) and Patagonian (C) types of geographic range

Table 3.27. Rodent species with different types of geographic range in the fauna of the Atacama province

Type of geographic range	Species
Endemic	*Phyllotis amicus, P. gerbillus, Octodon degus, Octodon lunatus, Abrocoma bennetti*
Subendemic	*Phyllotis darwini*
Patagonian	*Phyllotis xanthopygus, Lagidium viscacia*

Table 3.28. Rodent species with different types of geographic range in the fauna of the Monte province

Type of geographic range	Species
Endemic	*Andalgalomys olrogi, Graomys edithae, Octomys mimax, Ctenomys latro, C. occultus, C. saltarius*
Subendemic	*Andalgalomys roigi, Calomys callosus, C. musculinus, Eligmodontia moreni, Graomys griseoflavus, Salinomys delicatus, Lagostomus maximus, Dolichotis salinicola*
Atacama	*Phyllotis darwini*
Patagonian	*Phyllotis xanthopygus, Reithrodon physodes, Acodon molinae, Lagidium viscacia, Dolichotis patagonum, Microcavia australus, Ctenomys mendocinus*

3.2.6.2
The Monte Province

According to our scheme, the Monte province occupies the northern half of the Monte desert. The fauna consists of 22 species belonging to five taxa. The most numerous is Sigmodontinae (12 species), while the rest are Chinchillidae (two species), Caviidae (three species), Octodontidae (one species) and Ctenomyidae (four species).

All three Neotropical desert types of geographic range can be found in the fauna (Table 3.28). The Monte type of geographic range is represented by more than 50% of species. Six Monte species are strict endemics, while the other Monte species are subendemics that have also entered the Patagonian province, but not the Atacama province. *Phyllotis darwini* entered the Monte province from the Atacama desert, while the Patagonian fauna provides the Monte with seven species (Table 3.28).

3.2.6.3
The Patagonian Province

Geographically, this province covers the southern portion of the Monte desert and the entire Patagonian arid region. The taxonomic composition of

Table 3.29. Rodent species with different types of geographic range in the fauna of the Patagonian province

Type of geographic range	Species
Endemic	*Eligmodontia morgani, Akodon iniscatus, Tympanoctomys barrerae, Ctenomys emilianus, C. sericeus*
Subendemic	*Phyllotis xanthopygus, Reithrodon physodes, Acodon molinae, Lagidium viscacia, Dolichotis patagonum, Microcavia australus, Ctenomys mendocinus*
Monte	*Andalgalomys roigi, Calomys callosus, C. musculinus, Eligmodontia moreni, Graomys griseoflavus, Salinomys delicatus, Lagostomus maximus, Dolichotis salinicola*

the fauna is the same as of that of the Monte province. The only difference is that both Sigmodontinae and Ctenomyidae contain one less species in Patagonia.

Species of Atacama origin are completely absent from Patagonia. The Patagonian species compose 65% of the fauna. Five of them are endemic (Table 3.29). Other Patagonian rodents also occur in the neighbouring biogeographic regions, while eight species of the Monte origin have invaded the Patagonian province (Table 3.29).

CHAPTER 4

Geographic Patterns of Faunistic Composition

4.1 Taxonomic Composition

The zoogeographic subdivision of the desert areas of the world reflects the variation in the taxonomic composition of their rodent faunas. In the context of this chapter, we analyze and consider rodent distribution at the family level. The only exception is the Muridae which will be considered at the level of subfamily. This is because of the extremely high generic and species richness of murids and the distinct morphological and ecological specialization of most subfamilies. In addition, the taxonomic rank of the murid subfamilies is contestable and most of them are often considered as separate families within the superfamily Muroidea (see review in Carleton and Musser 1984).

The taxonomic composition of rodent faunas is presented in Fig. 4.1 for the Old World and Fig. 4.2 for the New World. It is possible to distinguish two groups among the rodent taxa inhabiting desert areas. One group consists of families, species of which occur entirely, or at least mainly, in deserts (for example, Heteromyidae, Dipodidae, Gerbillinae, Ctenodactylidae). Another group is basically non-desert with only some representatives being desert-adapted (for example, Cricetinae, Murinae, Sigmodontinae). Thus, the fraction of species belonging to the "desert taxa" in a given fauna reflects its specific position within the system of desert biogeographic units. Simultaneously, the fraction of species of "non-desert taxa" in a given fauna mirrors the influence of the non-desert surroundings and links the fauna to a certain biogeographic region. Each taxon demonstrates a specific pattern of geographic distribution. We begin with the description of "desert families".

4.1.1 Heteromyidae

This is a compact family both taxonomically and geographically. Some taxonomists divide Heteromyidae into three subfamilies (e.g. Patton in Wilson and Reeder 1992), although this division reflects the very intensive study of heteromyid taxonomy and evolution that has taken place during the last half-

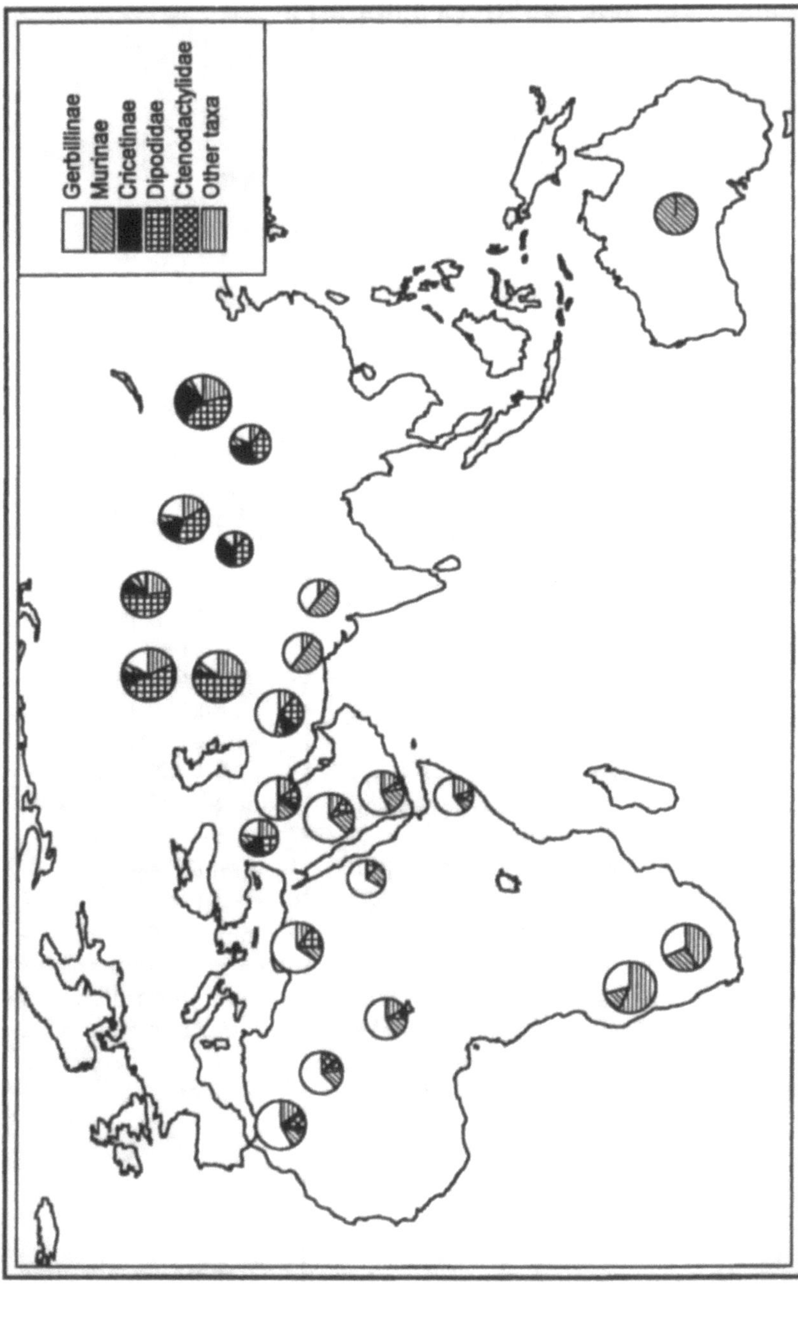

Fig. 4.1. Taxonomic composition of the provincial desert faunas of the Old World. The size of a pie corresponds to overall species richness in a province

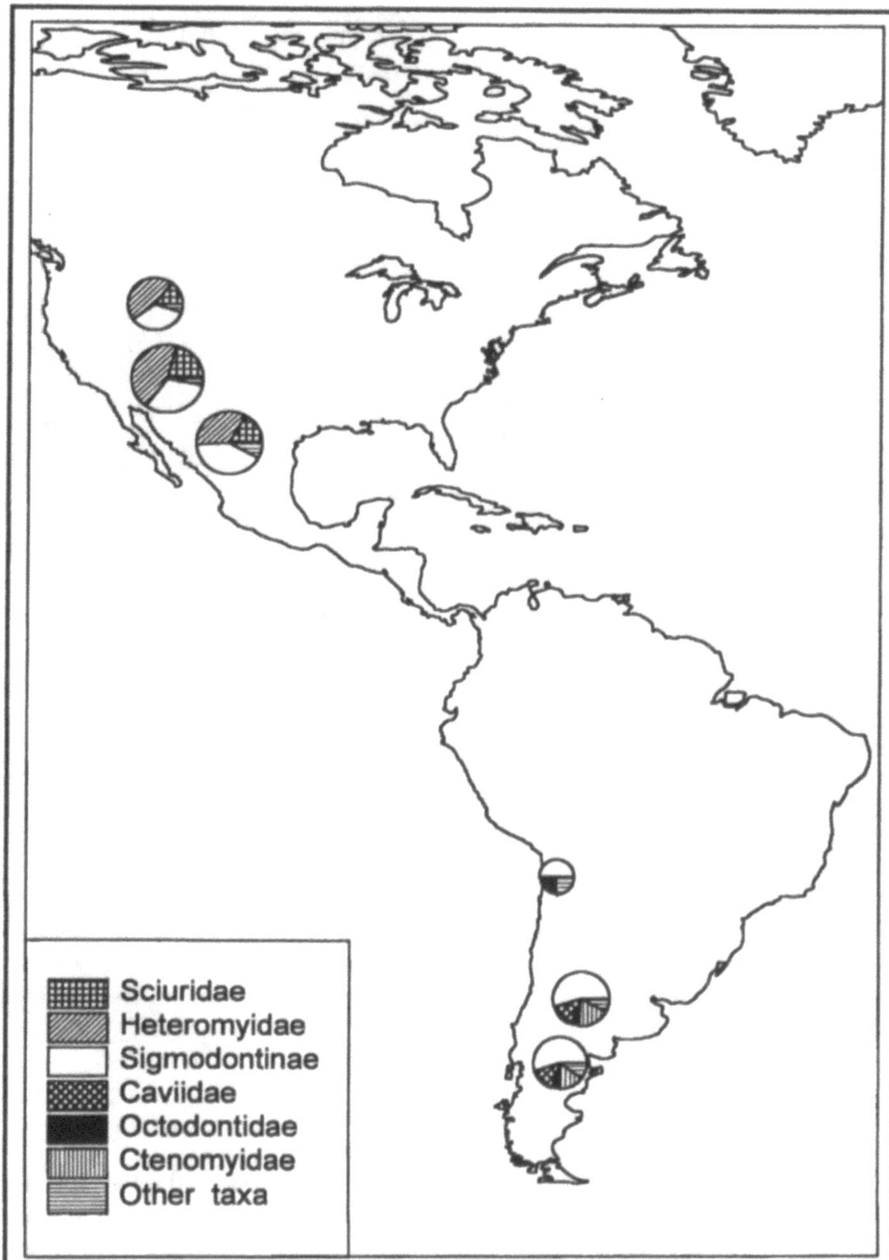

Fig. 4.2. Taxonomic composition of the provincial desert faunas of the New World. The size of a pie corresponds to overall species richness in a province

century, and the use of modern cladistic concepts rather than real evolutionary diversification (see Hafner 1993 for review). This taxon evolved and radiated in North America (see Schmidly et al. 1993; Wahlert 1993 for reviews). Four of six recent genera are desert-adapted and occupy a dominant position in each of the three North American provinces.

4.1.2
Dipodidae

This is a heterogeneous taxon that is composed of seven subfamilies, some of which can be regarded as separate families (Shenbrot et al. 1995). Two subfamilies, namely Sicistinae and Zapodinae, are non-desert. The early stages of dipodid evolution are still unclear. However, there is strong paleontological evidence that the late evolution of desert Dipodidae took place in Asia and a few desert species entered North Africa relatively recently (Black and Krishtalka 1986; Tchernov 1992; Shenbrot et al. 1995). Consequently, the fraction of dipodids in local Palaearctic faunas decreases from east to west. Two of five desert subfamilies have narrow geographic distributions as follows. The monotypic Euchoreutinae is restricted to the three easternmost provinces of the Turano-Gobian subrealm, while the monotypic Paradipodinae occurs in the Turanian province only. Other subfamilies have broader distributions. Cardiocraniinae inhabits the Turano-Gobian and Irano-Sindian subrealms. In almost all faunas, this taxon is presented by one species only, excluding the Sungorian and Gobian provinces with two and three co-occuring species, respectively. Dipodinae have the broadest distribution among all the jerboa subfamilies, inhabiting almost all Palaearctic provinces, excluding those of the Sind and South Pakistan. From southwest to northeast, the genera of Dipodinae replace one another from *Jaculus* through *Eremodipus* to *Dipus* and *Stylodipus*. Allactaginae are distributed in the Asian part of the Palaearctic (being absent in the Sind, South Pakistan and Kashgharia) with a small isolated patch in the northeast Sahara (the Libyan province). The allactagines are richest in the Central Kazakhstan province (seven species) and decrease sharply to the east and the south.

4.1.3
Subfamily Gerbillinae of Family Muridae

Gerbillines are represented almost exclusively by desert-adapted species. Only three genera out of 13 are non-desert. One of these genera (*Taterillus*) is endemic to the African savannas and no species enter the true desert. *Tatera* species are mainly savanna animals, although two species inhabit desert regions (*Tatera indica, T. brantsii*). Finally, the monotypic *Brachiones* is distributed within the Gobi desert, but is found in intrazonal habitats (oases and river valleys). According to our criteria, this species should not be considered as a desert taxon. The origin and early stages of the radiation of the Gerbillinae took place in Africa; their Asian history is much shorter (Pavlinov et al. 1990). Consequently, the greatest diversity and species richness of gerbils and jirds is observed in North Africa, Arabia and the Horn of Africa. In these regions, gerbillines comprise the largest fraction of the entire rodent fauna (approximately 70%). This fraction decreases to the south and to the northeast, attaining 25% of all rodents in the South African deserts and less than 10% in the Gobi desert. It should be noted that the South African gerbillines are represented by the endemic genera *Gerbillurus* and *Desmodillus* only. A similar situation occurs in the Somalian fauna (consisting of the endemic genera *Ammodillus* and *Microdillus*). The Palaearctic genus *Gerbillus* also occurs in the Somalia, although it is represented here by endemic species. Within the Palaearctic desert belt, a distinct trend of genus turnover from the southwest to the northeast can be depicted (from *Desmodilliscus, Monodia, Gerbillus* through *Pachyuromys, Sekeetamys, Psammomys* to *Meriones* and *Rhombomys*).

4.1.4 Ctenodactylidae

This family is represented by four genera distributed in the Sub-Maghreb (*Ctenodactylus*), Central Saharan (*Felovia, Massoutiera*) and Somalian (*Pectinator*) provinces. These taxa are relicts of the ancient (pre-Miocene) rodent fauna (Wang 1997). The family was broadly distributed in Asia in the Palaeogene (Shevyreva 1989, 1994, 1996).

4.1.5 Pedetidae and Petromuridae

These are monotypic families resticted to the South African desert realm. *Pedetes capensis* occurs across South Africa both in deserts and savannas, while *Petromus typicus* is endemic to the Namib.

4.1.6 Sciuridae

This is basically a non-desert family represented in the Holarctic and Palaeotropic faunas. There are two different opinions regarding sciurid taxonomy. According to one system, there is no subdivision within the Sciuridae sensu stricto (excluding the Petauristinae), while the other system proposes division into six subfamilies (see Hoffmann and Anderson in Wilson and Reeder 1992). We prefer to adopt the latter system, thereby recognizing the high diversity of sciurids. Two of six subfamilies of Sciuridae (Marmotinae and Xerinae) inhabit deserts, demonstrating different distribution patterns. Two of five genera of Marmotinae are desert inhabitants. The range of *Ammospermophilus* is restricted to deserts. Three quarters of *Spermophilus* species occur outside of deserts, and only ten species can be considered as desert dwellers. Desert marmotines are distributed in the provinces of the Turano-Gobian subrealm and the North American desert realm. In addition, one species (*Spermophilus fulvus*) penetrates the Iranian province. Xerinae is basically an African taxon inhabiting the South African desert realm, the African Horn realm and the Sub-Maghrebian and the South Saharan provinces of the Saharo-Arabian subrealm. The range of monotypic *Spermophilopsis* is restricted to the Turanian province and part of the Central Kazakhstan province and thus forms an isolated part of the subfamily range. In most Old World faunas, sciurids compose a minor fraction (no more than 10%), whereas in the North American faunas this proportion is higher (15-20%). The cause of the mosaic distribution of desert sciurids seems to be that the family is mainly non-desertic and occupies inter-desert steppe (Marmotinae) or savanna (Xerinae) biomes.

4.1.7 Geomyidae

The pocket gophers constitute North America's family of fossorial rodents. Like the majority of rodents specializing in a subterranean mode of life, this taxon is characteristic of grasslands rather than deserts. Only two of 35 species belonging to the Geomyidae inhabit deserts.

4.1.8 Subfamily Arvicolinae of Family Muridae

Voles constitute one of the most species-rich groups, and are widely distributed across the Holarctic. There are 26 genera and about 150 species in the subfamily all over the world. Voles occur in practically all biomes of the Palaearctic and Neoarctic, and only rarely extend into desert areas. These rodents do not inhabit the American and North African deserts at all and are represented only in the north-eastern part of the Asian deserts (four provinces) by six species. Furthermore, there is only one arvicoline genus that can be regarded as a true desert form, *Eolagurus*. No more than three species co-occur in a fauna, so the voles only comprise a minor fraction in rodent assemblages if they are presented at all.

4.1.9 Subfamily Cricetinae of Family Muridae

Old World hamsters are represented in Europe and Asia by a relatively small number of species (24 species of eight genera). Only one genus (*Allocricetu-*

lus) is truly desert-dwelling, although desert species can be found in some other genera. Hamsters occur across all the faunas of Asia excluding those of the Sindian and South Pakistan provinces. However, there are one to two species in most provincial assemblages, while the only fauna relatively rich in hamster species is that of the Gobian province. Cricetines comprise 25% of all fauna there.

4.1.10
Subfamily Murinae of Family Muridae

Rats and mice are distributed all over the Old World. This group is the richest of all rodent taxa (more than 100 genera and more than 500 species) and is predominantly tropical. In all the deserts of Africa and Asia, excluding the Sind, two genera of mice and rats are recorded. These genera (*Mus* and *Acomys*) are represented by a large number of species (38 and 14, respectively), but only a small percentage of these can be considered to be true desert-dwellers (three and five species, respectively). The Sindian and the South Pakistan faunas contain many more murine species. Each of these provinces is inhabited by five species of Murinae. Besides the above-mentioned genera, there are also *Cremnomys* and *Millardia*. The latter are desert representatives of primarily tropical genera. The historical aspect of geographic distribution of *Mus musculus* is uncertain. Based on a reconstruction of the history of *Mus musculus* by Din et al. (1996), we can assume that only a small portion of the present geographic range of this species (the Sindian, South Pakistan, Iranian, Mesopotamian and North Arabian provinces) should be considered to be natural, the species having been dispersed by humans throughout the remainder of its range. Finally, it is noteworthy that the Australian fauna is completely composed of murines. This is a specific case, the causes of which were described above.

4.1.11
Subfamily Sigmodontinae of Family Muridae

The size of the subfamily Sigmodontinae is slightly less than that of Murinae (about 80 genera and more than 400 species). In contrast to Murinae, this taxon is distributed throughout the New World. Different species of sigmodontines take the ecological place of the main Old World murid groups (gerbils, hamsters, mice and voles). They occupy almost all the biomes of the Americas, excluding the boreal zone of North America where lemmings and true voles occur. Sigmodontines are widely distributed in deserts, although only one North American genus (*Onychomys*) and five South American genera (*Andalgalomys, Eligmodontia, Graomys, Rheithrodon* and *Salinomys*) can be regarded as strictly desert dwellers. Other desert sigmodontines belong to genera that are mainly grassland or forest dwellers. The percentage of sigmodontine species in the New World faunas is relatively high, reaching 35-40% of all rodent species in the Neoarctic provinces and 50-55% in the Neotropical provinces.

4.1.13
Small Subfamilies of Muridae

Three small murid subfamilies, namely Dendromurinae, Otomyinae and Petromyscinae are restricted to Africa south of the Sahara; the representatives of these taxa are found only in the Kalahari and Namibian faunas. A single strictly desert genus (*Parotomys*) belongs to Otomyinae. Both *Parotomys* species and *Malacothrix typica* (Dendromurinae) inhabit both South African provinces, while *Petromyscus collinus* (Petromyscinae) can be found in the Namib fauna only.

4.1.14
Myoxidae

Dormice are basically forest-adapted rodents represented by eight genera with 26 species. The only truly desert genus is the monotypic *Selevinia*. Two species of *Eliomys* enter the desert, but only one of them, *E. melanurus*, is desert-adapted. The main part of the geographic range of *Eliomys quercinus* lies outside desert. The same is true for the South African species *Graphiurus ocularis*. Dormice occur in faunas of the Sub-Maghrebian, Libyan, both Arabian, Mesopotamian, Syrian, Balkhash and both South African provinces. Furthermore, each of the faunas mentioned consists of one myoxid species only.

4.1.15
Hystricidae

This is a small Old World family consisting of three genera with 11 species. Three species of a single genus enter desert regions, although none of them is a true desert dweller. Moreover, porcupines are absent from the hyper-arid deserts (such as central Sahara). In other Old World deserts, no more than one hystricid species occurs per fauna (*Hystrix cristata* in North Africa, *H. indica* in the Asian deserts and H. *africaeaustralis* in South Africa).

4.1.16
Bathyergidae

This is a South African group with two of its 12 species entering arid regions. One of these, *Bathyergus janetta*, is restricted to the Namib desert, while *Cryptomys hottentotus* occurs in both provinces, as well as in the surrounding non-desert areas.

4.1.17
Hystricognath Families

The Abrocomidae, Octodontidae, Ctenomyidae, Caviidae and Chinchillidae are characteristic of the South American deserts. Octodontidae and Chinchillidae occur in all provinces, while Ctenomyidae and Caviidae are absent from the Atacama provinces and Abrocomidae from the Monte and Patagonian provinces. Desert-adapted hystricognaths belong to Octodontidae (*Octomys, Tympanoctomys*) and Caviidae (*Dolichotis*). If each of these families is considered separately, its representation in local faunas is not very high (up to 18%). However, the overall proportion of hystricognaths is substantial in the faunal composition of all the Neotropic provinces.

4.2
Taxonomic Diversity

Concluding our review of the taxonomic composition of the regional faunas, it should be noted that the provincial faunas vary in their taxonomic diversity at the generic level as well as at the level of family/subfamily. To evaluate the taxonomic (family/subfamily or generic) diversity of a given fauna we have used a measure analogous to Hill's N_2 index for community diversity,

$$N_2 = 1/\Sigma p_i^2 \text{ (Hill 1973)},$$

where p_i is the proportion of species belonging to family/subfamily or genus i, respectively, in the fauna. Taxonomic diversity indices for each of the provincial faunas are shown in Table 4.1 and Fig. 4.3. The most diverse faunas at the level of family/subfamily are those of the Turano-Gobian and South African provinces as well as the Syrian province, while the Australian, North African and Arabian faunas are the least diverse. Patterns of generic diversity demonstrate a similar trend although, in spite of the strong correlation between these two levels (Spearman $r = 0.73$, $p < 0.0001$), some provinces with relatively low family/subfamily diversity have a relatively high generic diversity (for example, the Monte and Patagonian provinces) and vice versa (for example, the Syrian province). A relatively low level of taxonomic diversity can be determined either by a low number of occurring taxa (the Australian fauna at the family/subfamily level) or by overrepresentation of a certain taxon (for example, Gerbillinae in the North African faunas) or both (for example, the Atacama fauna). In turn, a relatively high level of taxonomic diversity is a result of the equal representation of a relatively high number of taxa. This can occur either when a province is the center of origin and/or of the radiation of several rodent groups (as in the case, for example, of the Monte and Patagonian provinces; see Simpson 1980) or when a province is a

Table 4.1. Family/subfamily and generic diversity of different provinces

Province	Family/subfamily		Genera	
	Number	Diversity	Number	Diversity
Sub-Maghrebian	7	2.71	9	5.55
Libyan	6	2.22	9	4.65
Central Saharan	4	2.32	8	4.83
Nubian	3	1.98	6	3.86
South Saharan	5	2.57	7	4.24
North Arabian	5	2.27	9	5.08
South Arabian	5	2.58	8	4.26
Syrian	7	6.23	9	9.00
Mesopotamian	7	3.38	10	7.54
Iranian	8	3.75	12	8.76
South Pakistan	3	2.38	7	5.56
Sindian	3	2.38	7	5.56
Turanian	10	7.14	15	11.11
Central Kazakhstan	8	5.38	14	9.31
Balkhash	9	5.45	13	9.97
Sungorian	8	6.75	15	12.46
Gobian	9	6.86	17	13.71
Kashgharian	7	5.40	8	7.36
Tzaidam	8	6.37	10	9.31
Somalian	5	2.50	7	4.55
Namibian	11	7.00	15	11.31
Kalahari	9	5.67	14	12.57
Australian	1	1.00	6	4.50
Great Basin	4	2.86	10	8.33
Sonoran	4	2.92	11	9.17
Chihuahuan	4	3.08	12	9.67
Atacama	4	2.91	4	2.91
Monte	5	2.78	14	11.00
Patagonian	5	2.78	14	11.11

zone in which the elements of two or more faunas which have different origin, overlap (as is the case with the Syrian province).

Provincial desert faunas can be classified by prevailing taxa. In the Old World it is therefore possible to distinguish (1) gerbilline faunas (in the Somalian, Iranian, Mesopotamian and all Saharo-Arabian provinces), (2)

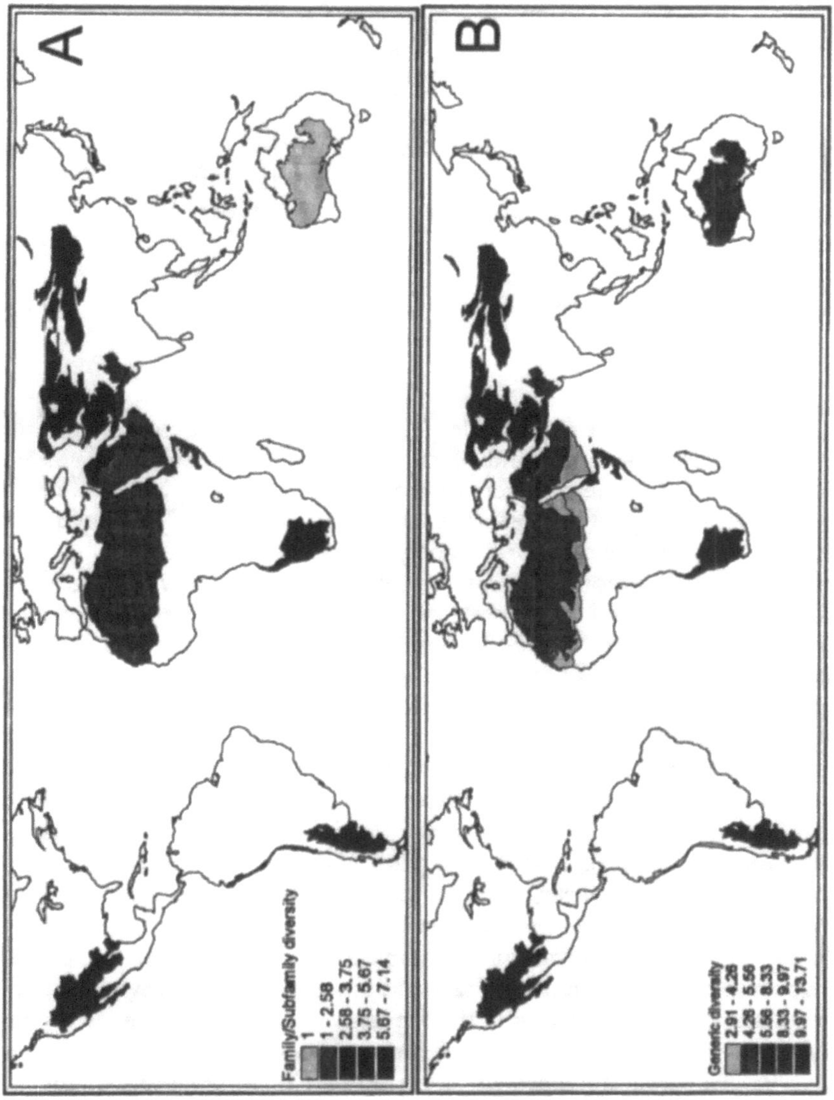

Fig. 4.3. Taxonomic diversity of the provincial desert faunas of the world. A Family/subfamily diversity, B generic diversity

dipodid faunas (in the Turanian, Central Kazakhstan and Balkhash provinces), (3) dipodid-cricetine faunas (in the Gobian superprovince), (4) murine faunas (in the Sindian, South Pakistan and Australian provinces) and (5) mixed murid faunas (in both South African provinces). Because of its transitional position, the fauna of the Syrian province must be excluded from this classification. All the desert faunas of North America can be regarded as heteromyid-sigmodontine faunas, while those of South America are sigmodontine-hystricognath faunas.

4.3 Species Richness

The species diversity of a region is largely determined by a combination of current environmental forces and of historical processes. Species richness provides a useful measure of diversity if the area considered is delimited in space and the constituent species are enumerated and identified (Magurran 1988). This is exactly the case for the zoogeographic units defined.

Patterns of distribution of species within biogeographically homogenous spatial units are reflected by the well-known species-area relationships (see e.g. Rosenzweig 1995 for review). These relationships are caused by increasing environmental diversity with increases in the area considered. They are not linear. Log-log transformation is necessary to linearize a species-area curve. For biogeographical units of the same rank, the resultant species-area regressions have the same slope (Rosenzweig 1995). However, their intercepts may differ as a result of historical reasons. Given that similar biomes (e.g. deserts) from different regions are considered, we assume that unidirectional environmental influences would minimize historically determined interregional differences. However, if the historical differences are deep enough, the minimizing effect of the environment will not be apparent.

We use log-log transformations to compare the number of rodent species among provincial desert faunas of the world in relation to the provincial area. The resultant trend demonstrates that species richness increases with the area of a province (Fig. 4.4), thus conforming with the statements of Rosenzweig (1995). However, some provinces are outliers in both positive and negative directions. The Central Saharan and Australian provinces contain fewer species than expected, while the Sonoran and Chihuahuan provinces have faunas richer than expected.

As stated above, the environmental component of species diversity is reflected by a linear trend, whereas the historical component is reflected by deviations of the provincial points from this trend. Each case of far outliers has its specific causes. For example, under-representation in the Australian fauna can be explained by the recent massive extinction of desert rodents. At least six species are recorded as becoming extinct over the last century (Stra-

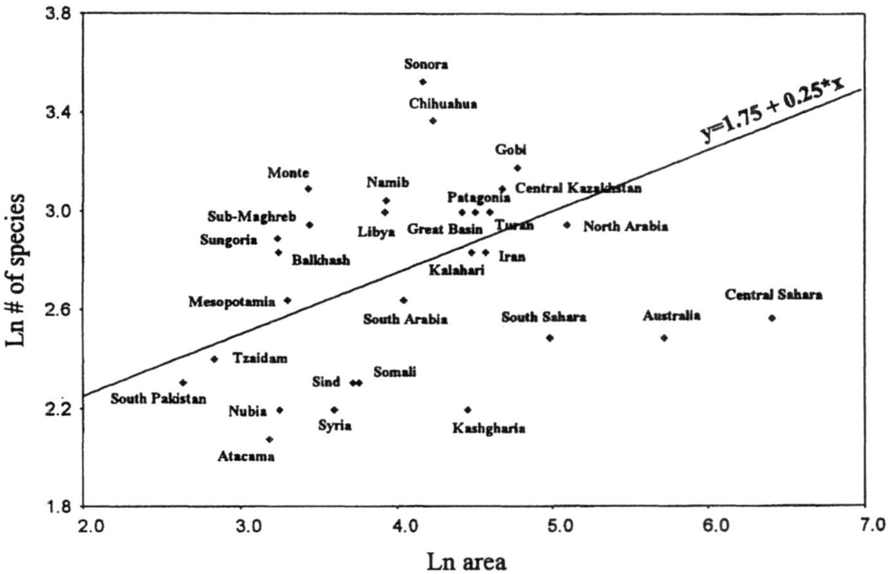

Fig. 4.4. Log-log plot of provincial species richness against the provincial area (units= 1000 km^2)

han 1983); it is highly probable that the number is actually higher. The faunal poverty of the Central Saharan province may be a result of the recent (Middle Holocene) increase in its desert area (Maley 1977; La Houérou 1997); the recent fauna was originally formed in a smaller area and then dispersed within its present boundaries.

The extreme richness of the two North American faunas can also be explained by historical reasons. The desert landscapes in North America are relatively young. They were formed during the Pleistocene (Axelrod 1958; Webb 1977; Van Devender and Spaulding 1979; Wells 1979; Thompson and Mead 1982). In addition, there were frequent and dramatic fluctuations in the size (both contractions and expansions) of these desert areas throughout the glacial-interglacial cycle. These were caused by deserts occurring in the valleys of a basin-and-range topography (see Schmidly et al. 1993 for review). During many glacial periods, desert habitats contracted considerably, persisting primarily in refugia further south than their present locations and sometimes being completely replaced by other vegetation types. The repeated restriction/expansion, fragmentation, and rejoining of the North American deserts during the Pleistocene, plus a topography with a great number of barriers, probably promoted speciation of the desert rodent fauna to a greater extent than in many other deserts (Shenbrot et al. 1994b). The above-mentioned interactions occurred to a lesser extent in the Great Basin province

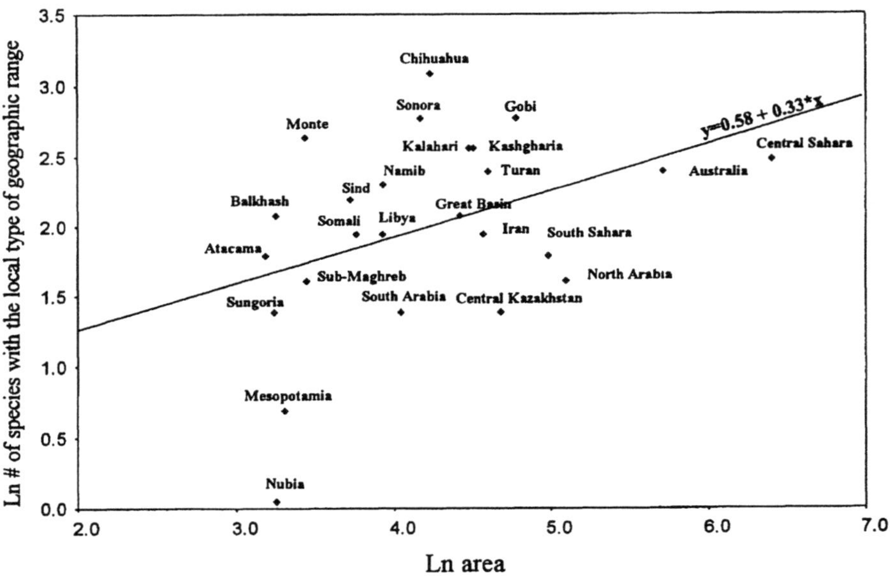

Fig. 4.5. Log-log plot of the number of species with a local type of geographic range against area of the respective province (units=1000 km^2)

because this lies far to the north and at higher elevation. It therefore seems probable that glaciation pressure on the fauna was stronger and that far fewer refugia were available.

A plot of the overall species richness of the faunas against the areas of the provinces reflects both the environmental and historical determinants of the faunal composition. Taking into account only those species that originated in a given province rather than the entire species pool, we can expect that the relation between this subset and the provincial areas would mainly reflect the historical component. Indeed, there is a strong positive correlation between the number of species with local origin and the area of a province (Fig. 4.5). It seems that the larger provinces support more intensive speciation. The only exception to this trend is the Nubian province which has fewer than "local" species expected (in fact, this is only one species). From this point of view, the Nubian province should not be considered as a true province but rather as a subprovince of the Central Saharan province. Indeed, the calculations show that the Nubian fauna lies between provincial and subprovincial rank.

The environmental component of species-area relationships is determined by an increase in environmental (=habitat) diversity with an increase in area (Rosenzweig 1995). We attempted to present this component by plotting species richness directly against environmental diversity. We used the number of major plant associations (=vegetation types) as a measure of the

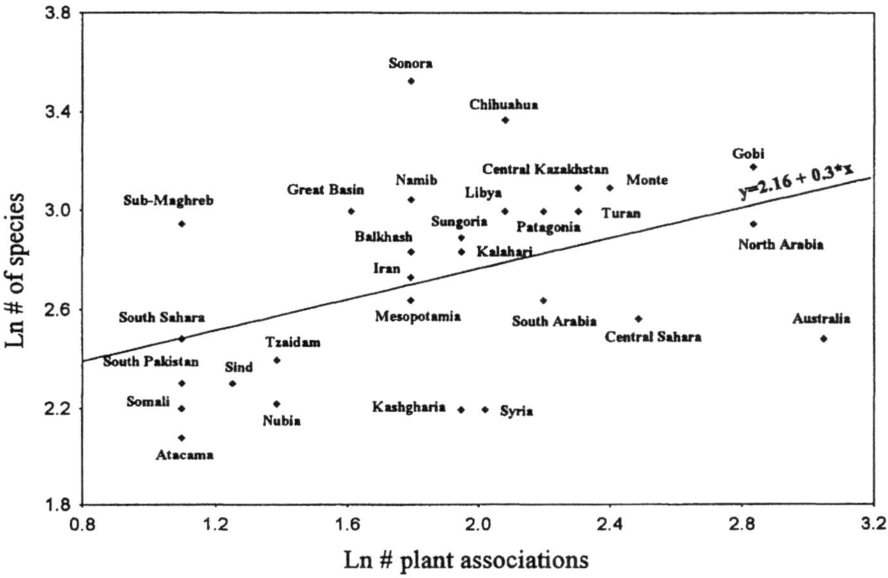

Fig. 4.6. Log-log plot of provincial species richness against number of plant associations

latter. Information on vegetation types was obtained from vegetation maps distributed by ESRI, Inc. (Environmental System Research Institute, Redlands, California USA) as online data for public use at their World Wide Web site (http://nutria.esri.com/geoserver). The positive relationship between species richness and environmental diversity is clear from Fig. 4.6. However, this correlation for the whole data set is not strong, albeit significant. This can be attributed to the different original sources of the vegetation maps. These sources seem to differ in their approaches to the degree of itemization in vegetation classification and mapping. Furthermore, if such a plot was restricted to a single biogeographic region, the correlation between species richness and environmental diversity would be more strongly expressed. This is exactly the case in the Palaearctic (Fig. 4.7). The number of the provinces in other regions is too small for such analysis.

Patterns of species diversity (=richness) considered at a lower spatial scale (within a biogeographically homogenous area among environmentally homogenous units) reflect current ecological relations and are free from the influence of historical processes. To understand spatial changes in species richness, areas of equal size should be compared. For this comparison we have chosen quadrats of 2° by 2° which follow lines of latitude and longitude, a method often used in similar studies (Schall and Pianka 1978; Currie and Paquin 1987; Meliadou and Troumbis 1997). We assume that this scale is a

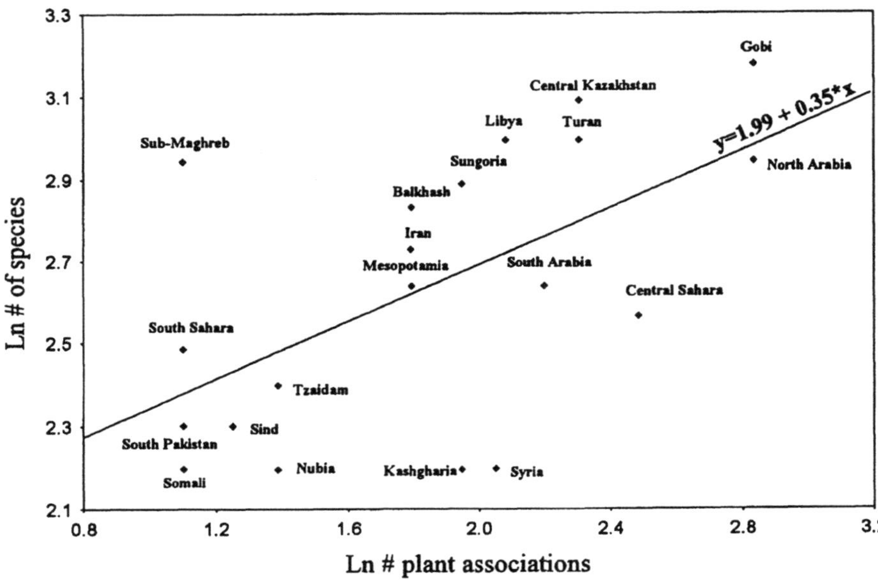

Fig. 4.7. Log-log plot of provincial species richness against number of plant associations for Palaeoarctic provinces

trade-off between two conditions, namely relative homogeneity of spatial units and the availability of faunistic data. For each quadrat, species richness of desert rodents was calculated by overlaying distribution maps of individual species on the grid map and by summing the number of species occurring in each quadrat.

The distribution of rodent species richness in different desert regions is presented in Figs. 4.8-4.12. There are several "hotspots" of high species richness in the Saharo-Gobian realm, whereas other realms have only one "hotspot" of species richness. A possible explanation of this pattern of distribution is that the area of the Saharo-Gobian realm is much larger that that of other realms; consequently environmental diversity is higher there. Furthermore, this realm takes the form of a long belt, which has repeatedly been divided into several parts and later unified (e.g. Matiushkin 1972; Tchernov 1988). As a result, different parts of the belt evolved as independent biogeographic units during certain historical phases. These processes have led to the present multiple "hotspot" distribution throughout the entire realm.

A well-known pattern of a latitudinal gradient of species diversity (see Rohde 1992; Rosenzweig 1995 for reviews) or Rapoport's rule (Stevens 1989) states that the number of species declines with increasing latitude. The New World deserts conform with this pattern, whereas the opposite is true of the Old World deserts (at least, those of the Palaearctic and Palaeotropic). How-

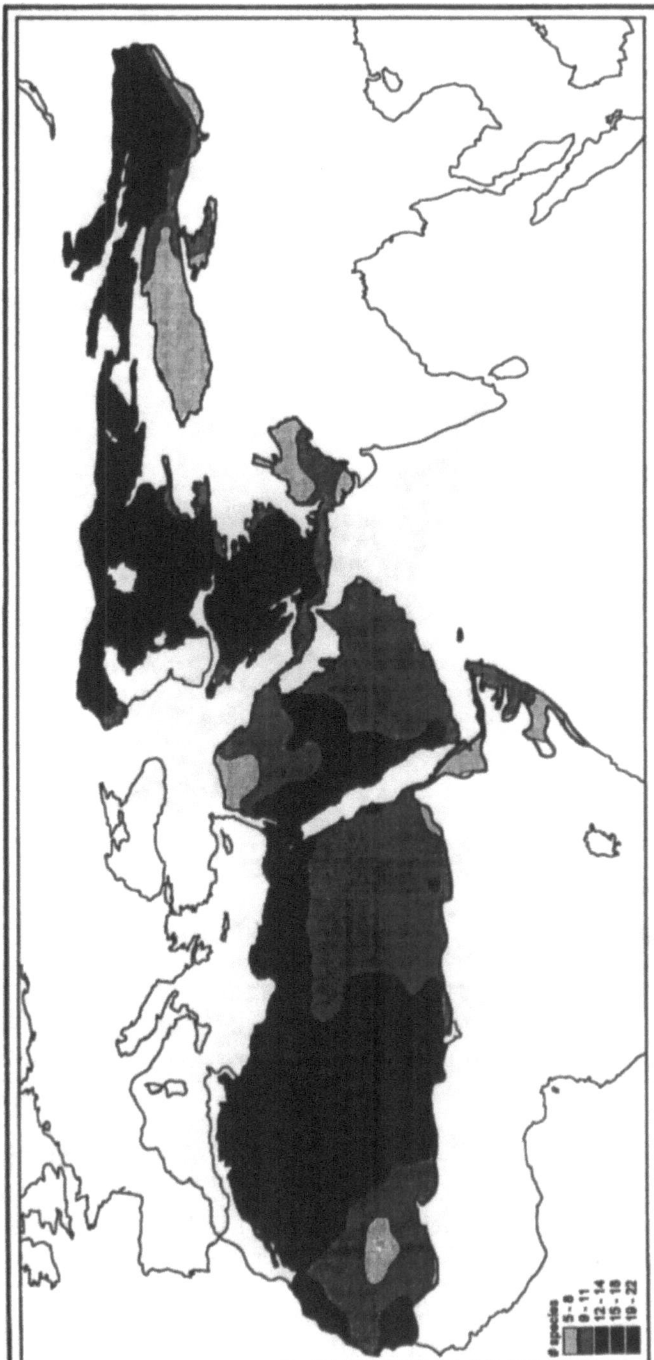

Fig. 4.8. Distribution of desert rodent species richness in the Saharo-Gobian desert belt

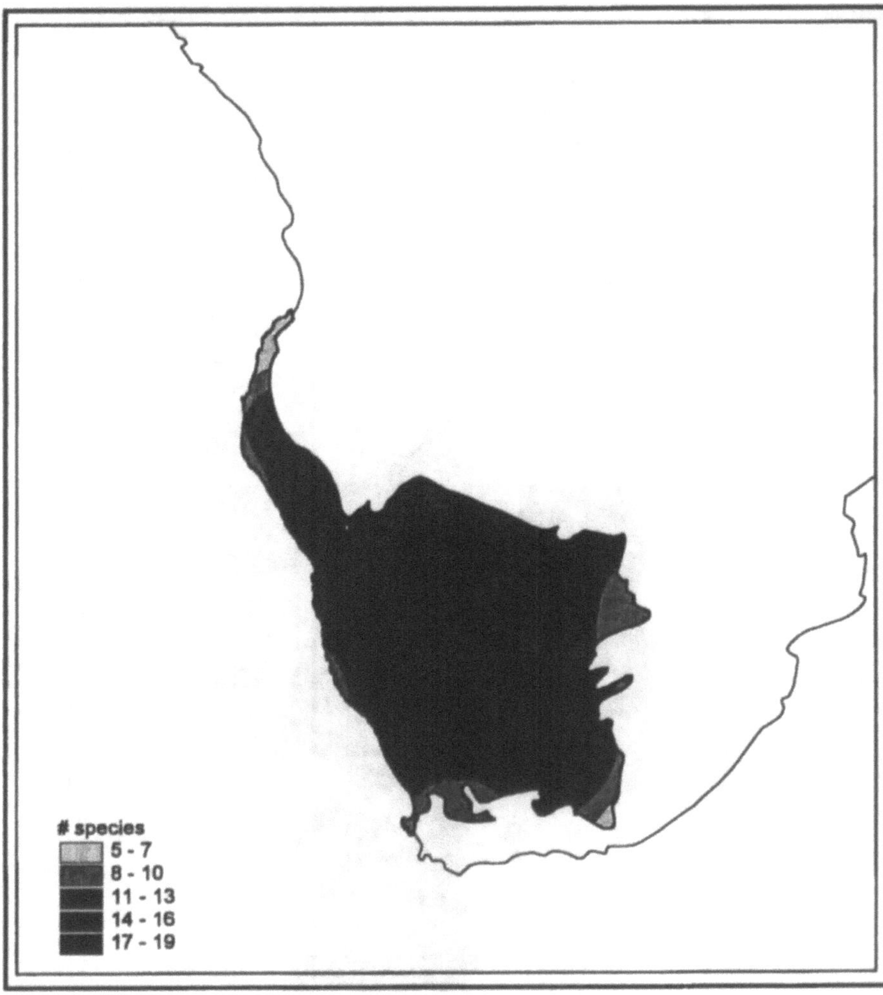

Fig. 4.9. Distribution of desert rodent species richness in South Africa

ever, it is difficult to expect a latitudinal pattern of species diversity to be manifested in a consistent manner. This is due to the influence of historical processes (e.g. Mönkkönen and Viro 1997) and the co-occurrence of interactions between species richness and other environmental parameters such as productivity.

Species richness, as a measure of the structural complexity of an ecosystem, is determined to a great extent by the level of productivity (Whittaker 1975; Odum 1983; Rosenzweig and Abramsky 1993). Two patterns in the re-

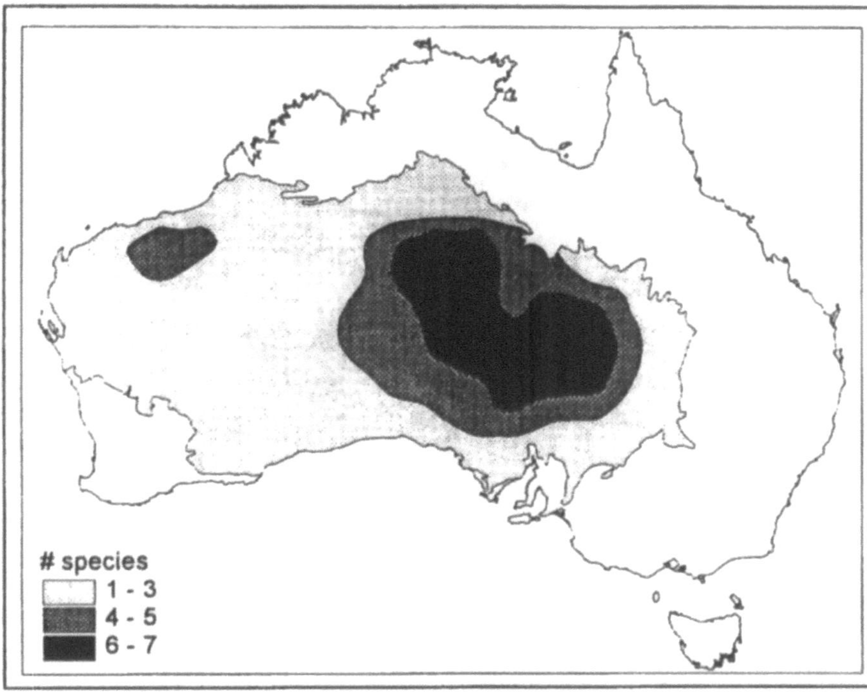

Fig. 4.10. Distribution of desert rodent species richness in Australia

lationship between species richness and the productivity of an ecosystem can be demonstrated. Both patterns have been shown to occur in deserts, namely linear and hump-shaped relationships. In the former, species richness simply increases monotonically with increasing productivity as a result of increased resource availability. The latter model is often ascribed to Tilman (1988) although it was also considered by Whittaker (1972), among others. In this model, species diversity plotted against a productivity gradient will increase from low productivity until some intermediate value is reached. Thereafter, competition between species occurs so that certain species outcompete others, thereby resulting in a reduction in species richness with a further increase in productivity, thus causing a hump-shaped curve. A linear relationship has been shown to be valid for the rodents of arid North America (Brown 1973; 1975), while Abramsky and Rosenzweig (1984) have demonstrated a hump-shaped curve for the rodent assemblages of the Negev Desert, Owen (1988; 1990) for the North American deserts, Shenbrot (1992) for the southern Turan Desert, and Rogovin et al. (1994) for the Thar Desert.

Productivity in deserts is directly determined by the level of annual precipitation (e.g. Southgate et al. 1996), so the rainfall can be a measure of gen-

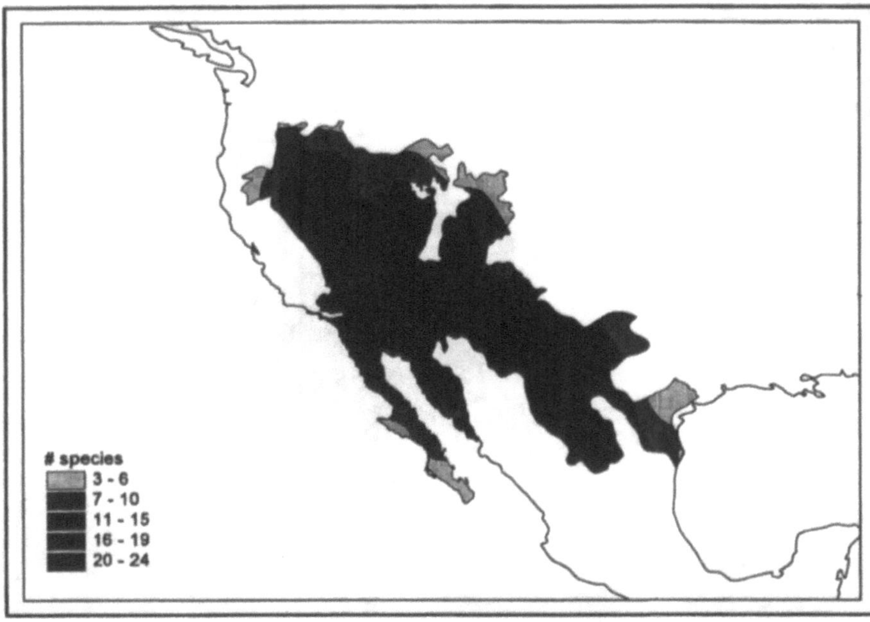

Fig. 4.11. Distribution of desert rodent species richness in North America

eral productivity. Most provincial faunas do not show any correlation between species richness and the level of precipitation[1]. This lack of pattern may be the result of a relatively narrow range in the variation of precipitation within the boundaries of these provinces. Other provinces demonstrate significant patterns of species richness along rainfall gradients (Figs. 4.13 and 4.14). Surprisingly, in most cases this correlation is negative; that is, the higher the annual precipitation, the poorer the desert rodent community. However, we have only taken into consideration species that are regarded as desert-adapted. High levels of precipitation cause a transition to non-desert environments and true desert species consequently drop out from the fauna as the ecosystem productivity increases, and are replaced by non-desert rodents. The total number of both desert and non-desert species sometimes increases toward the borders of desert zones (e.g. Abramsky and Rosenzweig 1984), but the number of desert species decreases.

The only exception is provided by the fauna of the Libyan province which displays an increase in the number of desert species with increasing produc-

[1]Information on annual preciptation was obtained from climatic maps distributed by ESRI, Inc. (Environmental System Research Institute, Redlands, California USA) as online data for public use at their World Wide Web site (http://nutria.esri.com/geoserver).

Fig. 4.12. Distribution of desert rodent species richness in South America

tivity. The Libyan province is the only one of all the desert provinces that has no contact with any terrestrial ecosystem at the upper end of the precipitation gradient because it is bounded on the north by the Mediterranean Sea.

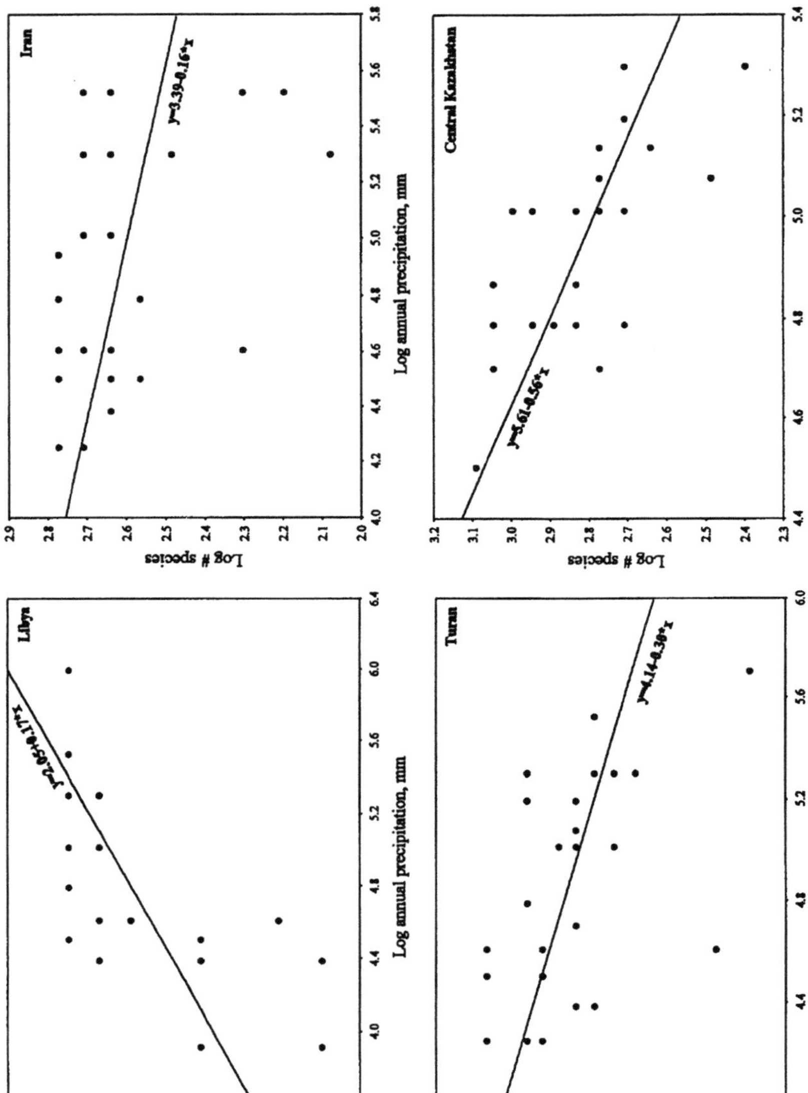

Fig. 4.13. Changes in species richness along the rainfall gradient for Libyan, Iranian, Turanian and Central Kazakhstan provinces

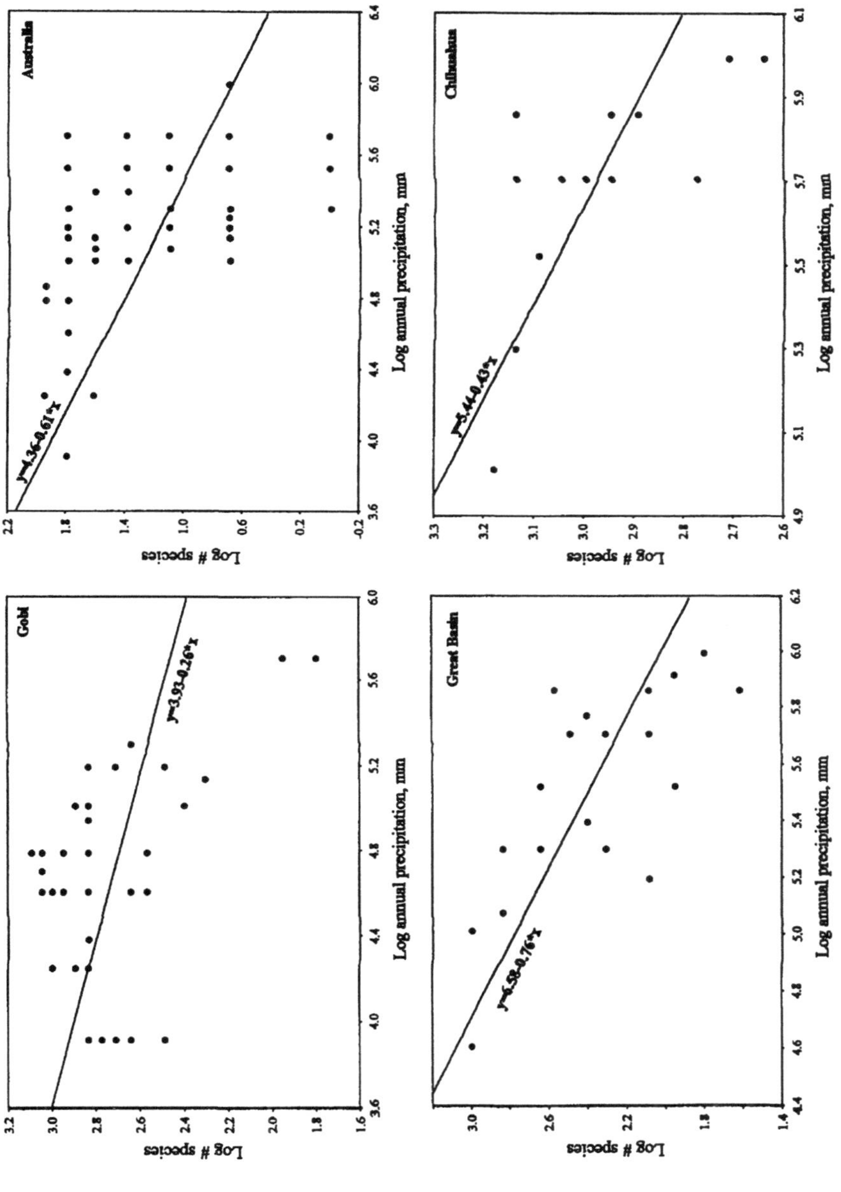

Fig. 4.14. Changes in species richness along the rainfall gradient for Gobian, Australian, Great Basin and Chihuahuan provinces

There are only two non-desert rodents in the province (*Spalax ehrenbergi* and *Microtus guentheri*), both of which are represented by small, isolated populations. The geographical position of the province provided specific conditions for the formation of its fauna when desert rodents evolved to use free ecological space that in other cases is occupied by non-desert species. Native Libyan species inhabiting the relatively highly productive coastal desertified grasslands (e.g. *Gerbillus andersoni, Meriones shawi, Jaculus orientalis*) exemplify this pattern. The Libyan province emphasizes the role of historical processes in producing the spatial patterns of species diversity, even on a small geographic scale.

CHAPTER 5

Inter-Desert Comparisons of Rodent Faunas

5.1
Climatic, Substrate and Vegetation Features Important for Rodents

Deserts share a number of properties. These are characterized by the contrasts in their combinations of environmental factors. Insolation may be extremely high, whereas moisture is scarce. Extreme environmental factors do not make life easy and restrict many animal and plant species to short periods in certain seasons when the combination of limiting abiotic factors have least effect. Deserts are arenas where Liebig's law is manifested, and desert life is usually represented by species adapted to very specific conditions. Specific features of the desert environment promote adaptive responses in its inhabitants. Indeed, a broad spectrum of rodent adaptations to the harsh desert conditions can be recognized, from biochemical (e.g. Downs and Perrin 1991) through physiological (e.g. Weisserberg and Shkolnik 1994) to ecological (e.g. Goyal and Ghosh 1993) and behavioral (e.g. Randall 1993). In this chapter we will limit ourselves to consideration of ecological-behavioural adaptive responses only (the border between ecological and behavioural reactions is vague). The various biochemical and physiological adaptations of rodents to the desert environment have been thoroughly described and analyzed recently by Degen (1996) and we will therefore not consider these questions here.

In spite of the variability of deserts in terms of their origins, geographic locations, precipitation regimes and moisture sources, the main features of the climate of different deserts are similar (chap. 2). Most important biologically are high temperatures (at least seasonally) and irregular, unpredictable and usually low precipitation. Desert mammals are therefore faced with the serious problem of maintaining constant body temperature while minimizing water use for evaporative cooling. This task can be achieved by selection of an appropriate microclimatic regime, both in time (i.e. by choosing an appropriate temporal pattern of activity) and space (i.e. by choosing habitats with an appropriate complex of environmental parameters).

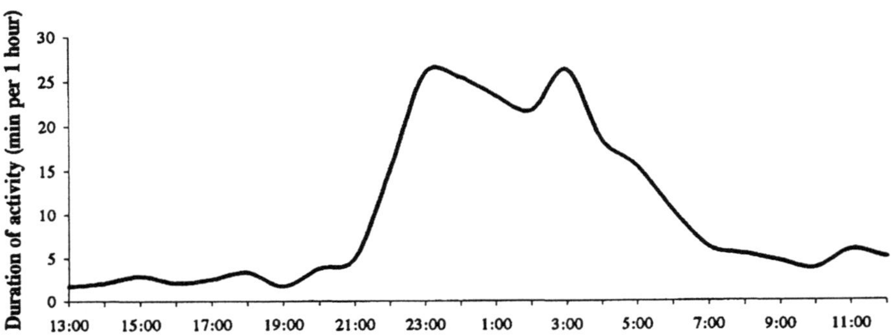

Fig. 5.1. Pattern of daily activity of *Cricetulus migratorius*. (Based on the data of Maizelis 1940, cited by Kalabukhov 1969)

5.1.1 Temporal Patterns of Activity.

Most rodents avoid the harsh daytime temperature of the desert by nocturnal activity. Jerboas usually leave their burrows at dusk and return just before sunrise (Shenbrot et al. 1995). This period of nocturnal activity is rarely continuous, but shows a polyphasic pattern. Granivorous and insectivorous species usually exhibit one long peak of activity (e.g., the gray hamster *Cricetulus migratorius*, Maizelis 1940, cited by Kalabukhov 1969; Fig. 5.1), whereas species with mixed diets demonstrate two peaks of activity, one just after sunset, the other just before sunrise (e.g., the thick-tailed three-toed jerboa *Stylodipus telum*, Lobachev and Shenbrot 1973; Fig. 5.2A) while herbivorous species show several short peaks of activity per night (e.g., the fat-tailed jerboa *Pygeretmus pumilio*, Rogovin 1981; Fig. 5.2B). In cloudy weather, nocturnal rodents sometimes begin their activity approximately an hour before dusk and terminate it shortly after sunrise (e.g. the fat-tailed jerboa *Pygeretmus pumilio*, Shenbrot et al. 1995). Both the pattern and the duration of activity change seasonally. *Pygeretmus pumilio* demonstrates two or three peaks of activity in summer and one or two peaks in autumn, whereas the duration of the activity decreases from 9 to 5 h (Lobachev and Shenbrot 1973; Shenbrot et al. 1995). Some jirds that are nocturnal in summer, become diurnal in winter (the midday jird *Meriones meridianus* – Gambarian et al. 1960; the Libyan jird *Meriones lybicus* – Harrison and Bates 1991; Chabovsky and Popov 1994).

Most desert rodent communities contain diurnal species. These include all desert Sciuridae (*Spermophilus, Ammospermophilus, Spermatophilopsis, Xerus, Atlantoxerus*), three jirds (*Meriones hurrianae, Psammomys obesus, Rhombomys opimus*), two murids (*Parotomys brantsii, Rhabdomys pumilio*), two arvicolires (*Eolagurus luteus, Eolagurus przewalskii*), four South Ameri-

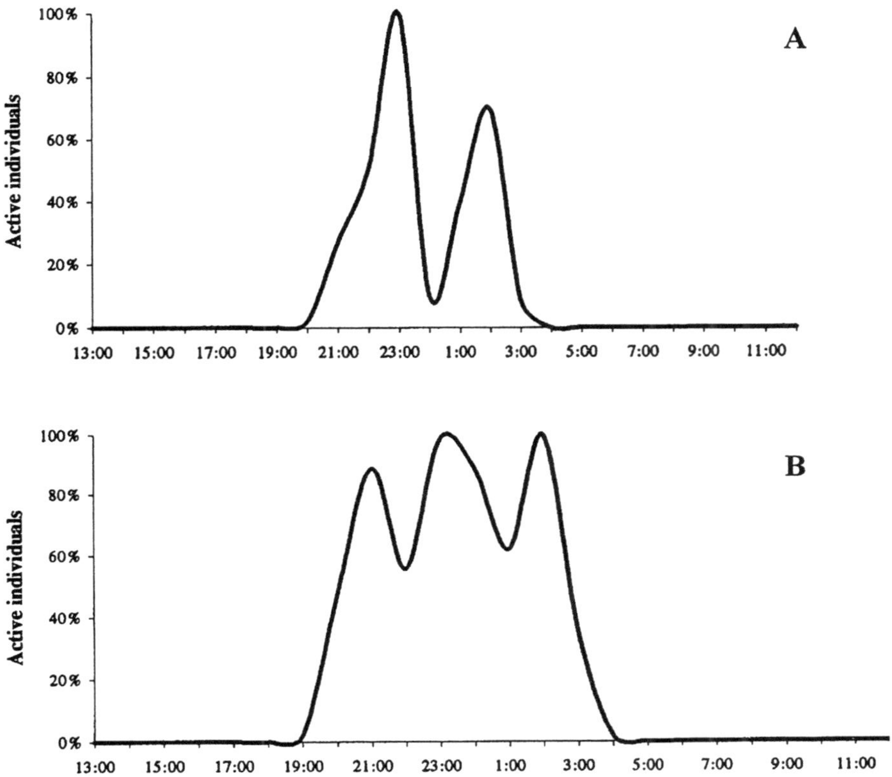

Fig. 5.2. Patterns of daily activity of *Stylodipus telum* (A, based on the data of Lobachev and Shenbrot 1973) and *Pygeretmus pumilio* (B, based on the data of Rogovin 1981).

can hystricognaths (*Octodon degus, Dolichotis patagonum, Dolichotis salinicola, Microcavia australus*), one petromurid (*Petromus typicus*), and all ctenodactylids (*Ctenodactylus gundi, Ctenodactylus vali, Massoutiera mzabi, Felovia vae, Pectinator spekei*). In summer, these species exhibit biphasic activity (morning and evening) avoiding the hottest part of the day, whereas in winter they are also active at midday. For example, biphasic activity has been described in the great gerbil *Rhombomys opimus* (Shekhanov 1952, cited by Sokolov and Kuznetzov 1978; Fig. 5.3). In addition, some diurnal species may also be active on moonlit nights, for example, Brants' whistling rat, *Parotomys brantsii* (Nel and Rautenbach 1974).

Diurnal desert species have specific morphological adaptations which enable them to endure high solar radiation. For example, they have a dense cover of hair on the soles of their feet which minimizes the effects of contact with the hot substrate. Another example of morphological adaptation is demonstrated by the diurnal long-clawed ground squirrel, *Spermophilopsis leptodactylus*. The skin of this desert species is much more pigmented than that of other species of ground squirrels (*Spermophilus undulatus, Spermophilus suslicus*) which inhabit more mesic regions (Sokolov 1973). Sokolov (1973) also notes that the number of melanocytes in the skin of this species is much higher than that of nocturnal rodents. Melanocytes are extremely abundant in the deep layers of the epidermis and form the black colouration of the skin. Moreover, the pigmentation of the skin is evenly distributed across the body, excluding the scrotum, the skin of which is less heavily pigmented. This could result from the necessity of preventing the overheating of the testes and consequent disturbance of the process of spermatogenesis (see Bronson 1989). The skin of the day active great gerbil *Rhombomys opimus* is not as highly pigmented as that of *Spermophilopsis leptodactylus*. However, the distribution of skin pigmentation is specific, namely large melanocytes are present in the dermis only and are absent from the epidermal layer (Sokolov 1973). Nevertheless, desert rodents do not possess specific skin structures that curb water loss; water balance is regulated by physiological mechanisms (see Degen 1996 for further references).

Diurnal rodents are either omnivorous (desert sciurids, the striped mouse *Rhabdomys pumilio*) or folivorous (all others). There are no strictly granivorous diurnal species. Degen (1996) points out that the absence of granivores can be explained by consideration of their water balance. Preformed water increases at night due to the hygroscopic nature of seeds, so strict granivores are forced into nocturnal foraging. The presumed cause of the absence of strict insectivory among diurnal species is that most insects also avoid the heat (see Cloudsley-Thompson 1991 for further references).

Some desert rodents demonstrate facultative diurnal activity. The classical example is the golden spiny mouse *Acomys russatus* which has often been observed to be active during the hottest hours of the day (Wassif and Hoogstraal 1954; Shkolnik and Borut 1969; Atallah 1970; Shkolnik 1971; Kronfeld et al. 1994). The diurnal activity of *A. russatus* is believed to be caused by the influence of its competitor the Egyptian spiny mouse *Acomys cahirinus* (Shkolnik and Borut 1969; Kronfeld et al. 1994). Where the density of the latter species decreases, *A. russatus* becomes nocturnal (Shkolnik 1971). In addition, the daily rhythms of metabolic rate and oxygen consumption of *A. russatus* in the absence of *A. cahirinus* are typical of those of a nocturnal animal (Haim and Zisapel 1995; Haim and Fluxman 1996). Our observations in the Central Negev desert show that at a constant low density of both species, *A. russatus* mainly demonstrates nocturnal activity.

Finally, subterranean species (*Ellobius, Thomomys, Pappogeomys, Ctenomys*) demonstrate polyphasis day and night activity. The same is true of two voles, *Alticola streltzovi* and *Alticola barakschin*.

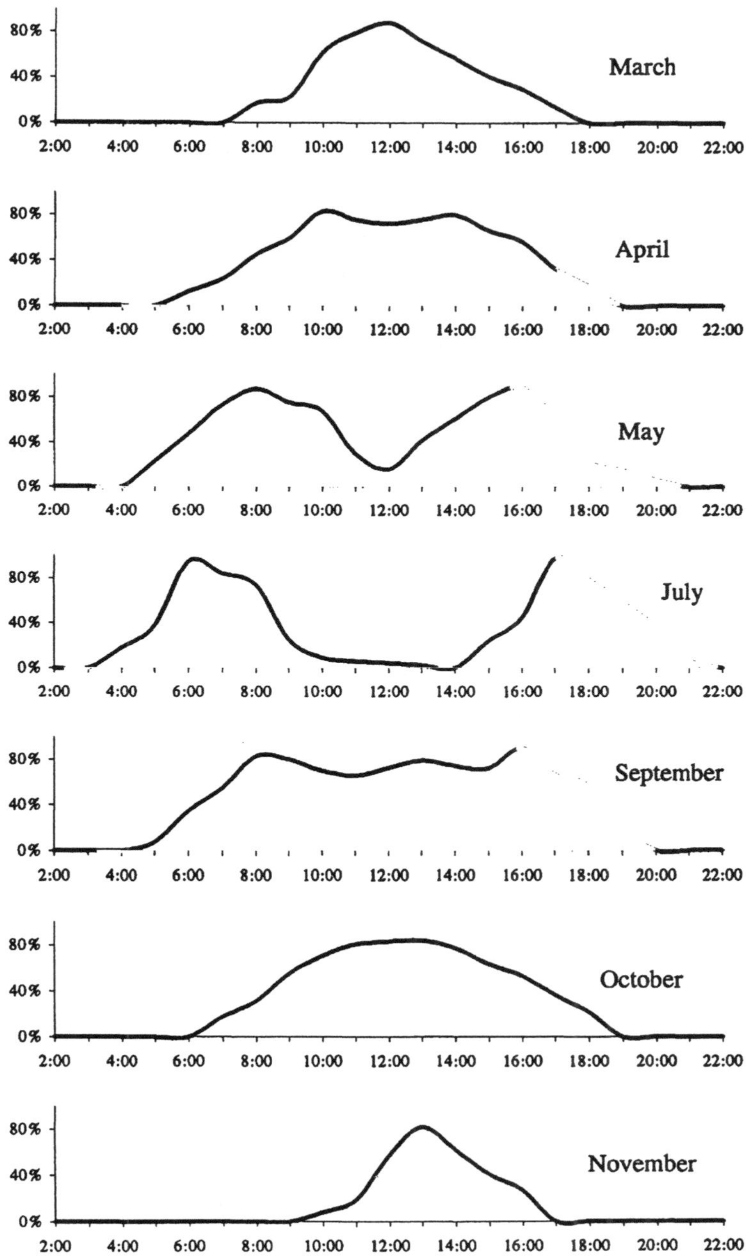

Fig. 5.3. Seasonal changes in pattern of daily activity of *Rhombomys opimus* (based on the data of Shekhanov 1952, cited by Sokolov and Kuznetzov 1978). Vertical axes are percentages of active individuals

5.1.2
The Desert Substrate: Problems of Burrowing and Locomotion

Desert soils differ from those in humid zones in some significant respects, all of which are related to scant moisture and exposure to evaporation. The biologically active layer of the soil in deserts is either extremely thin or often even absent (Petrov 1976; Cooke et al. 1993), so that surface properties are determined mainly by the geological substrate and/or aeolian processes. As a result, the types of substrates in deserts differ sharply, have abrupt boundaries and form a spatial mosaic. Furthermore, they vary from extremely hard (rocks) to extremely soft (sand).

All the above-mentioned features suggest that burrowing in desert substrates must be constrained, and that desert rodents live in harsher conditions, from this point of view, than do those in humid zones. Some desert rodents do not dig burrows, but construct aboveground dens. The dens of wood rats, *Neotoma* spp., are elaborate dwellings composed of foliage, stems, twigs, stones and any other available material (Bonaccorso and Brown 1972; Alvarez et al. 1988; Smith 1995). These dens are used, renovated and added to by several generations of *Neotoma* and may grow up to 2 m in both height and diameter (Alvarez et al. 1988; Nowak 1991; Smith 1995; Hrachovy et al. 1996). Stones and Hayward (1968) recorded dens of the desert wood rat *Neotoma lepida* with heights of about 7.5 m.

Rock-dwelling species also have no genuine burrows because their habitat precludes the possibility of digging. Strict rock-dwellers, such as *Acomys russatus, Sekeetamys calurus*, and *Petromus typicus* make use of rock crevices and piles of boulders as shelter. Facultative rock-dwellers, such as Wagner's gerbil *Gerbillus dasyurus* and the large North African dipodil *Gerbillus campestris* are likewise unable to construct underground refuges in rocky areas, whereas in other habitats they build regular burrows (Hatough-Bouran 1990).

Most desert rodents construct burrows of variable complexity. These are usually deeper that those of their relatives in humid zones. For example, the depth of the wintering chambers (=hibernaculae) of the steppe ground squirrel *Spermophilus dauricus* varies from 65 to 175 cm (Bibikov 1952, cited by Kucheruk 1983), whereas those of the desert ground squirrel *Spermophilus pygmaeus* have been found at depths from 89 to 259 cm (Rall and Demiasheva 1934, cited by Kucheruk 1983).

Soft desert substrates (e.g. loess soils, sandy-loam soils, hard-sand soils) do not require adaptations for digging that differ from those of non-desert rodents. However, two types of desert substrates, namely flat depressions with hard clay surfaces (=takyrs) and non-stabilized sands do demand specific behavior. The main problem encountered when burrowing in takyrs is to breach an extremely hard upper layer. The solution to this is called "vibration" burrowing. This behaviour is demonstrated by different species of the five-toed jerboas (Allactaginae), namely *Allactaga elater, Allactaga severtzovi, Pygeretmus pumilio*. These species have weakly arched, thin and relatively long lower incisors which they use as a main working tool (Zubtzova 1978). When burrowing, the lower incisors of five-toed jerboas perform quick percussive movements, thereby creating cracks in the substrate (Gambarian et al. 1978).

The problem of burrowing in sand is that sand mobility causes the walls and roof of the burrow to cave in, so animals need to strengthen these parts of their burrows. Most sand-dwellers, as well as opportunist species inhabiting sand patches, avoid this by burrowing under bushes or grass clumps where the sand is stabilized by plant roots [e.g. the three-toed dwarf jerboa *Salpingotus crassicauda* (Shenbrot et al. 1995)], Sundevall's jird, *Meriones crassus* (Krasnov et al. 1996b)). Seely (1977) noted that some South African gerbils in the genus *Gerbillurus* burrow where the sand has been consolidated by the urine of ungulates. Some psammophiles strengthen the walls and the roof of the burrow using their heads. The head of the animal performs vibrating movements which tamp the sand (I.M. Fokin, pers. comm.).

Another problem of sand-dwelling is locomotion; this is difficult on account of the softness of sand. The feet of most sand specialists have broad soles increased by brushes of long and coarse hairs. In some three-toed jerboas (*Dipus, Paradipus, Eremodipus*) this brush helps its owner to jump from the sandy surface (see below). The hairs of the brush of these rodents are oriented perpendicularly to the lower plane of each toe. When an animal takes off quickly this brush prevents the so-called bulldozer effect, i.e. the shifting of the sand under the jerboa's foot and the creation of a wall of sand (Fokin 1978).

5.1.3
Desert Vegetation and Rodents

The structure and composition of desert vegetation differs sharply from that of other biomes. Tall trees are almost absent from deserts, so arboreal rodents are extremely rare (with the notable exceptions of the desert dormouse *Selevinia*, in Central Kazakhstan, the garden dormice *Eliomys* in North Africa and Arabia, and the wood rats *Neotoma* in North American deserts, which, however, should be regarded as "climbing" rather than "arboreal" animals). The grass cover is seldom continuous in deserts and is mostly ephemeral, whilst the proportion of graminaceous plants is generally extremely low, except in the Gobi, Namib and Australian deserts. This explains the absence of both vole-like species in desert rodent faunas (except for steppe lemmings of the genus *Eolagurus* in the Gobi Desert) and species constructing aboveground nests of grass [with the exception of Littledale's whistling rat, *Parotomys littledalei*, in the Kalahari Desert (De Graaff 1981) and the desert mouse *Pseudomys desertor* in central Australia (Strahan 1983)].

Succulent vegetation is frequently abundant in deserts, encouraging the existence of folivorous species (e.g. *Psammomys, Rhombomys, Parotomys, Neotoma, Paradipus, Microcavia* spp.). Exceptions are *Euphorbia* and *Mesembryanthemum* in Africa which are succulent but chemically well-defended. Feeding on succulents provides rodents with water in the absence of free water sources (see below). At the same time, folivores frequently have a problem with salt excretion on account of the high salt level in many of these plants (see review in Degen 1996; Mares et al. 1997).

Plant cover in deserts occurs in a mosaic of vegetated patches alternating with areas of bare ground. This distribution is reflected in the dichotomy of locomotion found in desert rodents. Quadrupedal species mainly occupy covered patches and bipedal species mainly use open spaces (see below). Thus, the plant mosaic exacerbates habitat divergence determined by soil diversity in deserts.

5.2
Distribution of Body Sizes Within Regional Faunas

Variation in body mass is associated with a variety of physiological, behavioral and ecological responses to the desert habitat such as metabolic rate, litter size, longevity, home range size, etc. (McMahon 1973; McMahon and Bonner 1983, Calder 1984; Schmidt-Nielsen 1984; Reiss 1989; Charnov 1993 etc.). In fact, Peters (1983) related 88 physiological functions to animal body mass. In his recent book "Macroecology" Brown (1995) explicitly discussed the distribution of body sizes among species of different taxa on different scales. His work was based both on his own data (Brown and Nicoletto 1991; Brown et al. 1993 etc.) and on data from other studies (Hutchinson and MacArthur 1959; Morse et al. 1988; Holling 1992 etc.). The main rules of body size distribution are: (1) when considered on a large (regional) scale, size distributions are highly skewed, so that there are many more small species than large or extremely small ones; (2) when considered on a small (local) scale, these distributions become uniform and (3) the optimal size for mammals is approximately 100 g.

Several hypotheses have been proposed to explain the observed distributions. For example, the explanation of the skewness of body size distribution toward small species (on a large spatial scale) by Hutchinson and MacArthur (1959) is that the modal-sized species are more specialized and thus they can divide resources and space more finely. Contrastingly, Eisenberg (1981) stated that these species are the most generalized, and can therefore use the broadest range of resources. The uniform body size distributions on a local scale are explained by the more rapid spatial turnover of modal-sized species compared with larger or smaller species. Species of modal size differ among the small-scale samples, whereas the larger/smaller species occur repeatedly in different samples (Brown 1995). The hypothetical optimal mammalian size of 100 g, which closely matches the observed size frequency distribution, has been calculated according to a model of allometric energetic constraints on fitness (Brown et al. 1993).

Figures 5.4-5.6 demonstrate the distributions of body size of desert rodents on a logarithmic scale in different desert realms or subrealms. These distributions are limited to the same range of sizes across deserts, from pygmy granivorous or insectivorous species to relatively large omnivores and folivores. Except for the Atacama and Monte-Patagonian realms, the distributions are highly skewed. The uniform distribution of body size in the Atacama can be accounted for by the poverty of the fauna and by the relatively small size of this realm. In this case, the level of consideration lies somewhere in between those of regional and local scales. It should be also noted that species of the genus *Phyllotis* (the leaf-eared mice) in the Atacama realm are strictly allopatric (see above) and their spatial distributions can be regarded as being analogous to species turnover on a local scale. The case of the Monte-Patagonian realm is more complicated. It seems that here we are dealing with two independent, barely overlapping distributions of different taxonomic and trophic taxa, namely small and medium-sized omnivorous Sigmodontinae and large folivorous Hystricognatha (Fig. 5.7). Superimposing these distributions produces a distribution that appears to be uniform, although in reality it is not.

There are pronounced differences in modal body size among different desert biogeographic units. The mode is at 11-20 g body mass for the Saharo-Arabian subrealm, at 41-80 g for the Irano-Sindian subrealm, and at 21-40 g for the Turano-Gobian and Australian subrealms. The distribution of body mass among North American desert rodents is bimodal with one mode occurring at 11-20 g body mass and the other at 81-160 g body mass. In contrast, this distribution has a broad mode at 21-80 g body mass for the South African desert rodent assemblage. The difference in the positions of the modal values appears to contradict Brown et al. (1993), who predicted a unimodal distribution of body mass of mammals with the mode at 100 g. This is understandable, because rodents can be considered as a subsample of the class Mammalia and are represented mainly by small and medium-sized species (at least, in recent times). The rodent modal class is thus shifted to small body size values.

The differences described require some explanation, which can be differences in the distribution of rodents with different feeding modes in different desert regions. The details of feeding mode distributions will be described below. Nevertheless, we need to remember that rodents demonstrate a correlation between dietary habits and body size. Small species with high energy expenditure per unit body mass must use energy-rich food (seeds or invertebrates), whereas species that feed on food of low calorific value (folivores) must have greater body masses to be able to digest large amounts of food and

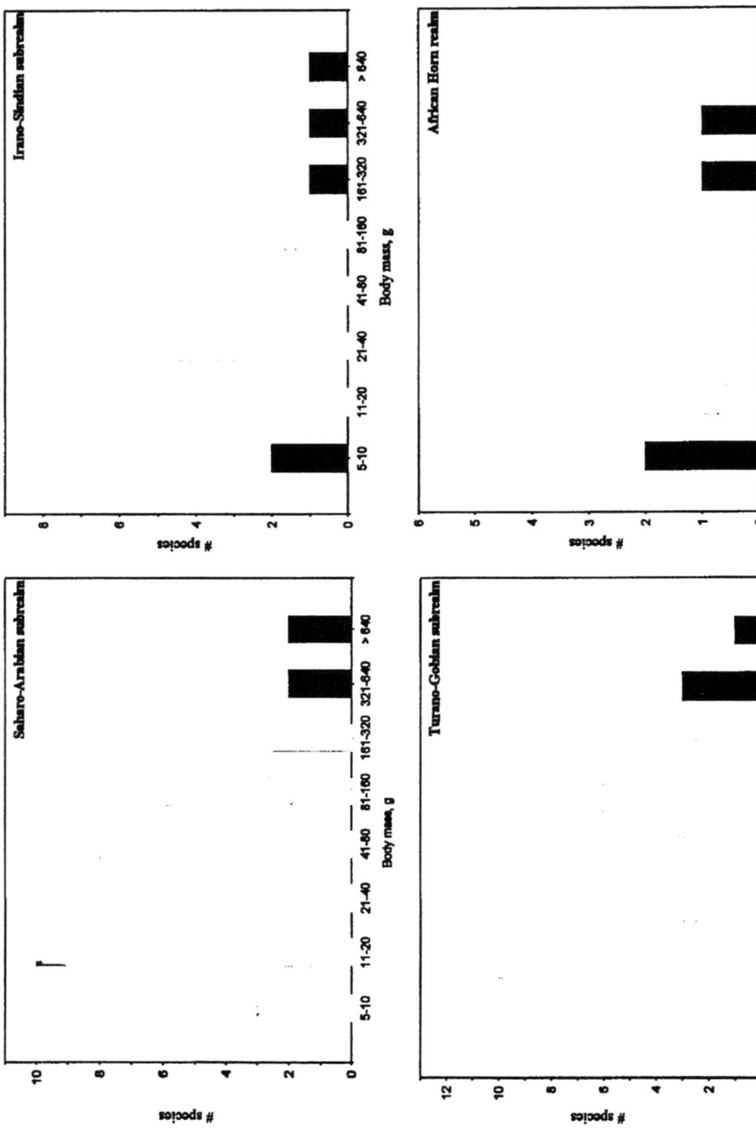

Fig. 5.4. Frequency distributions of number of species with different body masses for the faunas of the Palaeoarctic realms/subrealms

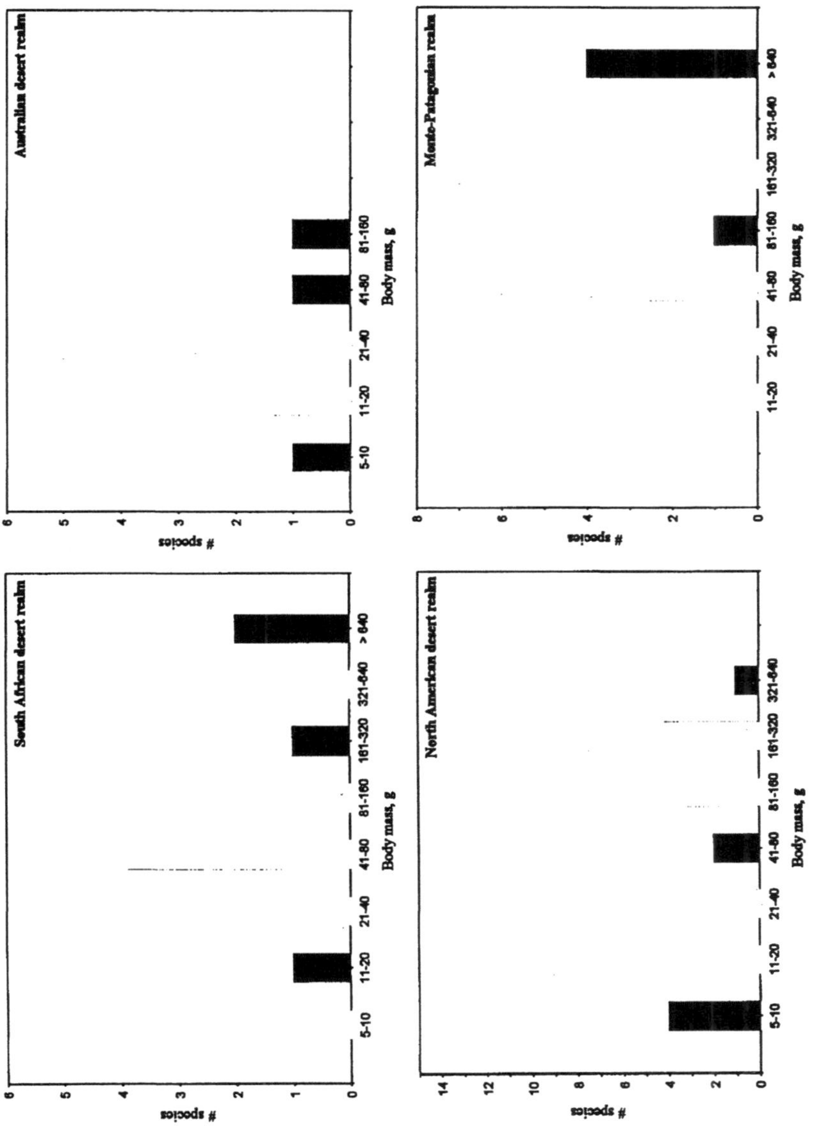

Fig. 5.5. Frequency distributions of number of species with different body masses for the faunas of the South African, Australian, North American and Monte-Patagonian realms

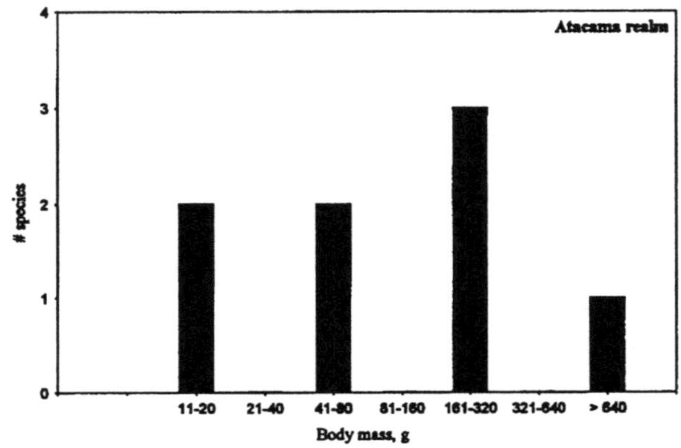

Fig. 5.6. Frequency distributions of number of species with different body masses for the fauna of the Atacama realm

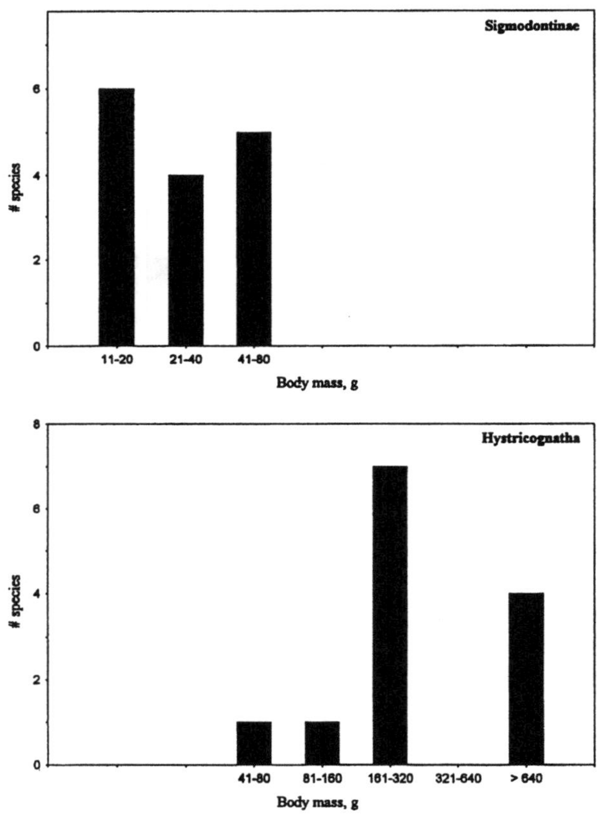

Fig. 5.7. Frequency distributions of number of species of Sigmodontinae and Hystricognatha with different body masses for the fauna of the Monte-Patagonian realm

not spend as much energy per unit body mass as do species with small body masses. These rules are true both on a very general level (with respect to the

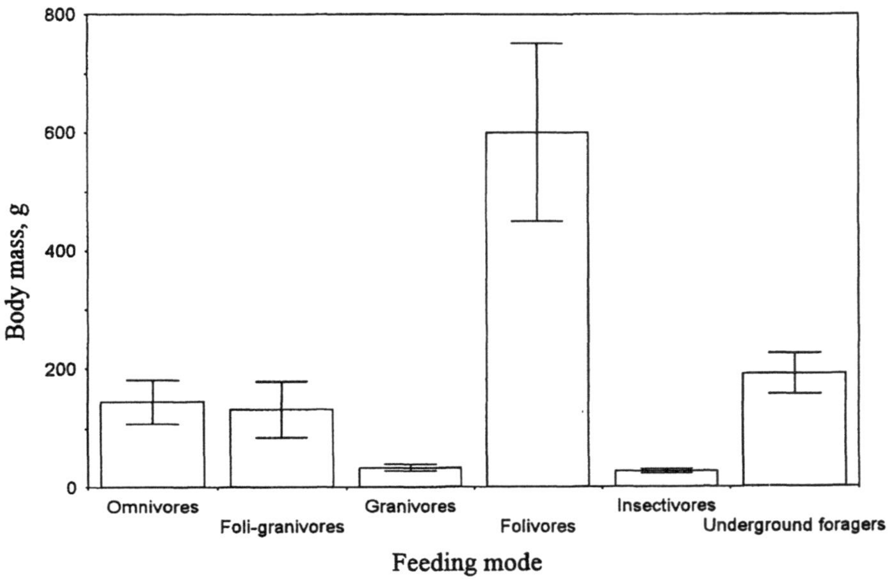

Fig. 5.8. Body masses of species with different feeding modes for the whole rodent fauna of all deserts

whole desert rodent fauna of the world, Fig. 5.8) and also when we consider species within biogeographic realms (Figs. 5.9, 5.10). In all cases, granivorous and insectivorous species are significantly smaller than folivores or foli-granivores. The only exception is the Australian rodent fauna which comprises relatively large granivores and relatively small foli-granivores. The variance in the body mass of omnivores is high, reflecting the mixed composition of this trophic category. The ratio of different components in the diet of an omnivorous species appears to play a primary role in the determination of its body mass.

These patterns explain the differences in modal body size among different desert regions. The large granivorous component of the Saharo-Arabian fauna (see below) is reflected by the shift to smaller body size, whereas the decrease of this component in the Irano-Sindian fauna shifts the mode towards that of larger species. The bimodal pattern of the distribution of feeding modes in North America is determined on the one hand, by small granivorous heteromyids and sigmodontines and, on the other, by omnivorous heteromyids and sciurids and folivorous wood rats (*Neotoma*).

The distributions of body sizes of rodent species occurring together at the local level (per 1-ha sampling site) are uniform (Fig. 5.11). This supports the findings of Brown and Nicoletto (1991). Local habitat patches comprise a similar number of species belonging to each logarithmic size category. In

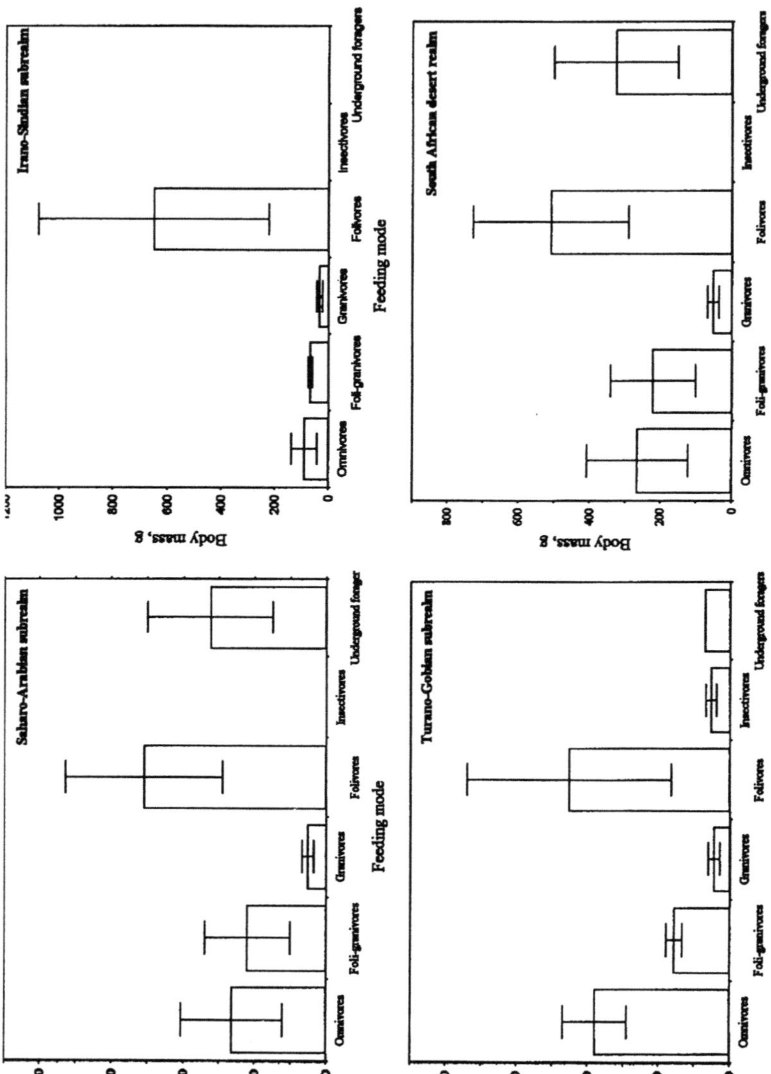

Fig. 5.9. Body masses of species with different feeding modes for the faunas of the Palaeoarctic and Palaeotropic realms/subrealms

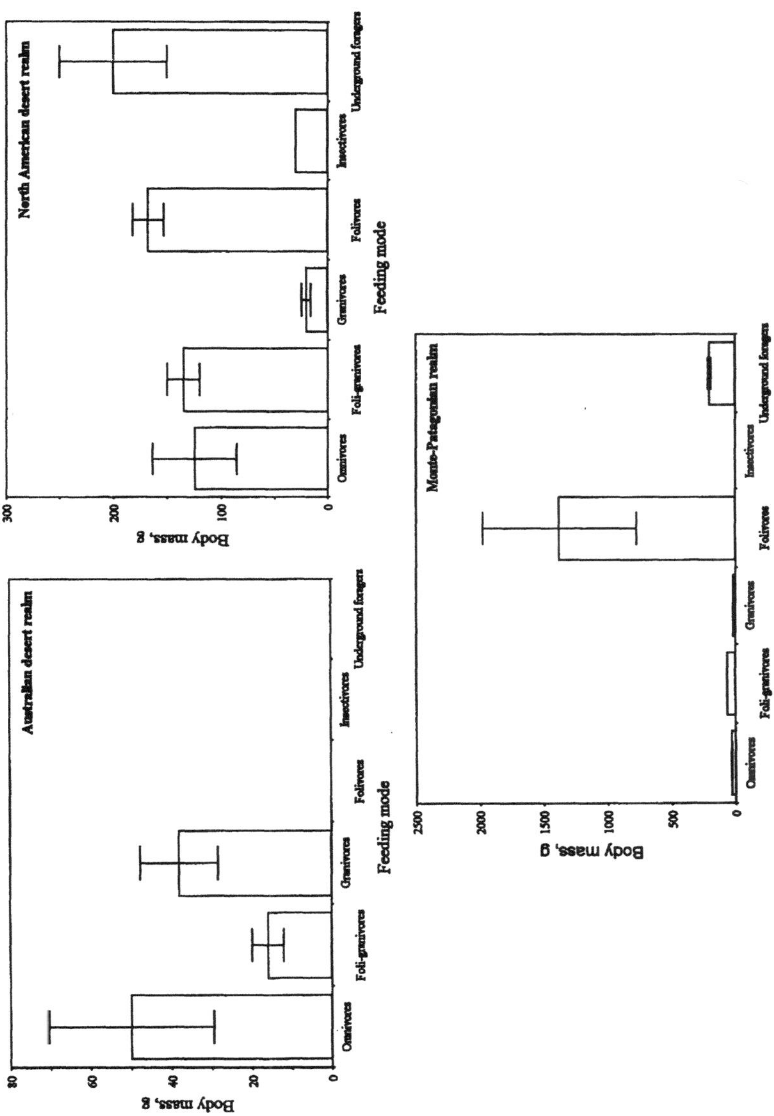

Fig. 5.10. Body masses of species with different feeding modes for the faunas of the Australian, North American and Monte-Patagonian realms

other words, a given community comprises less species of similar body size than it does species of different body size, and the diversity of species in relation to body size tends to be high. Brown and Nicoletto (1991) explained the observed distribution of rodents with different body sizes by competitive exclusion of species of similar size within local habitats (but see Abrams 1990a,b). There are a number of suggested mechanisms for the co-existence of rodents with different body size. For granivores, it was initially proposed that the size of seeds collected is positively correlated with body size (Brown and Lieberman 1973). Later field observations and experimental testing has not, however, supported this hypothesis unambiguously (Smigel and Rosenzweig 1974; Reichman 1975; Mares and Williams 1977; Lemen 1978; M'Closkey 1980, 1983; Price 1983). Later, a new hypothesis was proposed which predicted the different use of seed with spatial distribution by rodents of differing sizes. Larger species select clumped seed aggregations whereas the small species use dispersed seeds (Reichman and Oberstein 1977). The new hypothesis was supported by both experimental (Hutto 1978; Price 1978) and field (Bowers 1982; Thompson 1982a; Price and Brown 1983) data. Belovsky (1997) has proposed an explanation of the above mentioned pattern for herbivores using optimal foraging models. These include a forager's physiological and morphological abilities, so that plants with different characteristics are selected by herbivores of different body sizes. Allometric relationships in digestive physiology and nutritional demands cause the selection of plants having different characteristics by animals of different body size. The result is that a herbivore of each body size can utilize some plants exclusively while sharing other plants with other species. The above-cited hypotheses are not all mutually exclusive but rather are complementary. Some of them explain the mechanism of exclusion of species of the same size within local communities, whereas others explain the mechanism of co-existence of species of different size. Using computer-simulation models, Kelt (1997) argued that the assembly rules of local communities are governed mainly by the existence of a modal body size mediated by competitive character displacement. In his models, species attempt to attain this optimal body size which seems to reflect maximal fitness but are displaced from this point by competitive interactions manifested as character displacement.

5.3 Distribution of Locomotory Modes Within Regional Faunas

The appearance of any rodent seems to be determined to a great extent by the predominant type of locomotion used. There are three basic features of the desert landscape which influence the behavior of mammals of small body size: (1) the relative openness of the terrain, (2) the patchy distribution of feeding resources and shelters (desert shrub vegetation), and (3) the poverty of the

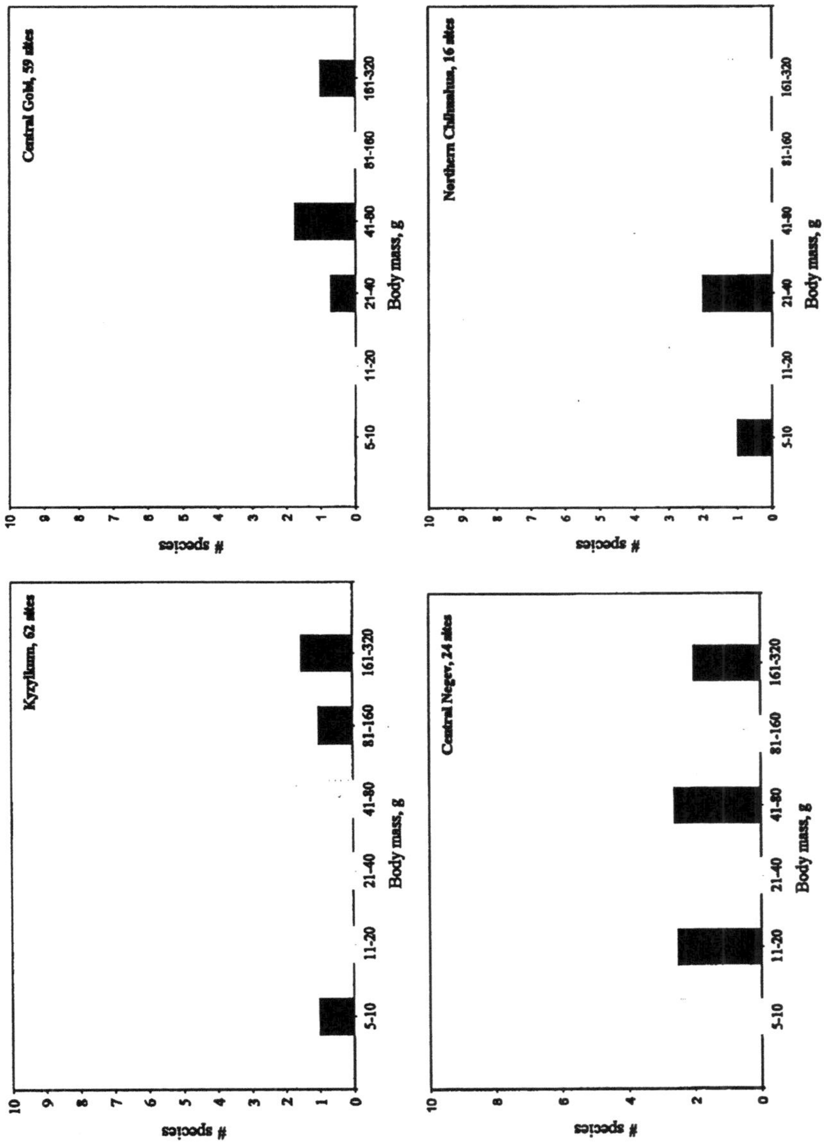

Fig. 5.11. Frequency distributions of number of species with different body masses for four local communities of desert rodents

resource base. Terrestrial desert rodents are forced to look for sparse and/or patchily distributed food and to escape predatory attacks under conditions of low and heterogeneous distribution of shelter. Both predator pressure and competition for food resources affect the behavior of desert rodents and, thus, may influence the evolution of locomotory mode (Price and Brown 1983).

Among rodent species that inhabit deserts, there are fossorial (=subterranean), terrestrial and even semi-arboreal (=climbing) forms, whose locomotion does not differ much from that of fossorial, terrestrial or semi-arboreal rodents in non-desert areas. However, there are also bipedal rodents whose locomotion appears to be specific to deserts. Bipedality has evolved independently in at least four rodent taxa: Dipodidae (Palaearctic), Heteromyidae (Neoarctic), Pedetidae (Palaeotropic) and Muridae (Australia) (Berman 1985; Mares 1993a). In addition, there are (or were) bipedal forms of other groups of mammals which converge on bipedal rodents. The Australian deserts are inhabited by bipedal marsupials of the families Dasyuridae, Peramelidae, Potoroidae and Macropodidae. There used to be bipedal marsupials of the family Argirolagidae in the desert and grassland areas of South America up to the early Pleistocene (Simpson 1980). The African order, Macroscelidea, also contains facultatively bipedal species.

The bipedal locomotory mode includes different gaits which are characteristic of a taxon-specific stage of specialization for long distance and fast running (Fokin 1963). Palaearctic three-toed (Dipodinae) and five-toed (Allactaginae) jerboas include the most advanced obligate bipeds (Fokin1978). In contrast, the North American kangaroo rats are facultative bipeds, and use their forelimbs when moving slowly (Eisenberg 1963).

Both the comparative morphological and biomechanical analysis of the locomotory system of recent desert quadrupeds and bipeds permits us to understand the specificity of rodents' bipedal gaits and to reconstruct the genesis of the bipedality to some extent. Gambarian (1972) deduced the bipedal saltatory gait from the "primitive quadrupedal ricochet jump". Ricochet jumping appears to be a primitive form of quadrupedal gallop in which the hind feet work synchronously and fulfil the main role in the initial push from the substrate surface. The input of forelimbs is much less. They serve as shock absorbers and balance the animal's body on the substrate until its hind feet are carried ahead of them (this gait is exemplified by some pocket mice, e.g. *Perognathus, Chaetodipus,* and gerbils, e.g. *Gerbillus, Meriones*; Fig. 5.12a). An increase in jump length and velocity is provided by carrying the hind feet so far forwards that the forelimbs lose their function. This type of gait can be observed in primitive bipeds such as dwarf jerboas (*Cardiocranius, Salpingotus,* Fig. 5.12b; Fokin 1963) and bipedal heteromyids (*Microdipodops, Dipodomys*; Eisenberg, 1963). In contrast to the heteromyids or cardiocraniines, most jerboas are more specialized for bipedal locomotion according to the morphology of their hind feet. The fast-running forms of jerboas (Allactaginae, Dipodinae) can use two bipedal gaits, namely (1) a symmetrical bipe-

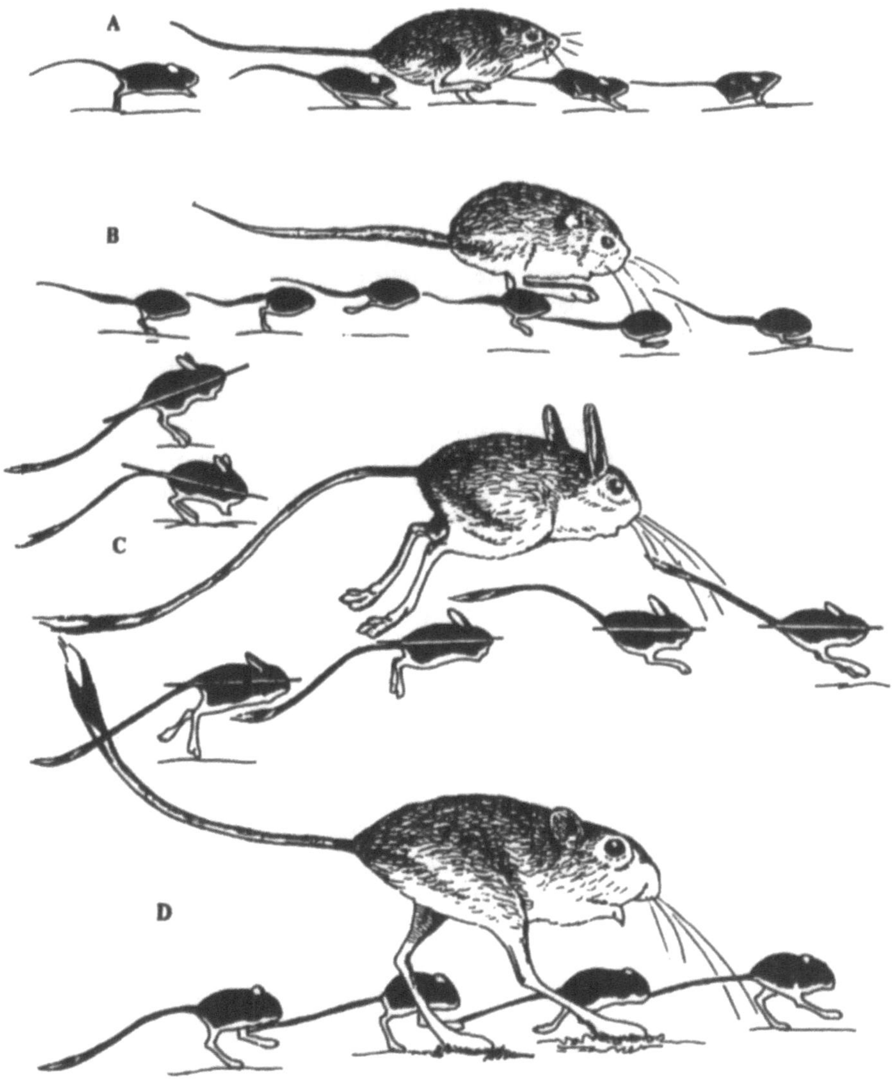

Fig. 5.12. The diversity of bipedal locomotory modes among the Asian Dipodidae: A Quadrupedal ricochet jump of *Sicista*. B Synchronous ricochet of *Cardiocranius*. C Nonsynchronous ricochet of *Pygeretmus*. D Symmetrical trot of *Dipus*

dal run, a trot pace with an alternate right and left foothold and the absence of a phase of "free flight" (Fig.5.12c); and (2) an asymmetrical saltatory pace during which a push is performed initially by one and then by the other hind foot, thus providing additional acceleration (Fig. 5.12d; Fokin 1963, 1978).

Quadrupedal and bipedal locomotory modes are traditionally considered to be qualitatively different but uniform within themselves. This point of view can be easily accepted in discussing the morphological divergence in the North American heteromyids. The foraging economics (Reichman and Oberstein 1977; Reichman 1981; Reichman and Price 1993 etc.), predators (Webster 1962; Webster and Webster 1971; Kotler 1984a,b; Price and Brown 1983; Price 1986; Kotler et al.1988, 1991, 1993) and the interaction of both factors (Price and Brown 1983; Price 1986) have been considered to cause this dichotomy which, in turn, is a keystone of microhabitat selection. Bipedal kangaroo rats usually prefer open patches, whereas quadrupedal pocket mice are associated with dense shrub vegetation (see Price and Brown 1983; Price 1986; Reichman and Price 1993 for reviews). However, a number of exceptions have been reported (Wondolleck 1978; Thompson 1982a; Kotler 1984a, 1985) and the dichotomy is not absolute. The reason for this seems to be a relatively low degree of specialization of heteromyids for fast bipedal running and, consequently, their dependence on the proximity of shelter. Considering the entire diversity of desert rodents, it becomes clear that the quadrupedal/bipedal contradistinction really represents opposite ends of the continuum of locomotor adaptation. For example, Asian hamsters (*Phodopus, Cricetulus*) use quadrupedal locomotion exclusively. The next stage on the continuum is exemplified by gerbils and jirds which use bipedal ricochet jumps when they begin to speed up (Fokin 1978). Dipodoidea demonstrate almost the entire evolutionary sequence of locomotion from the quadrupedal gait of the Sminthidae through the primitive ricochet jump of the Cardiocraniinae to the asymmetrical saltatory pace and symmetrical bipedal run of the Allactaginae, Dipodinae and Paradipodinae (see above). The type of locomotion is connected with escape tactics (Fig. 5.13). The best runners among jerboas (e.g. *Paradipus ctenodactylus*) tend to escape a danger by running away. Small and slow species (e.g. *Pygeretmus platyurus, Cardiocranius paradoxus*) usually hide using the relief features of the ground. Species of medium size that have a moderate running speed use both escape tactics. Either they hide under shrubs (e.g. *Eremodipus lichtensteinii*) or use an extensive system of shelter burrows. They may construct these burrows themselves (*Pygeretmus pumilio, Pygeretmus zhitkovi*) or make use of the burrows of other rodents (*Stylodipus telum*).

The distribution of locomotory modes in the main desert regions of the world as reflected by the distribution of the main morpho-ecological types of rodents is shown in Fig.5.14. When considered superficially, these distributions look similar. There are, however, clear distinctions reflecting major ecological differences between different faunas. Terrestrial quadrupeds, which predominantly use a ricochet gallop, dominate in the desert areas of all continents. Bipedal rodents are highly diverse in the Palaearctic region (except for the Horn of Africa). They are poorly represented in South Africa and Australia (where a rich fauna of bipedal marsupials occurs) and are com-

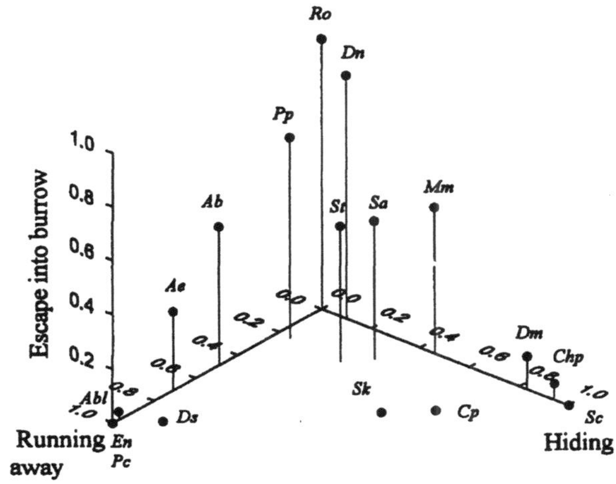

Fig. 5.13. Proportion of individuals demonstrating patterns of escape behaviour in some desert rodents. Animal responses to the human approach were registered in the field during night excursions with the searchlight. Numbers in the parentheses after species names are ratios of hind foot length/head-and-body length. Species names are as follows: Ab *Allactaga bullata* (0.44), En *Euchoreutes naso* (0.48), Cp *Cardiocranius paradoxus* (0.44), Sc *Salpingotus crassicauda* (0.42), Sk *S. kozlovi* (0.44), Ae *Allactaga elater* (0.48), Abl *Al. balikunica* (0.47), Pp *Pygeretmus pumilio* (0.44), Ds *Dipus sagitta* (0.48), St *Stilodipus telum* (0.42), Sa *S. andrewsi* (0.43), Pc *Paradipus ctenodactylus* (0.52), Mm *Meriones meridianus* (0.25), Ro *Rhombomys opimus* (0.24), Chp *Chaetodipus penicillatus* (0.30), Dm *Dipodomys merriami* (0.37), Dn *D. nelsoni* (0.34)

pletely absent from the South American deserts. Among quadrupeds, there are species that are able to climb. These usually have movable mouse-like tails and can climb small desert shrubs to feed. However, true arboreal forms are absent from deserts as a consequence of the scarcity of trees. Some desert regions are inhabited by subterranean rodents. These are most diverse in South Africa and North America.

Different distributions of rodents with different locomotory modes among deserts are also illustrated by comparison of the number of species with different types of locomotion collected at randomly-distributed sampling sites (Table 5.1). All regions differ significantly in the relative abundance of representatives of each functional group per sampling site (one-way ANOVA for non-climbing quadrupeds: $F = 129.1, p < 0.001$; for climbing quadrupeds: $F = 200.1$, $p < 0.001$; for bipeds: $F = 54.9$, $p < 0.001$). Only in two cases did pairwise comparisons between regions not show significant differences: (1) between the South Asian and Australian deserts as regards bipedal species, and (2) between the Australian and North American deserts with respect to climbing quadrupedal species (t -test for independent groups and Mann-Whitney U-test).

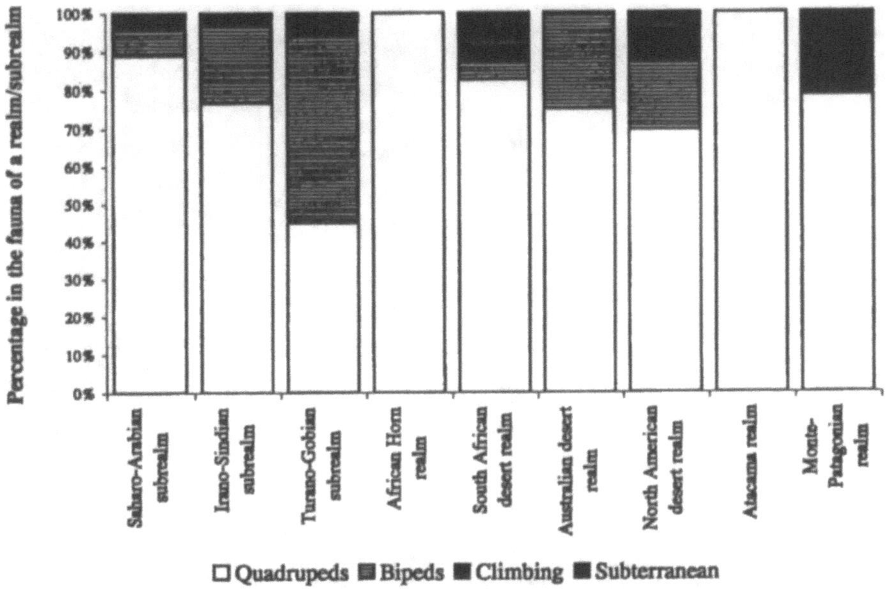

Fig. 5.14. Distribution of rodent species with different locomotory modes in different desert realms/subrealms

Despite a number of the observations of the local coexistence of bipedal and quadrupedal rodents in the North American (Brown and Kurzius 1987), Australian (Morton et al. 1994) and Asian (Kelt et al. 1996) deserts, there is considerable unevenness in the available information between North America and the other continents. Very intensive studies have been conducted in the North American deserts (see Price and Brown 1983; Price 1986; Kotler and Brown 1988; Brown and Harney 1993; Reichman and Price 1993 for reviews). A few studies have been performed in the Asian deserts (Rogovin 1988; Rogovin and Shenbrot 1995; Shenbrot and Rogovin 1995), but practically no work has been carried out on the deserts of Africa and Australia. The mechanisms of rodent coexistence within one habitat type will be discussed below. Here, we would like to emphasize that, to date, there is no common opinion on the co-evolution of bipedal and quadrupedal rodents. Given that co-evolution has occurred, interspecific competition for resources would be a likely cause.

It has been suggested that bipeds and quadrupeds may select different microhabitats. This suggestion is based on the difference in food distribution between vegetated and open patches, and on the differences in foraging energetics between bipedal and quadrupedal species (Reichman and Oberstein 1977; Reichman 1981). This may be explained by the difference in cost between bipedal and quadrupedal locomotion at high speed (Dawson and

Table 5.1. Mean number (<±> SE) of nocturnal species of different functional groups collected per sampling site in the Asian, Australian and North American deserts. Each pool of species assemblages at a number of sites within desert region, was represented by the randomly distributed field observations. Number of sampling sites for each region in brackets

Region	Non-climbing quadrupeds	Climbing quadrupeds	Bipeds	Data source
Northern Asia (169)	1.21 <±> 0.08	0.06 <±> 0.02	2.24 <±> 0.01	Our data
Southern Asia (43)	2.06 <±> 0.17	0.58 <±> 0.16	0.18 <±> 0.07	Our data
Australia (244)	0.01 <±> 0.01	1.07 <±> 0.06	0.25 <±> 0.03	Morton et al. (1994)
North America (201)	1.64 <±> 0.07	1.26 <±> 0.07	1.06 <±> 0.07	Brown and Kurzius (1987)

Taylor 1973; Dawson 1976; Hoyt and Tailor 1981; but see Thompson et al. 1980; Thompson 1985). If bipeds expend less energy at high speed than do quadrupeds of similar size, they would specialize on clumped, widely spaced seed resources to a greater extent than would quadrupedal species.

It has been shown that combinations of bipedal and quadrupedal heteromyids within local assemblages are not random (Brown and Harney 1993). In this study, 61 two-species combinations within local communities were analyzed by determining the frequency of coexistence of pairs of species in three functional groups: bipedal heteromyids (*Dipodomys* and *Microdipodops*), quadrupedal heteromyids (*Perognathus* and *Chaetodipus*) and quadrupedal sigmodontines (*Peromyscus* and *Reithrodontomys*). Pairs of species belonging to different functional groups occurred together more frequently than would be expected by chance, while pairs of quadrupedal heteromyids coexisted less frequently than expected. A similar pattern has been demonstrated in a test of the hypothesis of an assembly rule for functional groups developed by Fox (1989). North American desert rodent assemblages had significantly nonrandom compositions, consistent with the concept that species in different functional groups (based on mode of locomotion and/or trophic relations) are more likely to be able to coexist (Fox and Brown 1993). Indeed, rodent communities in the Asian deserts contain more species of a particular functional group (i.e. bipedal or quadrupedal) than expected by random samples drawn from the pool of species available. This has also been shown indirectly by comparison of the morphological organization of the Central Asian and North American desert rodent communities (Rogovin and Surov 1990). Both these findings support the role of interspecific competition, but the Asian desert rodent communities seem to be largely structured by past competitive interactions; consequently, the current community structure is based on strong habitat selection. Our observations in the North American and Central Asian deserts support this finding. The result of discriminant function analysis of the microhabitat characteristics at the sites where animals were recorded demonstrates that in the rodent community of the central Chihuahuan Desert, spatial niche overlap between bipedal and quadrupedal heteromyids is much greater than among only bipedal or quadrupedal species (Rogovin et al. 1991). In the rodent community of the Kyzylkum Desert there is no coincidence in the spatial preferences between bipedal and quadrupedal rodents (Shenbrot and Rogovin 1995).

When analyzing the differences between the Asian and North American rodent assemblages, we should take into account the difference in body size between coexisting bipedal and quadrupedal heteromyids. Harvest rate, travel speed and metabolic costs are allometric functions of body size, so the degree to which a particular species will specialize on the most profitable patches should depend on its size (Price 1981, 1983). Sophisticated statistical tests of null hypotheses, performed on the large data sets available from the North American deserts, have shown repeatedly that species coexisting in local habitats are highly nonrandom assemblages with respect to their sizes (Brown 1973; Bowers and Brown 1982; Hopf and Brown 1986). However, this is not the case for the Palaearctic deserts, where bipedal jerboas and quadrupedal jirds do not show pronounced size differences. Nevertheless, communities in the southern Palaearctic deserts formed mainly by granivorous gerbils appear to be well structured with respect to their body sizes (Yom-Tov 1991). Moreover, the northern Palaearctic deserts are widely inhabited by jerboas with diverse feeding adaptations. The distribution of body sizes within local assemblages of related jerboa species (Allactaginae) also appears not to be random (Rogovin 1995).

It must be emphasized that although bipedal locomotion evolved independently among different rodent lineages, the degree of specialization for fast bipedal running and the diversity of bipedal rodents differs drastically among regions. Factors such as the origin of the fauna, the phylogenetic inertia of different rodent groups and geographical, geomorphological, and climatological conditions during the formation of each desert region may have played a significant role in determining the diversity of bipedal and quadrupedal rodents within local assemblages.

5.4 Distribution of Feeding Modes Within Regional Faunas

Several of the characteristics of deserts influence the trophic specialization of small mammalian species. Short periods of plant growth, low biomass and low diversity of soil invertebrates, patchiness in the distribution of feeding resources, the sparse distribution of food due to overall low primary productivity and the low degree of predictability of food abundance in time and space (at local and regional spatial scales) appear to result from the desert climate (water deficit, high temperatures, extreme seasonality in both tropical and temperate deserts). Desert rodents can be readily classified into six tro-

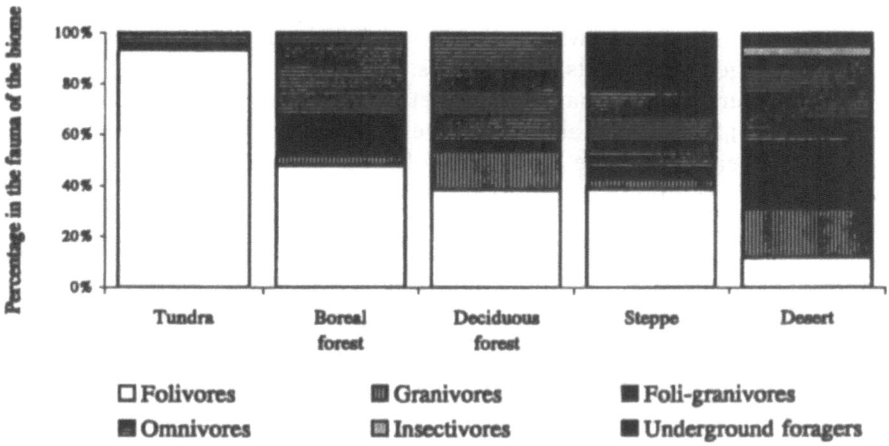

Fig. 5.15. Distribution of rodent species with different feeding modes in the biomes of the Northern Eurasia. Data from Gromov and Erbajeva 1995; Pavlinov et al. 1995

phic groups: (1) folivores, (2) granivores, (3) foli-granivores (species that feed on a mixed diet of seed and plant vegetative parts), (4) omnivores, whose diet includes both animal and plant components (seeds, green and underground parts), (5) insectivores (=carnivores that feed mainly on invertebrates, although they may sometimes eat small reptiles and mammals), and (6) underground foragers that feed on the subterranean parts of plants (roots, bulbs, rhizomes).

None of these categories is exclusive to deserts. However, the representation and distribution of feeding modes in the desert biome differs significantly from those of other biomes. For example, within the Eurasian continent, significant differences in the distribution of rodent species with different dietary habits occur between desert and tundra biomes (Kolmogorov-Smirnov two-sample test, $\lambda = 1.99$; $p < 0.001$), desert and boreal forest biomes ($\lambda = 1.56$; $p < 0.05$), and desert and steppe biomes ($\lambda = 1.45$; $p < 0.05$; Fig. 5.15). The fraction of folivores decreases from north to south, reaching a minimum in the desert zone. Deserts also contain the largest fraction of granivores, foli-granivores and insectivores. The high diversity of the last group in deserts can be correlated with the decrease in species diversity of small Insectivora (Soricidae) that inhabit the humid zones of the continent and depend greatly on the richness of soil invertebrates (Ivanter and Makarov 1994; Shvartz et al. 1997). There are two soricid species in the Eurasian tundra, nine in boreal forests, 11 in deciduous forests of the Far East, four in deciduous European forests, six in the steppe zone, and two species in deserts (Pavlinov et al. 1995). Marked seasonality, low productivity and the instability of feeding resources in the Eurasian deserts explain relatively high proportion of rodent species whose diet includes different foods, such as seeds and green vegetation, plants and animals.

Many desert rodents change their diet from season to season (Sludsky et al. 1977, 1978; Shenbrot et al. 1995) and even from year to year (Shenbrot et al. 1995) depending on the availability of food. Fresh green vegetation is a typical dietary component of most species during wet periods. Fresh vegetation provides animals not only with the water, but also with the green plant factor, 6-methoxybenzoxazolinone (6-MBOA), which has been shown to be important for successful reproduction (Negus and Prinier 1966; Field 1975; Negus and Berger 1977; Berger at al. 1987; Linn 1991). Consequently, an increase in population growth and reproduction of desert rodents often follows rainfall and subsequent flush of annual plants (Sludsky et al. 1977, 1978; Munger et al. 1983).

The proportion of seeds in the diet of desert rodents usually increases in the dry season and in years of unpredictable drought. For example, in Trans-Altai Gobi, where productivity is extremely low and precipitation unpredictable (Sokolov and Gunin 1986), in years of extreme drought jerboas and jirds forage on seeds to a greater extent than usual (Rogovin et al. 1986). In times of extreme drought, all species (even the predominantly granivorous *Cardiocranius paradoxus*) cease breeding. These and some other related patterns of response have been described repeatedly in the zoological literature and are reasonably considered as adaptations for surviving in harsh periods. Another example is that of Asian jerboas which never have more than one litter per year. They go into hibernation long before species that have two or three litters (Sludsky et al. 1977; Shenbrot et al. 1995). Accumulation of fat by these early hibernators is also associated with an increase of the percentage of the energy-rich food items in their diets. Other jerboas also reduce the number of litters in seasons of drought when only seeds are available (Shenbrot et al. 1995).

The low productivity and unpredictability of the desert environment do not favor narrow feeding specializations. Desert is not an exception from the general rule which predicts specialization only when there is a certain degree of abundance and predictability of the resource (Emlen 1966; Schoener 1971; MacArthur 1972; Charnov 1976). There are two main strategies among desert rodents to narrow a trophic niche, namely towards feeding on either energy-rich food (i.e. seeds and/or invertebrates) or on the green parts of plants.

An advantage of granivory is that seed banks in deserts seem to provide a relatively stable source of food (Reichman and Price 1993). However, feeding on an energy-rich diet such as seeds implies the ability of rodents to be independent of free water, at least periodically. The results of classic studies of the metabolic pathways and water regulatory affinities of desert rodents (Schmidt-Nielsen et al. 1948; Schmidt-Nielsen 1964, 1975a,b; MacMillen 1972; see Degen 1996 for review) provide us with information on one of the most remarkable mammalian adaptations to aridity – the ability of desert-adapted rodents to survive (without loss in body mass) when using only metabolic water. The "water-independent" species inhabit different desert regions where there is a long history of their adaptation to aridity. Examples of these can be found in the Australian deserts (MacMillen and Lee 1969; Baudinette 1972), Palaearctic (Winkelman and Getz 1962; Ghosh 1975; Kirmiz 1962) and Palaeotropic deserts (Christian 1978, 1979), and the South American deserts (Koford 1968, but see Mares 1977b). It is noteworthy, however, that Mares (1975b) analyzed the level of physiological adaptations (rates of water loss during complete water deprivation), natural history and distribution patterns of some South American and North American desert rodents, and found that the South American sigmodontines are not highly specialized for desert life (see also Mares 1975a, 1977a,b, 1980).

Because discussion of ecophysiological affinities lies outside the scope of this book, it is possible merely to note here that independence from free water seems to be strictly determined by body sizes among desert rodents (see MacMillen and Hinds 1983; Degen 1996). Jerboas that have diverse spectra of feeding adaptations, as well as predominantly granivorous heteromyids, demonstrate that diversity of diet increases linearly with increase in body mass (Fig. 5.16). An increase in diversity of diet is correlated with increase in body size, irrespective of the type of food consumed. This is universal pattern which has been reported in many different animal taxa (Freeland 1991, Brown 1995).

Relatively high predictability and abundance of food resources are considered to be conditions for the evolutionary development of folivory among desert rodents. Because the great abundance of ephemeral herbs in deserts is restricted to short periods of precipitation, specialization for feeding on green vegetation is possible only in species able to use plants that remain green during the dry season. Such plants usually have a high salt content. Numerous morphological, physiological and behavioral adaptations of desert rodents to overcome high salinity of food have been described (see Schmidt-Nielsen 1964, 1975a,b; Degen 1996 and references therein). For example, desert rodents are capable of producing highly concentrated urine, to reduce fecal and respiratory water loss. Elongated renal papillae may be an example of a morphological adaptation. Some desert folivores remove the most saline parts of plants or choose young leaves and sprouts which are less saline. The specialized folivorous jird *Psammomys obesus* of the Negev desert usually removes the hypersaline epidermal layer from the leaves or stems of its principal food plant *Atriplex halimus* (Degen et al. 1988; Kam and Degen 1989). Similar behaviour has been reported in the North American chisel-toothed kangaroo rat *Dipodomys microps* (Kenagy 1972, 1973) and the South American red vizcacha rat *Typmanoctomys barrerae* (Mares et al. 1997). The diurnal and folivorous Central Asian great gerbil *Rhombomys opimus* feeds on succulent sprouts of *Anabasis* or *Haloxylon* in midsummer, climbing over the plant to cut off twigs and feed on the youngest parts (Kucheruk et al.

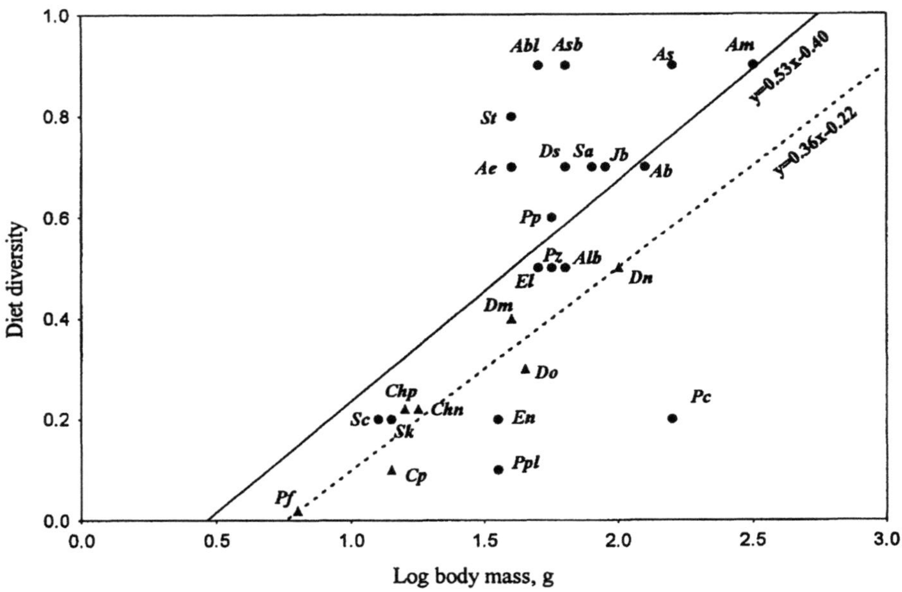

Fig.5.16. Diet diversity of dipodid (circles, solid line) and heteromyid (triangles, dashed line) rodents plotted against log body mass. Diet diversity was calculated based on stomach contents, using Shannon's entropy index. (Data from Rogovin 1985; Shenbrot et al. 1995). Species names are as follows: Am *Allactaga major*, As *A. severtzovi*, Asb *A. sibirica*, Ae *A. elater*, Ab *A. bullata*, Abl *A. balikunica*, Alb *Allactodipus bobrinskii*, Pp *Pygeretmus pumilio*, Pz *P. zhitkovi*, Ppl *P. platyurus*, Cp *Cardiocranius paradoxus*, Sc *Salpingotus crassicauda*, Sk *S. kozlovi*, Ds *Dipus sagitta*, St *Stylodipus telum*, Sa *S. andrewsi*, El *Eremodipus lichtensteinii*, Jb *Jaculus blanfordi*, Pc *Paradipus ctenodactylus*, Eu *Euchoreutes naso*, Dn *Dipodomys nelsoni*, Do *D. ordii*, Dm *D. merriami*, Pf *Perognathus flavus*, Chp *Chaetodipus penicillatus*, Chn *C. nelsoni*

1972). The desert wood rat *Neotoma lepida* in the Mojave Desert (which lacks succulents such as cactus, yucca, and agave), eats mature leaves and stems of the creosote bush (*Larrea tridentata*) during the late fall and early winter before the winter rains. After the winter rains it feeds on newly germinating annuals (Karasov 1989). The reason for this is that *Larrea* has a high resin content, so wood rats avoid eating it whenever possible.

In all cases, feeding specializations appear to be connected with strict behavioral adaptations. Asian Dipodidae, a taxon with diverse feeding modes, exhibits perhaps the most exotic strategies for food gathering among desert rodents. The dwarf jerboa *Salpingotus kozlovi* which inhabits sand dunes in the Central Asian deserts, is mainly granivorous although it feeds on insects too. It usually forages on the steep slopes of dunes. Its snout is surrounded by short dense bristles. There are also two bunches of long coarse vibrissae which are oriented perpendicularly to the body axis. Digging with its snout and forelimbs, the dwarf jerboa makes small slides of sand. By the use of its

Fig. 5.17. Foraging *Salpingotus crassicauda*

vibrissae it is able to feel the movements of insect larvae in the soft sand within a radius equal to its head-and-body length (Fig. 5.17). Searching for seeds, it erects its vibrissae, moves them down and "plows" the sand with them.

Another inhabitant of the Central Asian deserts, the long-eared jerboa *Euchoreutes naso*, can locate flying insects actively using its long mobile ears. It catches flying moths by quick vertical leaps (Fig. 5.18). This species also hunts lizards (*Phrynocephalus versicolor*) sleeping on the ground.

The specialist-folivore comb-toed jerboa *Paradipus ctenodactylus* inhabits sand deserts of Turan. It feeds on the young green twigs of succulent shrubs and annuals that grow on the slopes and ridges of sand dunes (Fig. 5.19). It seems surprising that, after cutting off a green twig, this rodent does not take the food into the shelter, but runs, carrying the twig, to the top of a sand dune, and eats it there. Extremely sensitive hearing and strong hind feet protect the animal from predators. Running across a moving sand dune, this species (body mass = 140 g) can reach a velocity of about 32 km/h.

How uniform are different desert regions with respect to the representation and relative abundance of different trophic groups of rodents? To what extent does the argument, developed in the tradition of inter-zone comparisons within one continent, apply to the situation on different continents? Deficiency of information limits our ability to make detailed intercontinental

Fig. 5.18. Foraging *Euchoreutes naso*

Fig. 5.19. Foraging *Paradipus ctenodactylus*

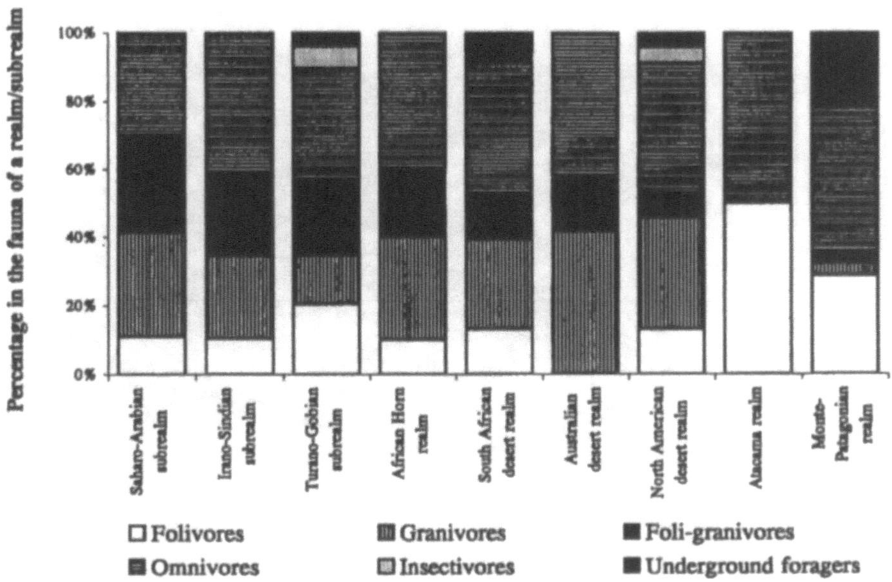

Fig. 5.20. Distribution of rodent species with different feeding modes in different desert realms/subrealms

comparisons. We can attempt to do that only in a very general way. Nonetheless, a relatively superficial classification of desert rodents into six trophic groups may help us to understand the degree to which we can justify broad-scale ecological extrapolations.

At first glance, the distribution of the dietary habits of rodents in different desert realms or subrealms seems similar (Fig 5.20). However, some differences should be noted. Faunas of most realms or subrealms contain species of four to five trophic modes, except for the Atacama realm where only two trophic groups are present. The cause of this seems to be the general poverty of the local fauna in the Atacama. The largest representation of folivores occurs in the Atacama and Monte-Patagonian realms. This group is only moderately represented in other regions, excluding Australia, where strict folivores are absent. One reason for the latter phenomenon may be that the extremely high diversity of lizards in the Australian deserts results in suitable for rodents niches being occupied. Granivores constitute a relatively high proportion of all faunas except in South America and the Turano-Gobian subrealm of the Palaearctic. Rodents that have a mixed (seed and green) diet are common only in the Old World, whereas insectivores and underground foragers are relatively infrequent everywhere.

Consideration of the trophic compositions of rodent faunas on a smaller scale provides us with additional information. Ordination of the provincial

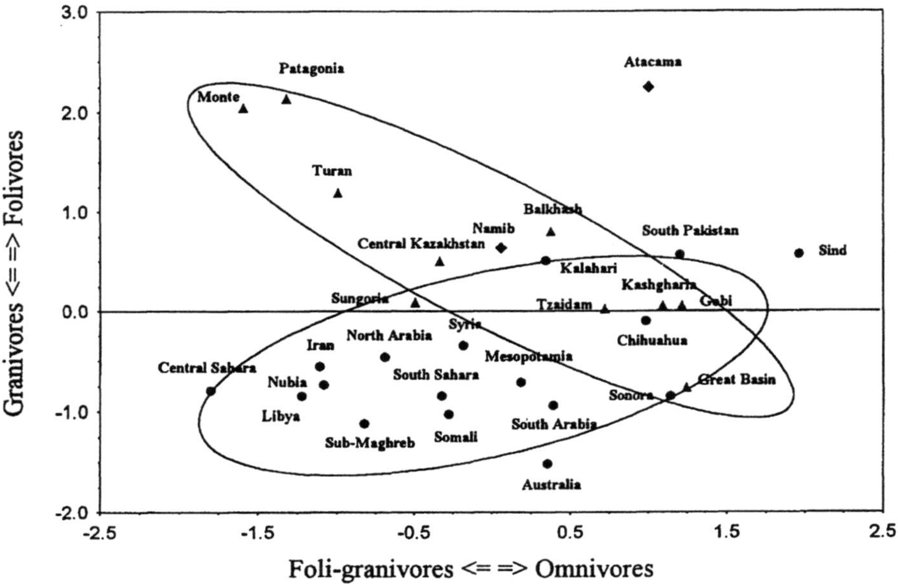

Fig. 5.21. Scatterplot of the ordination of provincial desert rodent faunas according to their composition of rodents with different feeding modes in the space of the two first principal component axes. Circles Subtropical provinces, triangles temperate provinces, diamonds coastal provinces

faunas by principal component analysis demonstrates that the faunas of the subtropical provinces contain more granivores, whereas those of temperate provinces contain more folivores (Fig. 5.21). Furthermore, there are two directions of trophic composition turnover, namely from the faunas containing mostly omnivores to those with a predominance of granivores (from the Gobian and Sindian superprovinces to the North Saharan and Central Saharan superprovinces) or to those with a predominance of folivores (from the North American desert realm to the Monte-Patagonian realm). Plotting the diversity of species with different feeding modes against the taxonomic diversity of the faunas demonstrates a significant positive correlation between these parameters (Fig. 5.22). Moreover, the percentage of folivores increases with an increase of taxonomic diversity (Fig. 5.23a), whereas the opposite is the case for the percentage of granivores (Fig. 5.23b). The proportion of both foli-granivores and omnivores as well as insectivores do not depend on taxonomic diversity.

Significant inter-regional differences exist in the trophic diversity of the local rodent communities of the deserts of North and South America, Australia and Asia (Kelt et al. 1996). Data used in this analysis were presence/absence matrices of terrestrial mammal species with a mass of less

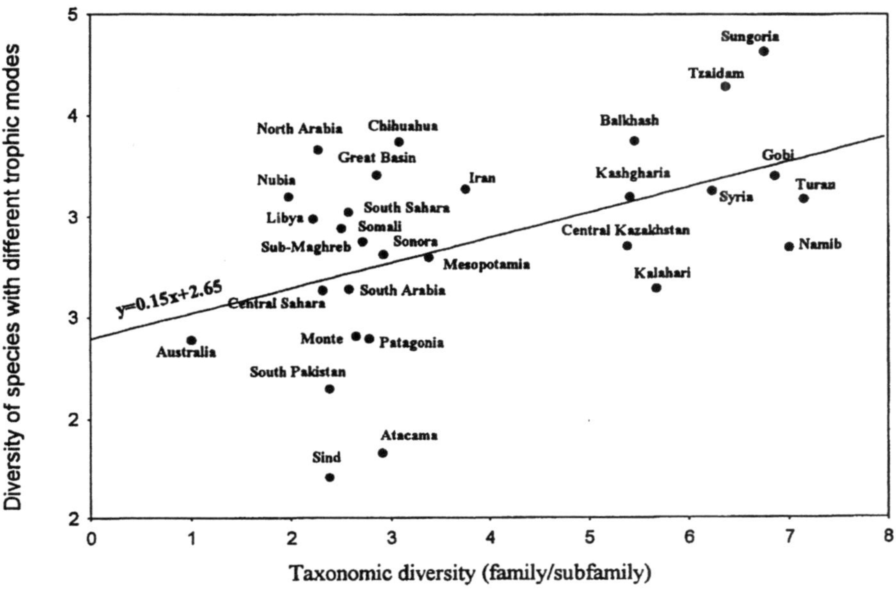

Fig. 5.22. The relationship between diversity of species with different feeding modes and taxonomic diversity of the provincial faunas. Indices of diversity are analogous to Hill's N2 diversity index (Hill 1973)

than ≈500 g which had been recorded at numerous sampling sites. The representation of species of four trophic categories, granivores, folivores, omnivores and carnivores was found to differ significantly among these deserts (Kelt et al. 1996). The highest percentage of granivores is in North America, although they are also numerous at sites in the Thar, Gobi, Negev, and Turkestan (Turan) deserts. Sites in the Turanian deserts contain significantly more folivorous species than sites in other deserts. The percentage of folivores does not differ between sites in the Gobi, Thar, Negev, and South American deserts, but was found to be significantly greater in these than in the deserts of North America or Australia. Omnivores are most abundant in the South American and Thar deserts.

These results demonstrate that in spite of a number of common biotic features and some remarkable examples of morphological similarities between species, desert regions demonstrate a relatively low degree of parallelism in terms of diversity and the existence of different feeding strategies. What are the reasons for the dissimilarities in the relative abundance of taxa exploiting different feeding resources across regions? These differences may be explained by differences in contemporary environments as well as by the different histories of the deserts and the mammalian lineages occurring in them

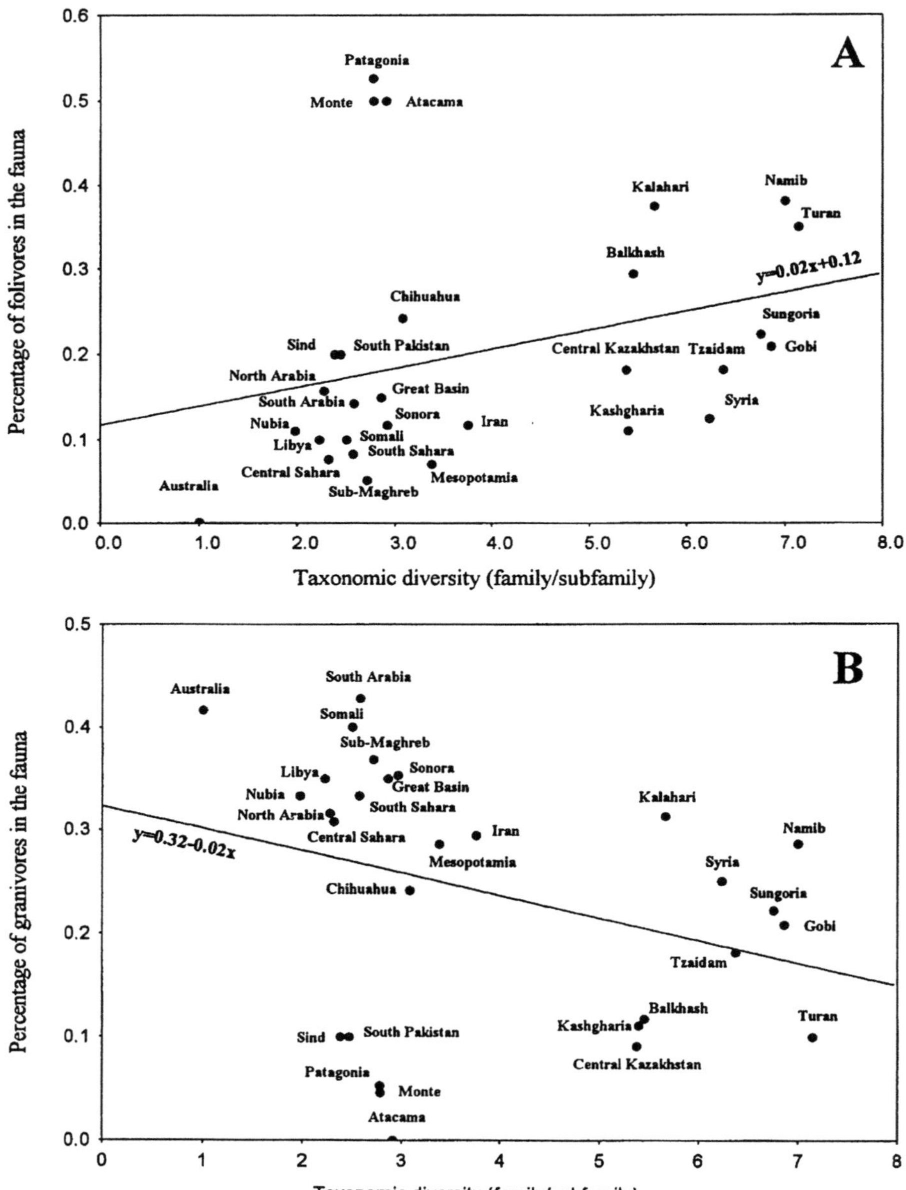

Fig. 5.23. The relationship between diversity of folivorous (A) and granivorous (B) species and taxonomic diversity of the provincial faunas. Indices of diversity are analogous to Hill's N2 diversity index (Hill 1973)

(Kelt et al. 1996). Three main features of the contemporary environment may influence the variability in the distribution of feeding modes among regional faunas. As mentioned above, desert regions differ drastically in: (1) latitude and elevation, (2) isolation from the ocean, and (3) topographic relief. These features strongly influence the climate, especially with regard to its predictability, and the daily and seasonal fluctuations in temperature and precipitation. For example, the deserts of the Turano-Gobian subrealm, the Great Basin of North America and the deserts of the Monte-Patagonian realm occur at temperate latitudes, whereas the North African, Irano-Sindian, South African and Australian deserts are essentially subtropical. The Namib and Atacama deserts occur near oceanic coastlines, whereas the Gobi and Great Basin deserts are located at relatively high elevations. Topographic relief also varies greatly across deserts. The southwestern part of North America presents alternating mesic and xeric conditions as a result of numerous mountain ridges oriented in a north-south direction. Mountain ridges in the desert belt of Asia are generally oriented in an east-west direction. The vast desert lands of Asia are characterized by extensive and continuous habitats, whereas the much more restricted desert basins of North America provide smaller habitats. The broad, open expanses with low topographic complexity of the Australian deserts provide vast areas of monotonous desert vegetation, whereas the vegetation in the coastal deserts of South America is extremely sparse.

Both the amount and predictability of precipitation vary greatly among deserts. This may affect the predictability of certain food resources and, thereby, the presence of different trophic strategies. The absence of strict granivores in Australia may partly reflect the low predictability of annual precipitation and, therefore, of a reliable seed crop (Morton 1993). However, the number of granivorous taxa does not depend much on either mean annual precipitation or on its coefficient of variation, suggesting that this cannot explain granivore species abundance in North America.

Another reason for differences in the distribution of rodent feeding adaptations among deserts can lie in the existence of historically-distinct faunal lineages. As previously mentioned, Australian desert rodents are represented by murids. This taxon retains its evolutionary ancestral omnivory (see Vorontsov 1967; Naumova 1981 for further references). The North African and Asian deserts are widely inhabited by Gerbillinae and Dipodidae. Gerbils (Gerbillinae) appear to be ancestrally granivorous and retain strict or partial granivory as a dominant feeding strategy. Jerboas (Dipodidae) demonstrate a wide range of feeding adaptations but species with the most primitive bipedal gaits and hind feet morphology retain a diet of seeds and insects (Fokin 1978). The North American family Heteromyidae is unique in the possession of external fur-lined cheek pouches, an evolutionary innovation that may have been central to its subsequent specialization on seeds. Cricetine rodents comprise another group inhabiting arid lands of both North and South

America. Sigmodontines of the North American desert are represented by omnivorous, folivorous and insectivorous species.

Rodents of South America illustrate the roles of both the age of the deserts and the evolutionary trail of the taxa within a given desert region. The latter may be decisive in determining the diversity of trophic strategies. The absence of granivorous sigmodontines in South America may be a consequence of the limited time available for adaptation to arid conditions as a result of the late invasion from the north. The absence of granivorous sigmodontines may also be connected with their origin in the humid tropical rain forests (Simpson 1980; Mares 1980, 1993a,b). The high proportion of folivores can be explained by the presence of "caviomorphs" (=South American hystricognaths), the ancient inhabitants of South America. It seems that the invading sigmodontines have displaced all the trophic generalist caviomorphs, whilst the relatively young specialist folivores remain. It should be remembered that folivory is considered to be the latest stage of the trophic evolution among rodents (Vorontsov 1967).

Although the Miocene was a period of global aridization (Alpers and Brimhall 1988; Singh 1988; Rundel et al. 1991), the extent and stability of desert areas differed significantly. For example, the deserts of the central Asia are known to have persisted since the Cretaceous, but modern landscapes began appear in the Miocene (Sinitzin 1962). The extensive area of the central Asian deserts and the uninterrupted process of desert formation during the Neogene provided conditions for the progressive divergence of the Dipodidae. These were originally omnivorous. This divergence resulted in the development of a wide spectrum of feeding adaptations. The North American deserts appeared as early as the late Miocene (Riddle 1995), but underwent significant fluctuations in the Pleistocene (Axelrod 1958; Webb 1977; Van Devender and Spaulding 1979; Thompson and Mead 1982). The contraction and expansion of arid intermountain basins with each pluvial/interpluvial cycle (Schmidly et al. 1993) in southwestern North America may have increased geographical isolation among related taxa, thereby resulting in greater opportunity for allopatric speciation than has occurred in the Asian deserts (Shenbrot et al. 1994b). However, the essential conditions for ecological divergence did not occur. North American Heteromyidae are derived from savanna-dwelling ancestors in the late Oligocene (Hafner and Hafner 1983; Hafner 1993; Wahlert 1993), but have not as yet diverged much with respect to their feeding strategies.

It would be easy to continue this list of the possible explanations of trophic diversity in desert rodents, but we will stop here, restating that both similarities and differences across geographically isolated desert faunas reflect the contemporary geography and physiognomy of the deserts, as well as the phenomenon that Brown (1995) has termed the "history of place" and the "history of lineage". The variation in rodent trophic strategies among desert regions signifies the low degree of convergence at the level of multispecies communities. An emphasis on local processes alone seems insufficient to provide reasonable explanations of large-scale patterns (Reichman 1991). Only careful integration of contemporary influences and the dynamics of history can help to explain geographic variation in trophic strategies among desert rodents.

CHAPTER 6

Life Forms of Desert Rodents and Convergent Evolution Among the Species of Different Faunas

6.1 Life Forms of Desert Rodents

A *life form* can be defined as a group of species that share a common life history strategy, are similar in general appearance, and have similar adaptations to a particular environment. A life form should therefore be considered as a set of adaptations to a specific environment, so that a species appertaining to a particular life form is the result of certain evolutionary processes. A classification of life forms should therefore generally coincide with a taxonomic classification because the latter (at least, an ideal one!) is also a reflection of evolution. A typical example is demonstrated by ground squirrels. Desert species of this group are very similar in general appearance, have similar biological features, life history traits and are closely related taxonomically. In contrast, there are some species that originate from different phyletic lineages but which appear to enjoy the same life form; such species are usually aboriginals of different geographic regions and are known as "ecological equivalents" or "ecological analogues" (Cloudsley-Thompson 1996). For example, Mares (1993b) noted a profound similarity between *Gerbillus* and *Perognathus*, although the former genus belongs to the Old World Gerbillinae and the latter to the New World Heteromyidae. The existence of ecological equivalents demonstrates that the pathways of adaptation have similar evolutionary directions in similar environments, even though the geographical positions of the milieus and primary gene pools of the specializing groups are quite different (Cloudsley-Thompson 1993, 1996).

One of the earliest efforts to classify the life forms of desert animals was undertaken by Formozov (1929). He based his classification on general biological features, namely food habits and the vertical stratification of environmental use. Later, Formozov proposed the theory of life forms (Formozov 1956, 1964, 1976). He noted that the initial concept of life forms was elucidated as early as in the middle of the nineteenth century by Severtzov (1855). Khodasheva (1953) has attempted a classification of desert rodents based on the same features that Formozov (1929) used. The resulting classifications of life forms partly overlapped with taxonomic classification. This is not surprising because the faunas under consideration were local (Mongolia and Kazakhstan, respectively). Moreover, only two parameters were taken into account and this had the result of combining quite different species (both ecologically and taxonomically) into the same life form, for example jerboas (Dipodidae) and mice (Murinae; Khodasheva 1953).

Mares (1973, 1980, 1993b) used the above-mentioned parameters in addition to body size and proportions (the locomotory mode), habitat preference, certain physiological and anatomical features, daily and seasonal activity and pattern of sociality to classify desert rodent life forms. In addition, he used the term "ecological niche" rather than "life form". Mares' (1993b) classification reflects the natural patterns of adaptation much better than those of Formozov (1929) and Khodasheva (1953). Nevertheless, it too is flawed. The flaws are caused primarily by two factors: (1) the procedure of classification was not so much formalized as intuitive, and, consequently, not repeatable; (2) Mares used well-defined ecological types of North American rodents and tried to incorporate the worldwide diversity of rodents into this Procrustean bed. Additionally, some of the data he used on Old World rodents were not correct. For example, one of the Tables in Mares (1993b) indicates that *Jaculus* is a granivore, *Allactaga* a herbivore, while *Salpingotus, Pygeretmus* and *Dipus* are omnivores. However, *Jaculus* has a mixed diet of seeds and green vegetation, *Allactaga* is omnivorous, *Salpingotus* and *Dipus* are mainly granivores, and *Pygeretmus* is a herbivore (Shenbrot et al. 1995). As a result, *Microdipodops* was included in the same life form as *Jaculus*, yet the former is a perfect ecological equivalent of the dwarf jerboas (*Salpingotus, Cardiocranius*). Moreover, the dwarf jerboas and true jerboas (*Jaculus*) should not be combined in one group because they show distinct size differences and size-related biological features.

We present here our classification of the life forms of desert rodents. It is based on the formal analysis of the following features: body size, dietary habits, locomotory mode, vertical stratification of habitat use and nesting activity, pattern of daily activity, the presence of hibernation and/or aestivation and degree of sociality. Principal component ordination of more than 200 species of desert rodents shows that four main life forms can be distinguished (Fig. 6.1a,b). The life forms of the highest rank differ mainly in (1) pattern of daily activity, which is correlated with size (on average, diurnal species are larger than nocturnal ones), and (2) locomotory mode.

One of the life forms, namely fossorial species, is homogenous. Three other life forms can be subdivided into life forms of lower rank. Diurnal quadrupedal species are represented by two groups that differ drastically in dietary mode, which is correlated with level of sociality and also with body size (Fig. 6.2). The classification of nocturnal rodents results in a two-level hierarchical scheme (diet mode and type of shelters for quadrupeds [Fig. 6.3]; body size and diet mode for bipeds [Fig. 6.4]). The resultant classification is given below.

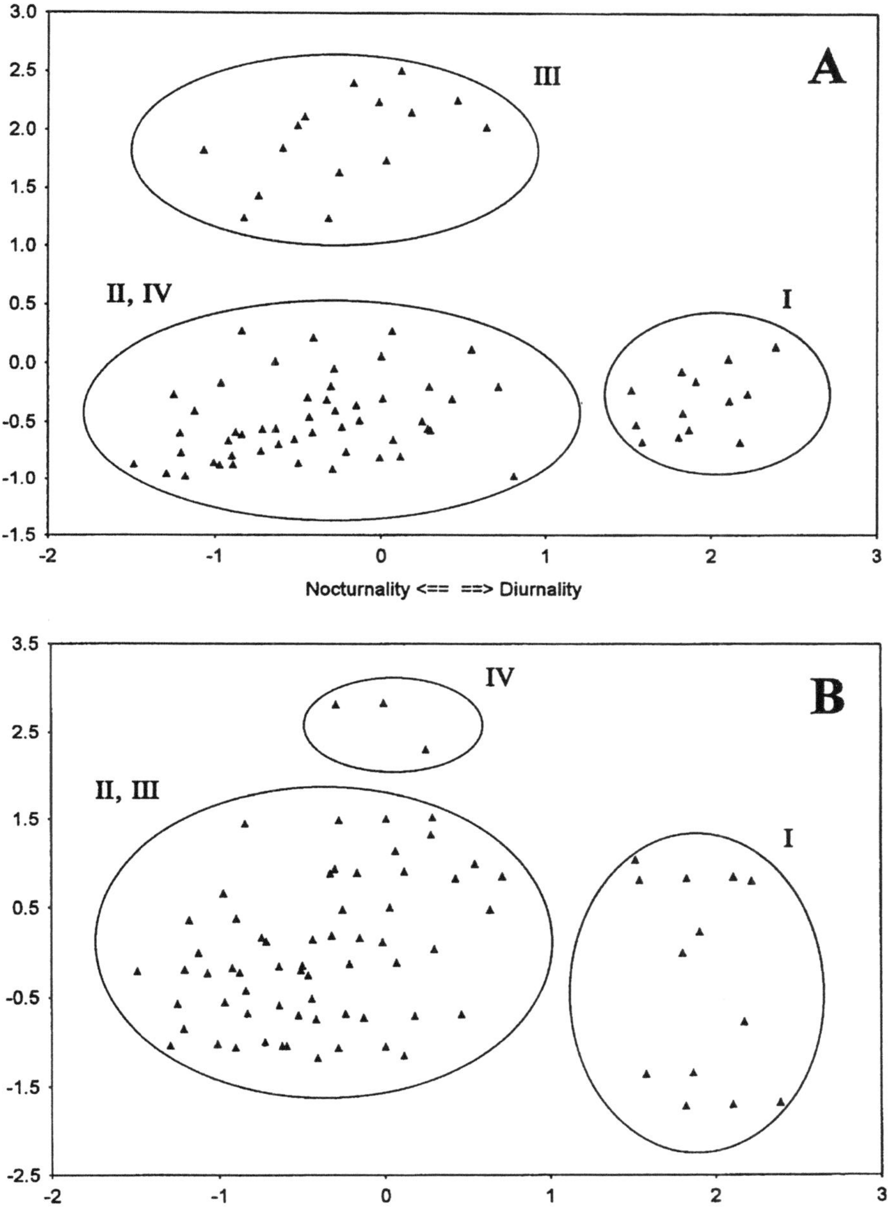

Fig. 6.1. Scatterplots of the ordination of the desert rodent species in the space of (A) the first and the second and (B) the first and the third principal component axes. I diurnal quadrupedal species, II nocturnal quadrupedal species, III bipedal species, IV fossorial species

I. **Diurnal quadrupeds**
IA. Omnivores (Sciuridae: *Atlantoxerus, Xerus, Spermatophilopsis, Spermophilus, Ammospermophilus*; Muridae: Murinae: *Rhabdomys pumilio*)
IB. Folivores (Muridae: Arvicolinae: *Eolagurus*; Gerbillinae: *Meriones rex, M. hurrianae, Psammomys, Rhombomys*; Otomyinae: *Parotomys*; Caviidae: *Dolichotis, Microcavia*; Chinchillidae: *Lagidium*; Octodontidae: *Octodon*, Petromuridae: *Petromus*; Ctenodactylidae: *Ctenodactylus, Felovia, Massoutiera, Pectinator*)

II. **Nocturnal quadrupeds**
IIA. Folivores
IIA1. Burrow shelterers (Muridae: Gerbillinae: *Meriones sacramenti*; Sigmodontinae: *Sigmodon, Reithrodon*; Hystricidae: *Hystrix*; Octodontidae: *Octomys, Tympanoctomys*; Abrocomidae: *Abrocoma*; Chinchillidae: *Lagostomus*)
IIA2. Rocky crevice shelterers (Muridae: Arvicolinae: *Alticola*)
IIA3. Aboveground nest shelterers (Muridae: Sigmodontinae: *Neotoma*)
IIB. Omnivores, foli-granivores, granivores and insectivores
IIB1. Burrow shelterers (Muridae: Murinae: *Aethomys, Leggadina, Millardia*, most species of *Mus, Rattus villosissimus*, most species of *Pseudomys, Zelotomys, Zyzomys*; Dendromurinae: *Malacothrix*; Cricetinae: *Allocricetulus, Calomyscus*, most species of *Cricetulus, Phodopus*; Gerbillinae: *Ammodillus, Gerbillurus, Gerbillus, Desmodilliscus, Desmodillus, Microdillus*, most species of *Meriones, Monodia, Pachyuromys, Tatera*; Sigmodontinae: *Andalgalomys, Acodon, Calomys, Eligmodontia, Onychomys, Peromyscus, Phyllotis, Reithrodon, Salinomys*; Heteromyidae: *Chaetodipus, Perognathus*)
IIB2. Rocky crevice shelterers (Muridae: Murinae: *Acomys, Cremnomys, Mus saxicola*; Petromyscinae: *Petromyscus*; Gerbillinae: *Sekeetamys*; Cricetinae: *Cricetulus longicaudatus*; Myoxidae: *Graphiurus*)
IIB3. Aboveground nest shelterers (Muridae: Murinae: *Pseudomys desertor*; Sigmodontinae: *Reithrodontomys, Graomys*; Myoxidae: *Eliomys, Selevinia*)

III. **Bipeds**
IIIA. Large (Pedetidae: *Pedetes*) (>500 g)
IIIB. Medium-sized (between 30 and 500 g)
IIIB1. Folivores (Dipodidae: Allactaginae: *Allactodipus, Pygeretmus*; Paradipodinae: *Paradipus*)
IIIB2. Omnivores (Dipodidae: Allactaginae: most species of *Allactaga*)
IIIB3. Foli-granivores (Dipodidae: Allactaginae: *Allactaga elater, A. vinogradovi*; Dipodinae: *Stylodipus, Eremodipus, Jaculus*; Heteromyidae: *Dipo-*

domys spectabilis, D. nelsoni, D. deserti)

IIIB4. Granivores (Dipodidae: Allactaginae: *Allactaga euphratica*; Dipodinae: *Dipus*; Heteromyidae: *Dipodomys merriami, D. ordii, D. microps*)

IIIC. Small (< 30 g)

IIIC1. Granivores (Dipodidae: Cardiocraniinae: *Cardiocranius, Salpingotus*; Heteromyidae: *Microdipodops*)

IIIC2. Insectivores (Dipodidae: Euchoreutinae: *Euchoreutes*)

IIIC3. Omnivores (Muridae: Murinae: *Notomys*)

IV. **Fossorials**
(Muridae: Arvicolinae: *Ellobius*; Geomyidae: *Pappogeomys, Thomomys*; Ctenomyidae: *Ctenomys*; Bathyergidae: *Bathyergus, Cryptomys*)

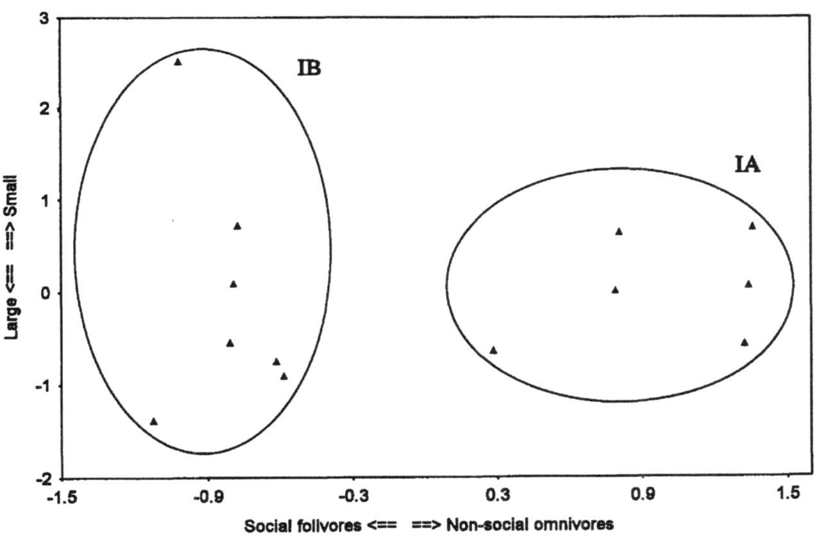

Fig. 6.2. Scatterplot of the ordination of the diurnal quadrupedal species in the space of the first two principal component axes. IA Omnivorous species, IB folivorous species

The geographic distribution of the four main life forms is shown in Figs. 6.5 and 6.6. Diurnal quadrupeds comprise 10-30% of all species in most local faunas. They are under-represented in the southern and eastern Sahara, Somali and Arabia, and are completely absent from Australia. Nocturnal quadrupeds form the main component in all local faunas (more than 50% of all species) excluding those of the northeastern portion of the Saharo-Gobian desert belt, where they comprise only 20-30% of all species. Bipedal species occur in most deserts excluding the Sind, the deserts of the Horn of African

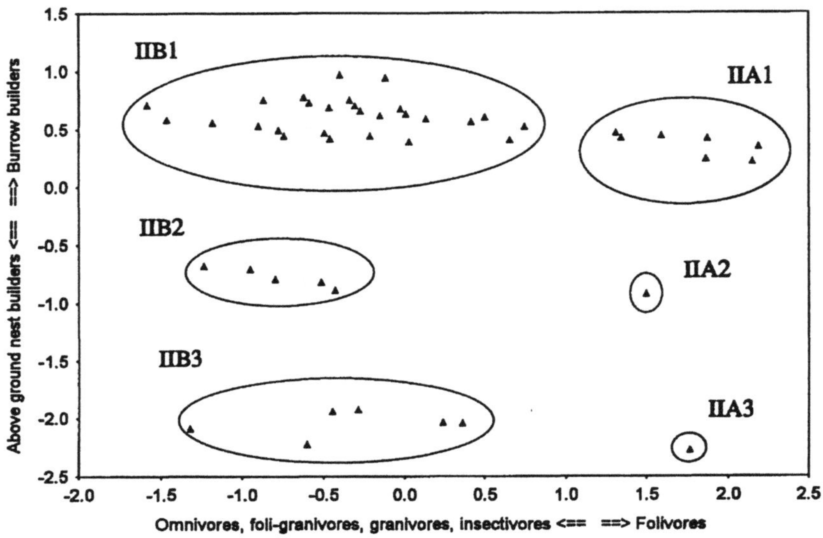

Fig. 6.3. Scatterplot of the ordination of the nocturnal quadrupedal species in the space of the first two principal component axes. IIA1 Folivorous species constructing burrows; IIA2 folivorous species sheltering in rocky crevices; IIA3 folivorous species constructing aboveground nests; IIB1 omnivorous, foli-granivorous, granivorous and insectivorous species constructing burrows; IIB2 omnivorous, foli-granivorous, granivorous and insectivorous species sheltering in rocky crevices; IIB3 omnivorous, foli-granivorous, granivorous and insectivorous species constructing aboveground nests.

and the South American deserts. Bipeds usually constitute a minor component of a fauna (up to 25% of all species). Only in the northeast of the Saharo-Gobian desert belt and in the Great Basin province does the proportion of bipedal species become considerable (30-60% of all species). Subterranean rodents are completely absent from the southwestern and central parts of the Saharo-Gobian desert belt and from Australia. In other deserts they comprise no more than 10% of each fauna. The only place with a high taxonomic diversity of subterranean species is the Argentinian desert in which a massive radiation of tuco-tucos (*Ctenomys*) has taken place. However, these species are all allopatric and there is only one species of *Ctenomys* in each local fauna (Redford and Eisenberg 1992).

Diurnal quadrupedal rodents are represented by two groups, namely trophic generalist species (Fig. 6.7a) and specialized folivores (Fig. 6.7b). As mentioned above, both these groups are absent from Australia. Generalists are absent from also the central and northeastern Sahara, Arabia and Mesopotamia, whereas specialists are lacking in the southern Sahara and the deserts of the Horn of Africa (Fig. 6.8). The distribution of these two groups in the New World is characterized by the presence of generalists only in

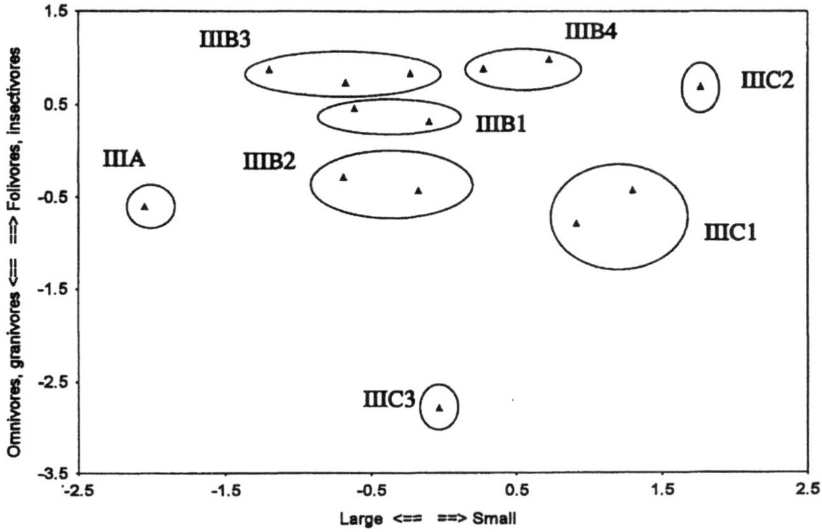

Fig. 6.4. Scatterplot of the ordination of the bipedal species in the space of the first two principal component axes. IIIA Large species, IIIB1 medium-sized folivorous species, IIIB2 medium-sized omnivorous species, IIIB3 medium-sized foli-granivorous species, IIIB4 medium-sized granivorous species, IIIC1 small granivorous species, IIIC2 small insectivorous species, IIIC3 small omnivorous species

North America and specialists only in South America (Fig. 6.9). This pattern of distribution is determined mainly by the taxonomic affiliations of the species involved. Most of the generalist species are Sciuridae and only one belongs to Muridae. On the other hand, the taxonomic composition of the group of specialist folivores is more diverse. The reason being that adaptations to folivory (both morphological and functional) have originated independently in most rodent taxa and represent the most specialized stage in the evolution of the digestive system in each phylum (Vorontsov 1967).

As with diurnal quadrupeds, nocturnal quadrupedal rodents can be subdivided into two main life forms, namely those with specialized folivory, and a complex group consisting of species with granivorous, insectivorous and omnivorous food habits. From an ecological point of view, granivores and insectivores are food specialists. However, these dietary modes are ancestral for rodents (Vorontsov 1967) and, from an evolutionary point of view, can be considered as generalized. Each of the two main life forms in turn is composed of three groups of lower rank, namely species excavating burrows, species using rocky crevices and niches as shelters, and species constructing aboveground nests. Generalized burrowing species are represented by Muridae and Heteromyidae (Fig. 6.10a) and comprise the bulk of nocturnal quadrupeds in all desert faunas (60-90% of all nocturnal quadrupedal species, Figs. 6.11, 6.12).

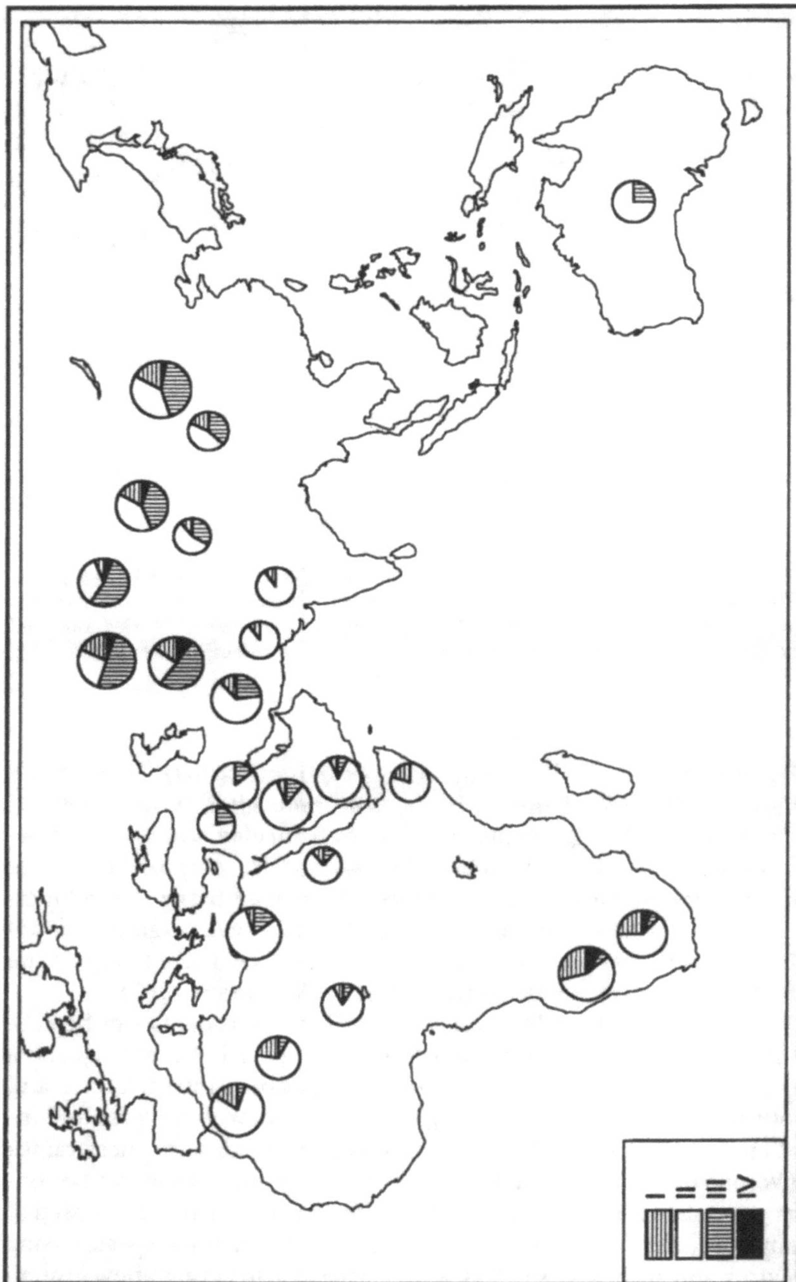

Fig. 6.5. Distribution of rodent life forms of the highest rank in the Old World desert provinces. The size of a pie corresponds to the overall species richness in a province. I Diurnal quadrupedal species, II nocturnal quadrupedal species, III bipedal species, IV fossorial species

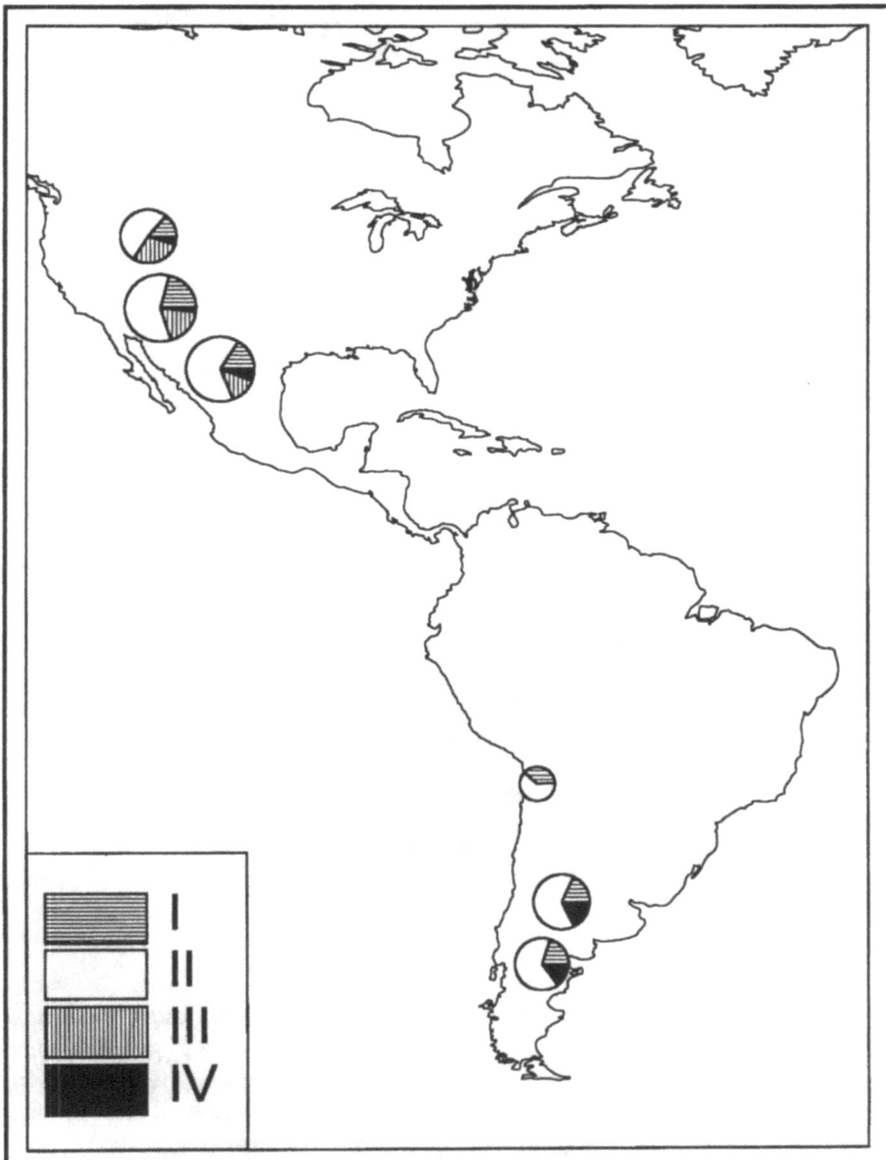

Fig. 6.6. Distribution of rodent life forms of the highest rank in the New World desert provinces. The size of a pie corresponds to the overall species richness in a province. I Diurnal quadrupedal species, II nocturnal quadrupedal species, III bipedal species, IV fossorial species

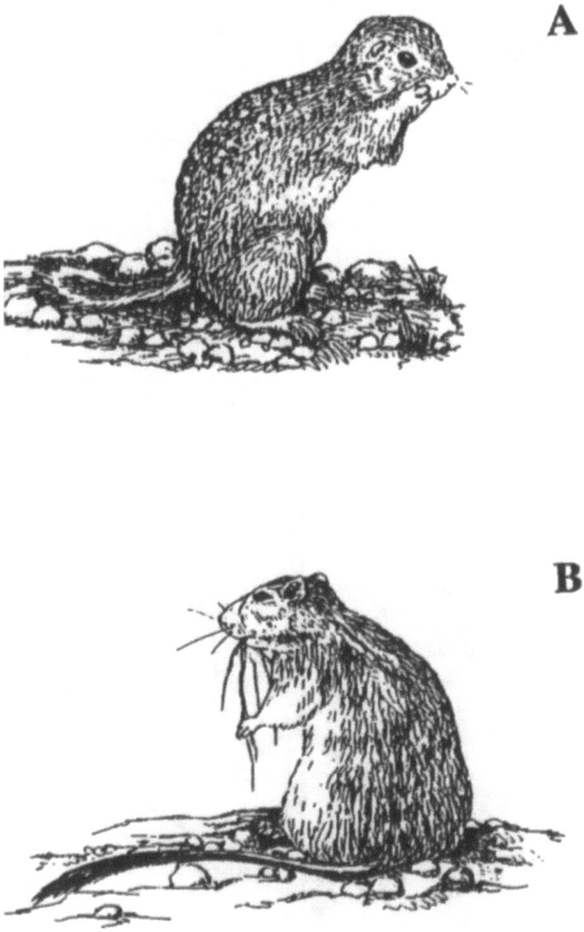

Fig. 6.7. Diurnal quadrupeds. A Omnivorous *Spermophilus spilosoma*, B folivorous *Rhombomys opimus*

Rock-dwelling (petrophily) is a particular adaptation to desert life. It has evolved in several different taxa (Muridae in Asia, North Africa and South America and Myoxidae in South Africa). It is not weil represented except in the African and Asian desert faunas and does not occur in the desert faunas of the Turan, Iran, Mesopotamia and the Northern Sahara (Fig. 6.10b).

Construction of aboveground nests is widespread among grassland and forest rodents (Fig. 6.10c). In deserts, this characteristic persists in taxa that are not typically desert dwellers but penetrate desert regions from adjacent grasslands (Muridae in Australia and South America and Myoxidae in Asia and North Africa). Species showing this life form are weakly, if ever represented in local faunas. The only exception is the Monte-Patagonian realm, where 20% of all nocturnal quadrupeds build aboveground nests. These

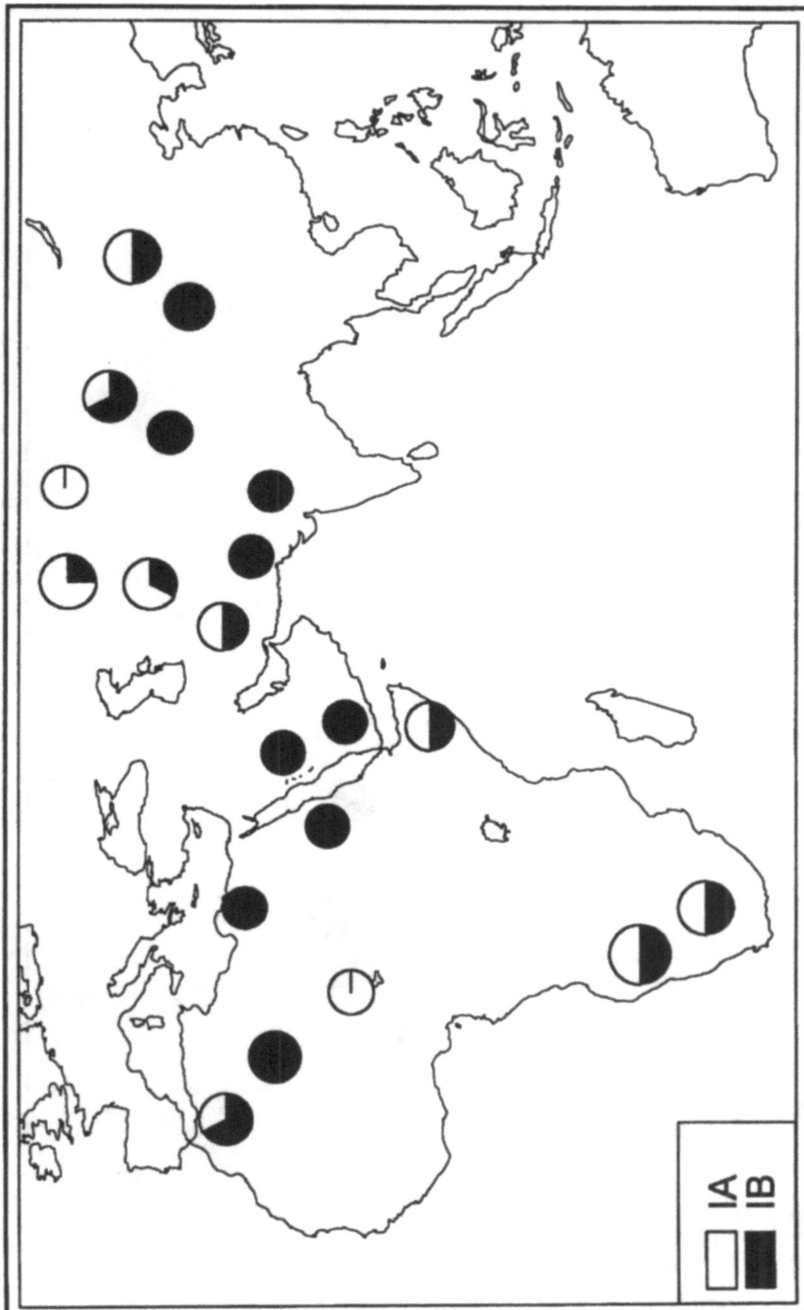

Fig. 6.8. Distribution of diurnal quadrupedal species in the Old World desert provinces. The size of a pie corresponds to the overall number of diurnal quadrupeds in a province. IA Omnivorous species, IB folivorous species

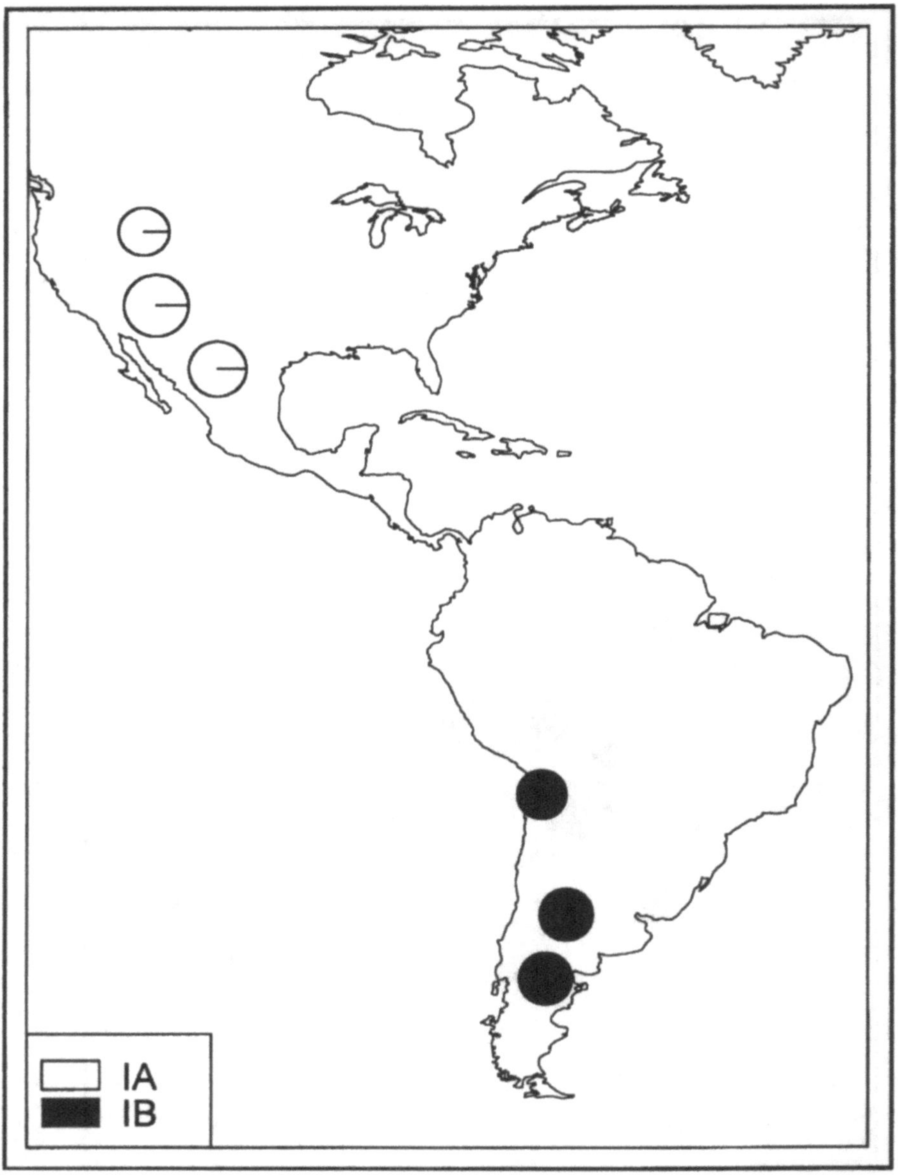

Fig. 6.9. Distribution of diurnal quadrupedal species in the New World desert provinces. The size of a pie corresponds to the overall number of diurnal quadrupeds in a province. IA Omnivorous species, IB folivorous species

Fig. 6.10. Nocturnal quadrupedal rodents. A *Meriones meridianus*, B *Acomys cahirinus*, C *Selevinia betpakdalensis*

rodents are absent from the most arid deserts such as the Gobi, the central Sahara, and the Namib.

The geographical distribution of burrowing folivores (Fig. 6.13a) is uneven (Figs. 6.11, 6.12). They are absent from Australia, the northeast of the Saharo-Gobian desert belt, the central Saharan, Namib, Great Basin and Sonoran deserts. However, they are over-represented (30% of all nocturnal quadrupeds) in the Monte-Patagonian deserts. The reason for this is that folivory is usually associated with diurnality, and this connection is only evolutionarily broken in some hystricognaths and murids. Each of the other two life forms of folivorous nocturnal quadrupeds (the rock-dwellers and builders of aboveground nests) is represented by a few species belonging to a single, originally non-desert genus. Most species of mountain voles of the genus *Alticola* (rock-dwellers, Fig. 6.13b) occur in high-altitude mountains, whereas wood rats, *Neotoma* (builders of aboveground nests, Fig. 6.13c) actually constitute a forest-adapted genus. Desert *Neotoma* spp. have no specific functional adaptations to arid environments (Schmidt-Nielsen 1964). Each of these genera has a restricted distribution, so nocturnal quadrupedal folivorous rock-dwellers are represented only in the Kazakhstan, Sungorian and Gobi deserts, while nocturnal quadrupedal folivorous builders of aboveground nests are found only in the North American deserts.

Bipedal rodents can be divided into three size classes, two of which consist of several life forms with different food habits. In general, the classification of bipedal life forms coincides with their taxonomic classification. The majority of species with bipedal locomotion belong to Dipodidae and only a few species to Pedetidae, Muridae and Heteromyidae. Because of this, the variety of biped life forms is highest in Central Asia (Fig. 6.14) where the dipodid fauna is the most diverse (Vinogradov 1937, Shenbrot et al. 1995). Large bipedal rodents are represented by only a single species (the folivorous springhare, *Pedetes capensis*, Fig. 6.15). Indeed, this is the only biped in the South African deserts.

Medium-sized bipeds comprise four trophic groups, two of which (folivores, Fig. 6.16a, and omnivores, Fig. 6.16b) consist of dipodids only. These groups are restricted geographically to the Turan, Kazakhstan and Gobi deserts. Two other groups of medium-sized bipeds (foli-granivores, Fig. 6.16c, and granivores, Fig. 6.16d) include both dipodid and heteromyid species. They occur either in all the Holarctic deserts (foli-granivores) or in the North American and the eastern Palaearctic deserts (granivores; Figs. 6.14, 6.17).

Small bipeds can be divided into three trophic groups. Two of these are monogeneric and restricted geographically either to the Gobi desert (the insectivorous long-eared jerboa *Euchoreutus*, Fig. 6.18b) or to Australian deserts (the omnivorous hopping mice *Notomys*). Small granivorous bipeds (Fig. 6.18a) are represented by one heteromyid and two dipodid genera and are confined to two disjunct regions, in Asia (the eastern part of the Saharo-

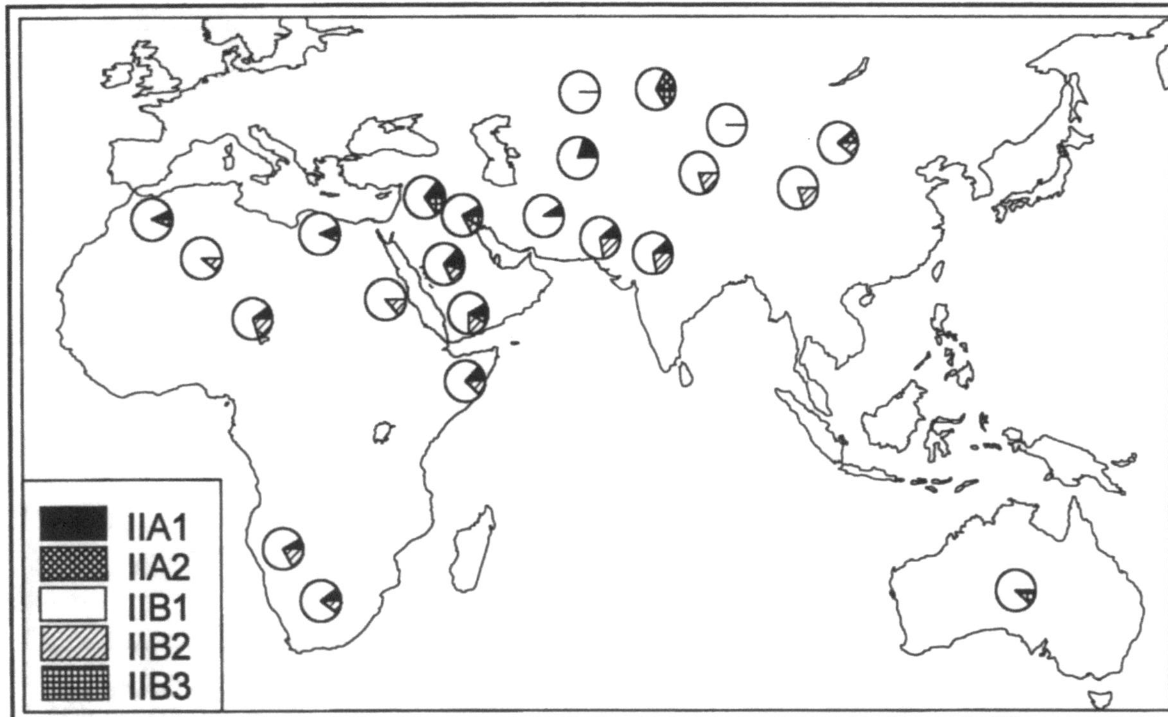

Fig. 6.11. Distribution of nocturnal quadrupedal species in the Old World desert provinces. The size of a pie corresponds to the overall number of nocturnal quadrupeds in a province. IIA1 Folivorous species constructing burrows, IIA2 folivorous species sheltering in rocky crevices, IIB1omnivorous, foli-granivorous, granivorous and insectivorous species constructing burrows, IIB2 omnivorous, foli-granivorous, granivorous and insectivorous species sheltering in rocky crevices, IIB3 omnivorous, foli-granivorous, granivorous and insectivorous species constructing aboveground nests

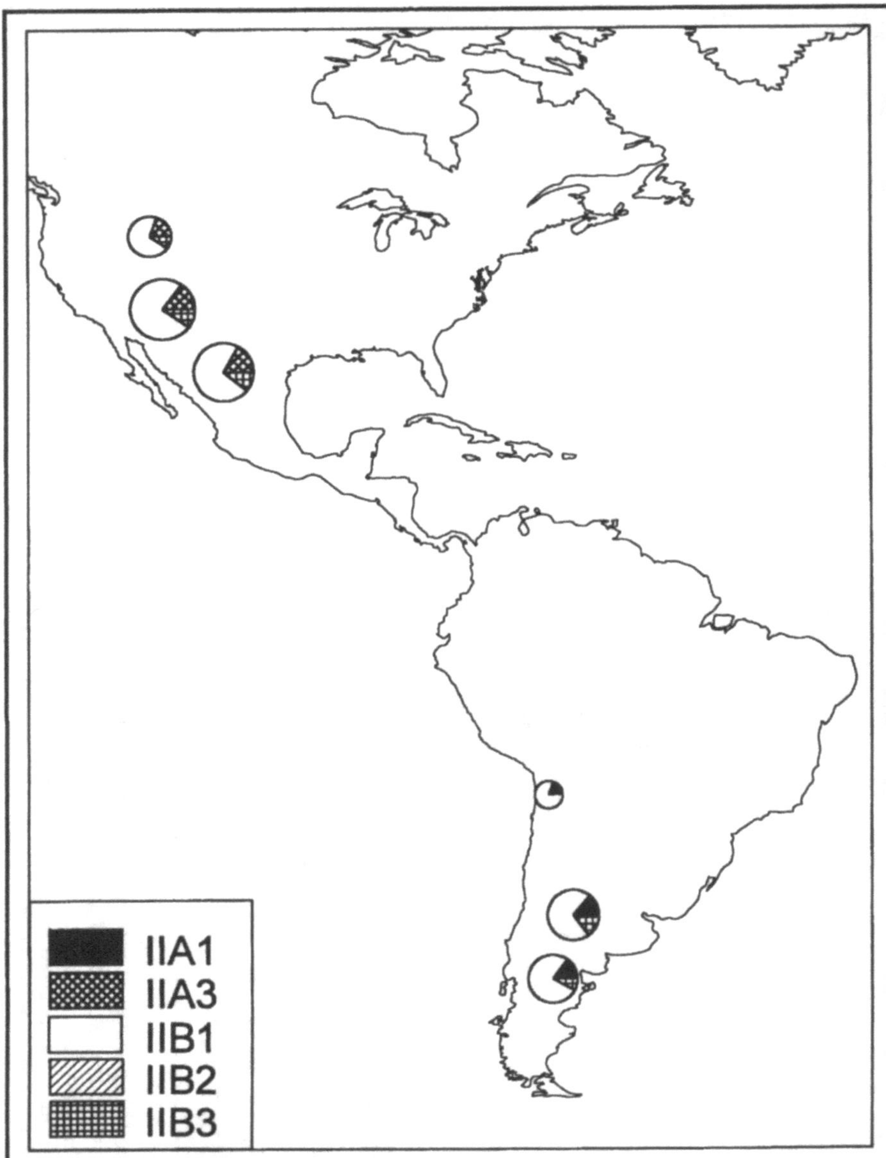

Fig. 6.12. Distribution of nocturnal quadrupedal species in the New World desert provinces. The size of a pie corresponds to the overall number of nocturnal quadrupeds in a province. IIA1 Folivorous species constructing burrows, IIA3 folivorous species constructing aboveground nests, IIB1 omnivorous, foli-granivorous, granivorous and insectivorous species constructing burrows, IIB2 omnivorous, foli-granivorous, granivorous and insectivorous species sheltering in rocky crevices, IIB3 omnivorous, foli-granivorous, granivorous and insectivorous species constructing aboveground nests

Fig. 6.13. Nocturnal folivorous quadrupeds. A *Sigmodon hispidus*, B *Alticola streltzovi*, C *Neotoma albigula*

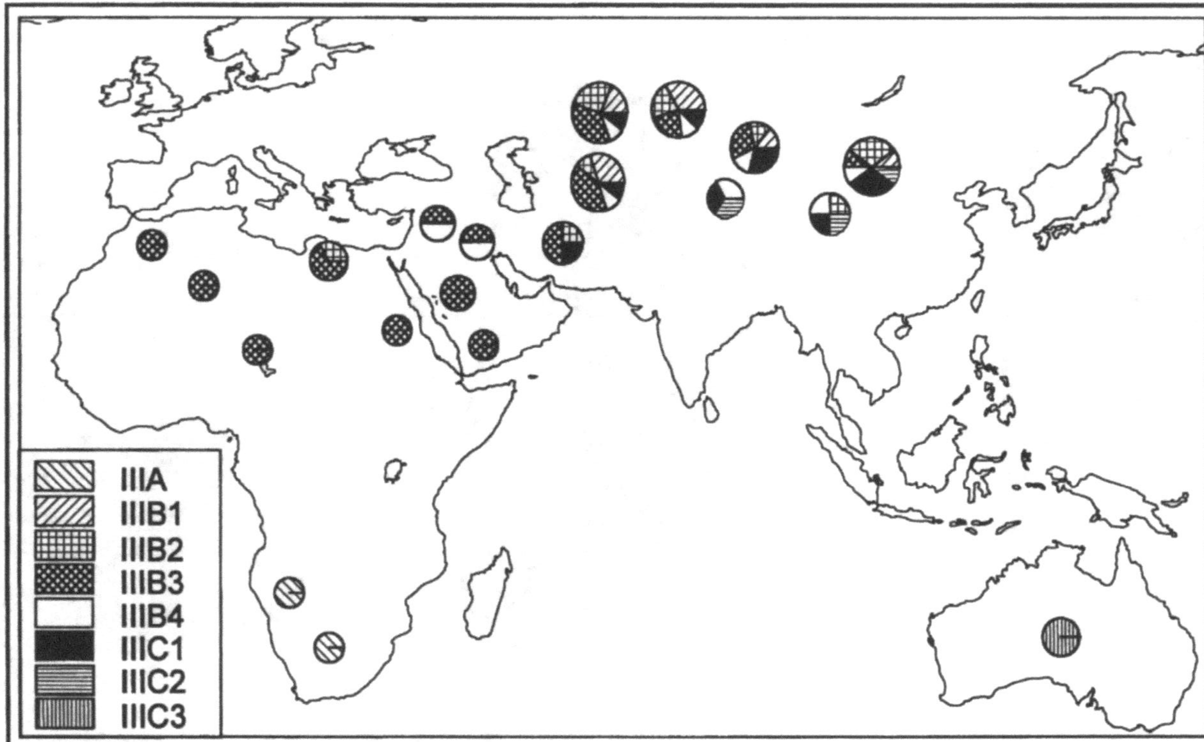

Fig. 6.14. Distribution of bipedal species in the Old World desert provinces. The size of a pie corresponds to the overall number of bipeds in a province. IIIA Large species, IIIB1 medium-sized folivorous species, IIIB2 medium-sized omnivorous species, IIIB3 medium-sized foli-granivorous species, IIIB4 medium-sized granivorous species, IIIC1 small granivorous species, IIIC2 small insectivorous species, IIIC3 small omnivorous species

Fig. 6.15. Large biped, *Pedetes capensis*

Gobian desert belt, – from Iran to the Gobi desert) and North America (the Great Basin desert; Fig. 6.14, 6.17).

The homogeneity of the fossorial life form (Fig. 6.18c) may be accounted for by the strong constraints on morphological features and by the ecology of a species imposed by its mode of life (Nevo 1979). All subterranean species therefore are similar in general appearance and ecological characteristics despite their different taxonomic affinities and geographic distributions; Muridae in Asia, Geomyidae in North America, Ctenomyidae in South America and Bathyergidae in South Africa. This convergent evolution of unrelated groups affords strong evidence for adaptation to specific environmental features in these rodents.

6.2 Convergent Evolution Among Species of Different Faunas

The subject of convergent evolution among desert rodents of different continents was discussed by Mares (1973, 1980, 1993a,b) who has shown that cases of convergent evolution are often observed, although some of the examples he cited in published papers are doubtful (see above).

It is clear that similar environments demand similar adaptations of their inhabitants. Unidirectional adaptations are concerned with a single function (e.g. bipedalism as a locomotory mode, yellowish "desert" coloration and enlarged tympanic bullae). They should be regarded either as parallel evolution if they evolve from the same morphological basis or as convergence if they evolve from different morphological bases (Cloudsley-Thompson 1993, 1996). When these adaptations evolve as a complex of features, different taxa are considered as belonging to the same life form category. In the present

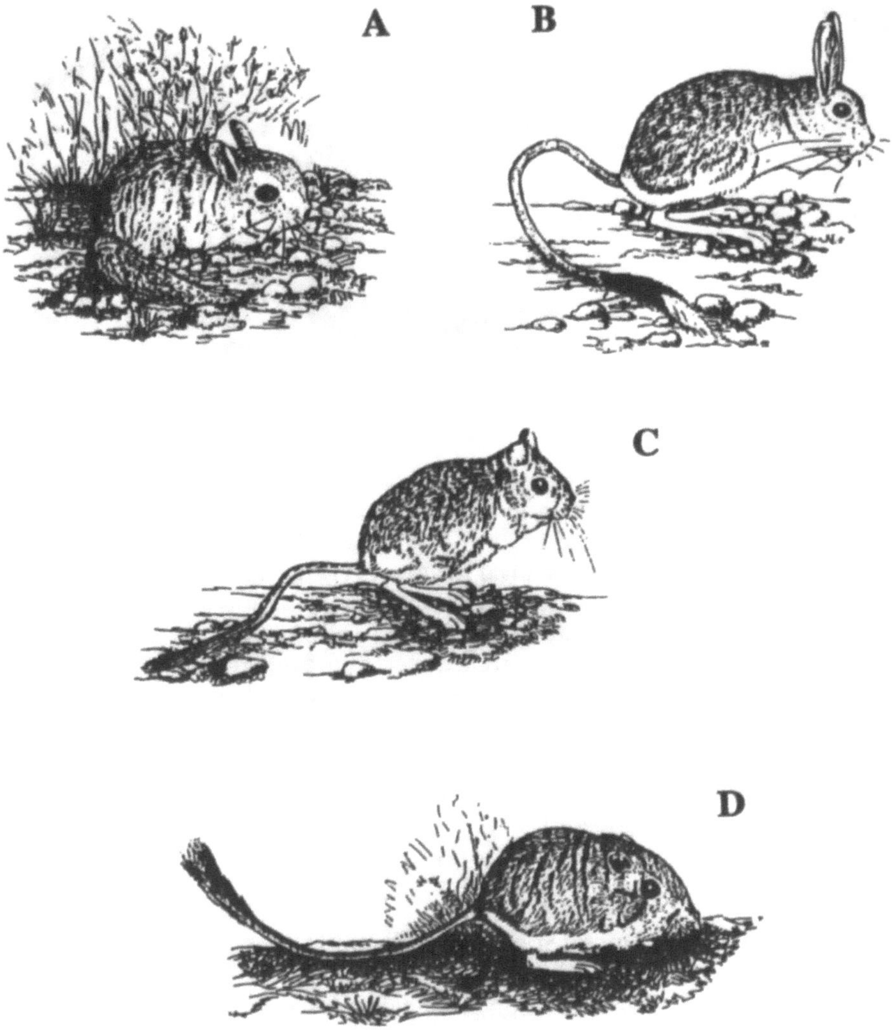

Fig. 6.16. Medium-sized bipeds. A *Pygeretmus platyurus*, B *Allactaga severtzovi*, C *Stylodipus telum*, D *Dipodomys merriami*

context we are interested in adaptational complexes rather than with the analysis of separate morphological, ecological or behavioural adaptations to either a particular habitat type (see below) or to the desert environment in general.

Generally speaking, the classification of life forms described above represents those rodent taxa which result from convergent evolution. However,

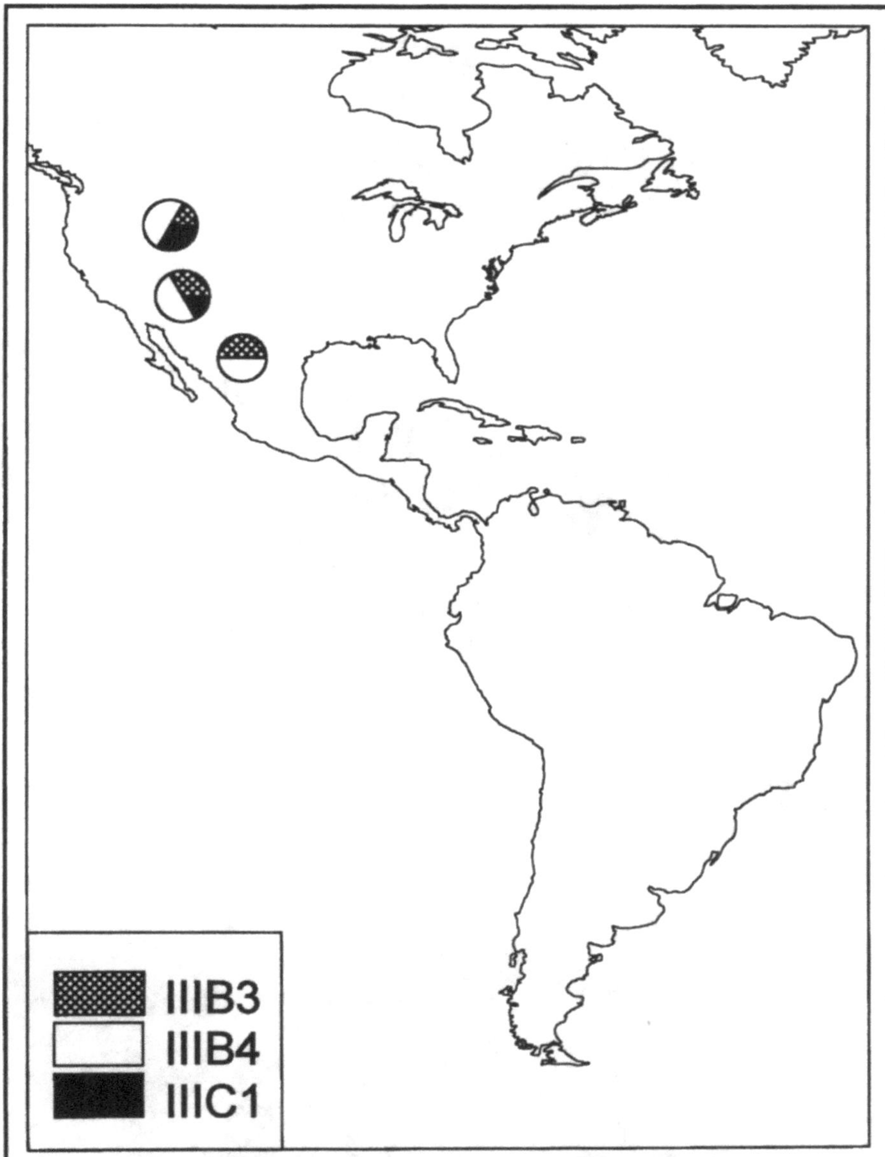

Fig. 6.17. Distribution of bipedal species in the New World desert provinces. The size of a pie corresponds to the overall number of bipeds in a province. IIIB3 Medium-sized foli-granivorous species, IIIB4 medium-sized granivorous species, IIIC1 small granivorous species

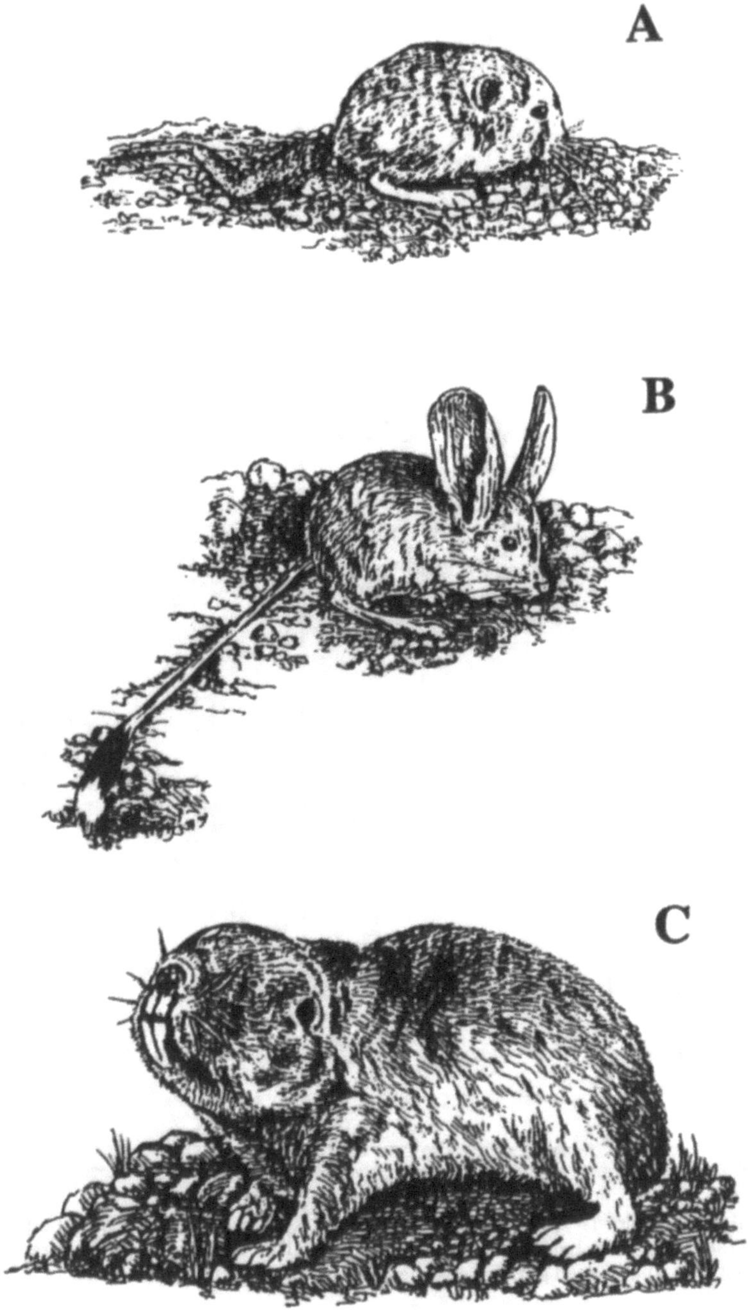

Fig. 6.18. Small bipeds, *Cardiocranius paradoxus* (A) and *Euchoreutes naso* (B) and fossorial *Cryptomys hottentotus* (C)

there is one important exception, namely nocturnal quadrupeds. Examples of surprising morpho-ecological similarities are to be found within this life form (*Onychomys* and *Allocricetulus, Chaetodipus penicillatus* and *Gerbillus dasyurus*). Nevertheless, most species belonging to this category are murids with the appearance of a mouse/rat archetype which have not evolved greatly (from a morpho-ecological point of view) from a generalized murid ancestor. This claim cannot be rejected on the grounds that common "desert" features are present in most such cases. For example, the only characters common to *Eligmodontia* and *Gerbillus* are enlarged tympanic bullae and pale coloration. These characters are parallelisms rather than convergent features. Among the few non-murid rodents belonging to this life form, some have the appearence of non-specialized mice/rats (e.g. *Octomys, Tympanoctomys, Abrocoma, Selevinia, Chaetodipus, Perognathus*) and the above comments should be applied to them too. Three other non-murid exceptions (porcupines, plain viscacha and dormice) were included in this category as a result of the formality of the analysis. Nevertheless, these rodents belong to principally non-desert taxa, do not possess specific desert adaptations and, consequently, can be considered as non-specialized species as far as deserts are concerned.

Other life forms display a number of examples of convergent evolution. This is expressed most clearly in two life forms, folivorous diurnal quadrupeds and both medium and small bipeds. Of course, subterranean rodents demonstrate one of the best-known examples of convergent evolution. However, this convergence has been repeatedly described and discussed in detail (e.g. Gambarian 1960; Nevo 1979 etc.) and, actually, has nothing to do with "desert" specialization. That is why we have not broached this subject in the present volume. Formally, ecological convergence also exists in the category of omnivorous diurnal quadrupeds. It can be seen in ground squirrels and *Rhabdomys*. Nevertheless, it is not sensible to discuss this as a case of ecological convergence because the resemblance is clearly superficial and does not affect any morphological or even external character.

Contrary to the claim of Mares (1993b), convergence is widely represented among folivorous diurnal quadrupeds although not in all desert regions. Examples of convergent evolution among folivorous diurnal species occur in the Palaearctic, Palaeotropic and Neotropic regions. Moreover, this convergence occurs not only among taxa of different regions but among taxa within each region as well. It seems that folivory is functionally connected with diurnal activity in desert rodents. All folivore species showing nocturnal activity belong either to principally non-desert genera (e.g. *Sigmodon, Neotoma, Alticola, Hystrix*) or else their folivory has evolved recently (e.g. in *Meriones sacramenti* according to the structure of its molars).

Convergent evolution among diurnal folivore rodents can be illustrated by a set of common morphological and behavioral features. The only exception is the cursorial mara *Dolichotis*, which occupies the ecological place of the leporids (see Mares 1993b). All other diurnal folivores share a set of common features, although in some species one or other character of this set may be absent or weakly expressed. The general appearance of diurnal folivores is exemplified by either great gerbil-like (Gerbillinae, Otomyinae, Petromuridae) or guinea pig-like (Hystrocognatha and Ctenodactylidae) or something in-between (*Eolagurus*). In all of them the eyes are elevated by being situated on the upper part of the head, thus increasing distance vision. The ears are usually small, both in burrow dwellers and in species that shelter in rocky crevices. The mountain viscacha *Lagidium viscacia* represents an exception to this general rule having relatively long hare-like ears.

Most diurnal folivores are social and live either in pairs or in colonies. For example, observation and radio-tracking of maras *Dolichotis patagonum* has demonstrated that they travel as monogamous pairs that breed either alone in solitary burrows or communally in settlements where up to 29 pairs inhabit shared warrens (Taber and Macdonald 1992). All species of Ctenodactylidae live in colonies of different complexity (George 1974). One of the classic examples of sociality is demonstrated by the great gerbil *Rhombomys opimus* which lives in large complex families (see Goltzman et al. 1977 for further references). Also noteworthy is the fact that the sociality in a very similar species, *Psammomys obesus*, is expressed much more weakly. This is mainly a solitary species (Rogovin 1996). Littledale's whistling rat, *Parotomys littledalei*, also shows an extremely low degree of sociality, usually living a solitary life in a single hummock (Coetzee 1969).

Individuals of social species communicate both by visual and acoustic means. Many of them (excluding gundis and some hystricognaths, such as *Microcavia*) are able to adopt a vertical posture, thus increasing their field of vision. The enlarged field of vision enables early identification of predators. Furthermore, the animal monitors the behavior of neighboring conspecifics, that may have already recognized a predator, and consequently is able to avoid predation by mimicking their escape behaviour (Shilov 1977). Most species have a rich repertoire of vocalizations and sound alarm calls which are elicited by the approach of predators (e.g. Speke's pectinator, *Pectinator speeki*; George 1974). These alarm calls are often not specific and are produced whenever animals are excited or disturbed. In such cases, ctenodactylids use footdrumming in addition to calls (George 1974). A similar response also occurs in the great gerbil *Rhombomys opimus* (Goltzman et al. 1977). Interestingly, the weakly (if at all) social fat sand rat *Psammomys obesus* scarcely vocalizes, although it does footdrum (our observations).

Bipedality is the most frequently cited example of convergent evolution specific to arid environments. As stated above, bipedal rodents occur in all desert regions of the world, excluding the Neotropics, the Horn of Africa and the Sind. However, a closer look at this category of life forms reveals that the patterns of convergence are not so simple, and differ among different ecological groups of bipeds. Generally, convergence has been shown to occur by

Mares (1993), but his conclusions are problematic due to some misinterpretations concerning the biology of non-American bipedal species (see above). We consider true examples of convergence only to be found in the medium- and small-sized foli-granivore and granivore bipeds. They are absent from the other groups of bipedal rodents. Thus, true convergent evolution among life forms of bipedal rodents (in the strict sense) took place only in the North American heteromyids and Palaearctic dipodids. Moreover, deep morphological and ecological convergence is exemplified only by a single pair of taxa (the heteromyid genus *Microdipodops* and the cardiocraniine genus *Salpingotus*). Species of both genera are of the same size and general appearance. They have very large heads comprising about the half the total head-and-body length, elongated hind feet with an undersurface bearing fine, soft hairs that form a "sandski", and long tails with tufts of hairs on their tips. Both genera are able to store fat in the anterior part of their tails and to use this during dormancy. Both genera have similarly enlarged and inflated tympanic bullae. The skeleton of the hind extremities is formed by elongated, but not consolidated, metatarsals. Both *Microdipodops* and *Salpingotus* prefer sandy habitats, and are primarily granivorous, although insects are also eaten. They build simple burrows, are solitary, territorial and highly aggressive (see O'Farrell and Blaustein 1974a,b for *Microdipodops*; Rogovin 1985 for *Salpingotus*). It is clear that there are also huge morphological differences between these genera due to their different origins and taxonomic affiliation. For example, the internal structures of the tympanic bullae are completely different, being formed from different elements (our data). Cheek pouches, very characteristic of heteromyids, are absent in dipodids, and so on.

Convergence among other species of bipeds is weakly expressed. The third member of the category of small-sized granivores (*Cardiocranius paradoxus*) differs markedly from both *Salpingotus* and *Microdipodops*. The "sandski" on the hind foot is absent, the tail is short and fat along its entire length, and the species is exclusively granivorous. Convergence among medium-sized foli-granivores and granivores is expressed even more weakly. There are distinct morpho-ecological differences between true jerboas (Allactaginae, Dipodinae) and heteromyids (*Dipodomys*). Some external similarity can be found between three-toed jerboas (Dipodinae) and *Dipodomys*. This includes similar shape, bicoloured tail banners, short and almost immovable necks, small ears and extremely large tympanic bullae. Five-toed jerboas (Allactaginae) appear unlike both Dipodinae and *Dipodomys* because of their long hare-like ears, relatively small tympanic bullae, mobile necks, and protruding incisors. Moreover, *Dipodomys* species demonstrate adaptations to bipedality to a lesser extent than jerboas do. The metatarsals of *Dipodomys* are elongated but separate, whereas, in the Dipodinae and Allactaginae, three central metatarsals are consolidated into a single bone (*os metatarsus*).

In conclusion, we can claim that, despite the fact that different taxa of rodents in the various desert regions of the world share similar ecological affinities and that several categories of life forms can be distinguished, cases of deep convergent evolution are rare (except for subterranean rodents). Among desert rodents, it seems that such cases occur with about the same frequency as in many other animal groups and/or biomes.

CHAPTER 7

Species-Habitat Relationships in Desert Environments

7.1 Methodology of Analysis: Niche Concept and Description of Niche Dimensions

Rodents are found in many different deserts and each species uses its environment in its own manner. In order, therefore, to extract some generalizations, a common approach that permits us to compare patterns of environmental resource utilization among the different species and regions is necessary. This approach can be based on the concept of the ecological niche of a species in terms of its characteristic resource utilization. The term "niche" was proposed as early as the beginning of the century (Grinnell 1917) for the description of the place of a species in an ecosystem, i.e. as a distributional unit of a species. Later, Elton (1927) defined niche as a functional concept for the description of the role of a species in an ecosystem in terms of its relations to food resources and predators, i.e. the position of a species in its trophic interactions. There are several explicit reviews of the history of the niche concept (e.g. Vandermeer 1972; Whittaker 1975; Pianka 1978; Giller 1984 etc.). Here we want only to emphasize that there are some differences in modern understanding of the term "niche" and these differences are derived from the initial differences between the definitions of Grinnell and Elton. The main difference between these two approaches is the sphere of application of the term: Grinnell's approach connects it with a species, whereas Elton's definition is connected with an ecological function. Consequently, it is correct to speak of "empty niches" within the frame of Elton's definition, but is impossible using Grinnell's concept. It is necessary to point out that usage of the term "niche" in this book is based mainly on Grinnell's approach.

The niche concept was defined quantitatively by Hutchinson (1957) who stated that the niche of a species is defined by the total range of environmental factors to which it responds and under which a population exists and replaces itself indefinitely. Every environmental factor can be conceived as a gradient along which species abundance changes and along which species replacements occur due to differences in the environmental optimum of each species (Whittaker 1972; Austin 1985). A gradient can be imagined as a dimension in space, so the niche can be described in terms of *n*-dimensional space, or hypervolume. Furthermore, Hutchinson proposed distinguishing between two states of the niche of a species. Following Hutchinson, the entire set of optimum conditions which a species can potentially occupy should be considered as a *fundamental* niche, whereas the *realized* niche is the actual spectrum of conditions in which it actually exists. Competitors and/or predators frequently prevent a species from occupying all its fundamental niche. Thus, the realized niche is a subset of the fundamental niche.

If we consider an ecological niche as a characteristic of resource utilization by a species' populations, it can be described along each dimension (axis) by the frequency distribution of utilization of different classes (or states) of a given resource. In most theoretical models, this distribution is approximated by a normal distribution (Levins 1968; May and MacArthur 1972; Southwood 1978; ter Braak and Prentice 1988). Following such an approach, a niche can be characterized along each separate axis by two parameters, namely the mean value (a niche center or position) and the variance or standard deviation (a niche breadth or width). This can easily be represented by a multidimensional model if the axes are considered as being orthogonal (=independent). A number of different measures of niche breadth have been proposed (Levins 1968; Van Valen 1968; Feinsigner et al. 1981; Yu and Orloci 1992, 1993). Of course, there are constraints on the usage of these measures. One of the main problems is that it is impossible to compare niche breadth between two or more species if they use a different spectrum of resources. Thus, for correct comparison, it is necessary to use the same resource axes and/or resource categories. Another problem is the necessity of taking the relative abundance of resources available into account in order to compare this with the observed pattern of resource use. It is easy to show with some simple examples that without such a comparison, estimations of niche breadth for a specialized species may be higher than for an opportunistic species, due to the inconsistency between innate resource preferences and real relative resource availability (see Petraitis 1979). Imagine that there are two habitats (A and B) and two species (X and Y). Species X is a specialist and occurs in habitat A only, while species Y is opportunistic and uses both habitats equally. In addition, densities of both species are such that the probability of recording species X in habitat A is 1 and in habitat B it is 0, while the probability of recording species Y in each habitat is 0.5. Now imagine that we sample the entire study area randomly without taking the relative representation of each type of habitat into account. If both habitat types are equally represented, we will record 50 individuals of each species per 100 samples, causing the estimated values of niche breadths to be equal. If habitat A occupies 75% of the area, we will record 75 individuals of species X and 50 individuals of species Y per 100 samples, thereby causing the estimated value of niche breadth for specialist X to be higher than that for opportunist Y. This problem can be solved by using the equation $1 - O_{ij}$ as an index of niche breadth, where O_{ij} is the value of overlap between the distributions of resource use and resource availability, (Schoener 1968; Feinsigner et al. 1981). In addition, it should be noted that some measures of niche breadth are density-dependent and special correction is needed to eliminate this effect (Rosenzweig and Abramsky 1985).

Following the theoretical ideas of multidimensional fundamental niche (Hutchinson 1957), the number of resource axes (factors) for each ecological niche is infinite. However, all differences in resource use among species can, in principle, be reduced to differences in the type of source of energy resource (trophic axis), the place (spatial axis) and the time (temporal axis) of resource use (Schoener 1974). To avoid interspecific competition, coexisting species have to divide the available resources in some way. The relative importance of different resource axes differs for different taxa. Available information demonstrates that coexisting species of non-carnivorous mammals divide the environmental resources mainly along a spatial axis, while the trophic axis is of lesser importance and temporal resource partitioning is relatively rare (Schoener 1974). Therefore, it is legitimate to consider spatial, trophic and temporal niches or niche components separately.

The resource state can be defined along each axis as either continual or categorical. For example, when studying the habitat distribution of desert rodents, we can either define two habitat types, for example "densely vegetated sand dune" and adjacent "open gravel plain", or else describe a composite environmental gradient determined by continual change in the relative abundance of particles of different size classes in soil composition and in the cover of different plant species. Both approaches are valid, and the application of either one or the other depends on the aims of the study. Furthermore, it is frequently difficult to determine the border between two habitats when there is a gradual transition between them. In a more broad sense, the distinction between discrete versus continual habitat boundaries is a problem of spatial scale. For example, a mixture of two habitat types with an increase in the first type towards one end of a transect and an increase in the second type towards the opposite end will, on a small scale, be represented as a gradient. On the other hand, a situation that appears on the small scale as a sharp ecotone between two habitat patches will on a large scale look like a gradient.

7.2 Habitat Structure and Types of Habitat Specialization

The term "habitat" is used in a variety of ways. Usually, it is used to describe an area of particular relief, vegetation and soil structure. It is commonly thought that the structure of an animal community is determined by the habitat structure of an area. Another concept considers habitat as related to a particular species or group of species. In the framework of this latter concept, a habitat is a patch with a particular set of environmental conditions and resources promoting the occupancy, survival and reproduction by individuals of a given species (Morrison et al. 1992). It is clear that the rates of reproduction and survival of a species vary in response to environmental conditions, thus reflecting the quality of a habitat (Van Horne 1983). Hence, the habitat structure of an area is not pre-defined, but is a complex of coevolved responses of organisms to abiotic and biotic factors (Rosenzweig 1995). In our opinion, this latter concept is more objective because it reflects the integrated results of the interactions of organisms with their environments. Examples of such an approach to habitat classification using cluster-analysis based on rodent species composition have been elucidated for the Mojave (Matson 1976) and the central Chihuahuan deserts (Rogovin et al. 1985).

It is clear that each spatial patch has a specific set of environmental features. Based on this set, one can build a hierarchical classification of spatial patches, either intuitively or using some mathematical algorithm. The level of detail of the classification and consequently the number of the resultant categories, is arbitrary. However, it is unknown whether the animals inhabiting an area "measure" the spatial structure of that area in the same way as would a researcher or his/her computer software. To determine the correct level of spatial subdivision, and to verify whether the obtained differences among spatial categories can be distinguished by animals, it is necessary to compare this classification with the classification of spatial patches based on the relative abundances of the species of animals. The result of this comparison is an a posteriori scheme of habitat types from an "animal's point of view".

The coincidence of the two (environmental and species composition) classifications should occur under at least two necessary conditions, namely (1) when landscape units are sharply distinct and have clear borders without smooth transitions, and (2) when, for each landscape unit, there is at least one species that clearly prefers and/or is specialized for it. Because there is usually a lack of strong correspondence between these two conditions and a real situation, the complete coincidence of the two classifications appears to be extremely rare. For example, the habitat structure of the desert of the Negev Highlands identified by environmental measurements is similar to that identified from rodent species composition (Fig. 7.1). There are, however, two main points that should be noted: (1) the sets of habitat categories of the two classifications are the same – excluding a "thin loess" category in the environmental classification that is absent in the "rodent" classification; and (2) the assignment of some sampling plots to a habitat type differs between classifications. The most sharply-delineated habitat types that are also easily distinguishable physiognomically are sands and rocks; they also have specific rodent assemblages. The same is true for "deep loess" habitats, whereas differences in rodent assemblages between hammadas, wadis and "thin loess" habitats are less pronounced.

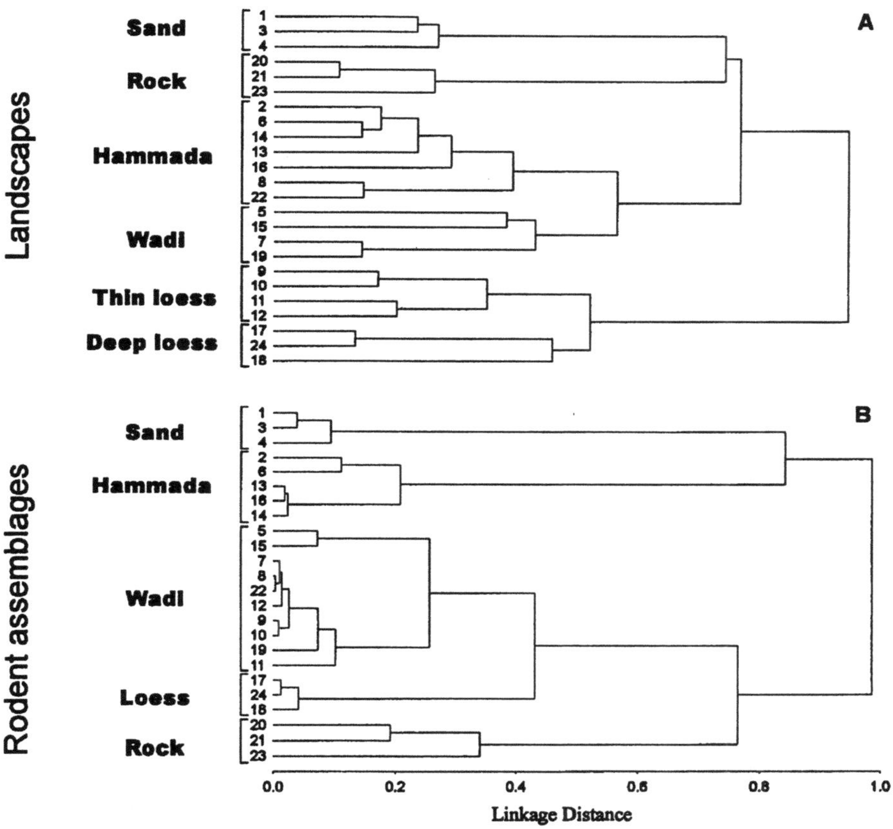

Fig. 7.1. Dendrogram showing the results of the cluster-analysis of sample plots in the Negev Highlands, based on their similarity in environmental parameters (A) and rodent species composition (B). Numerals indicate the numbers of sample plots

The comparison of the cluster-diagrams of the two classifications for the Thar Desert reveals a slightly different case (Fig. 7.2). The sand dune habitat type differs distinctly in its environmental characteristics and is perceived by rodents in like manner, while three other habitat types, accentuated by cluster-analysis based on relief, soil composition and vegetation variables, are split by rodents into two types according to the amount of vegetation present and without any relation either to soil structure or to relief.

The comparison of two classifications for the southern Kyzylkum Desert (Fig. 7.3) demonstrates results that are even fuzzier than in the previous case. There is not much in common in the manner that rodents respond to the landscape mosaic and the structure of this mosaic as revealed by environ-

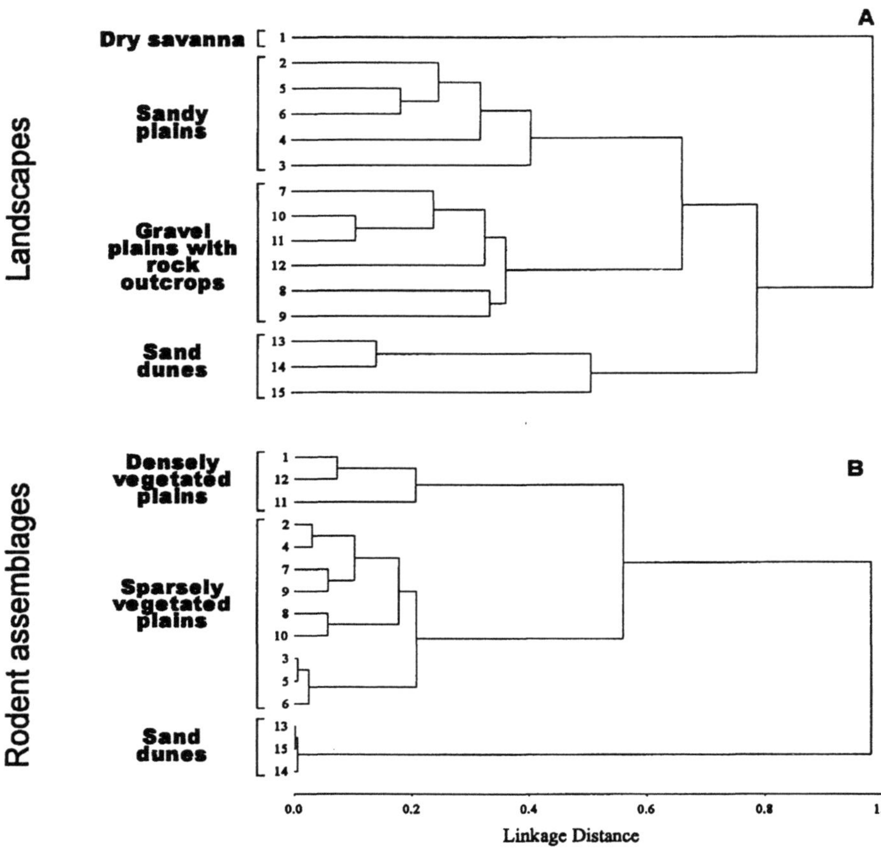

Fig. 7.2. Dendrogram showing the results of the cluster-analysis of sample plots in the Thar Desert, based on their similarity in environmental parameters (A) and rodent species composition (B). Numerals indicate the numbers of sample plots

mental similarity, other than the distinction between the two large habitat blocks, namely sands and non-sandy habitats. This means that, even in this case, rodent assemblages can be related to some landscape units, albeit in a rather poorly-defined manner.

A completely different phenomenon was exposed when cluster-analyses were applied to classifications of sampling plots in the northern Chihuahuan Desert (Fig. 7.4). Clusters of rodent assemblages have nothing in common with delineated landscape units. Consequently, we failed to distinguish habitat types and were forced to name rodent assemblages according to their taxonomic composition. The reasons for such a pattern may be found in the absence of strong environmental preferences among most North American

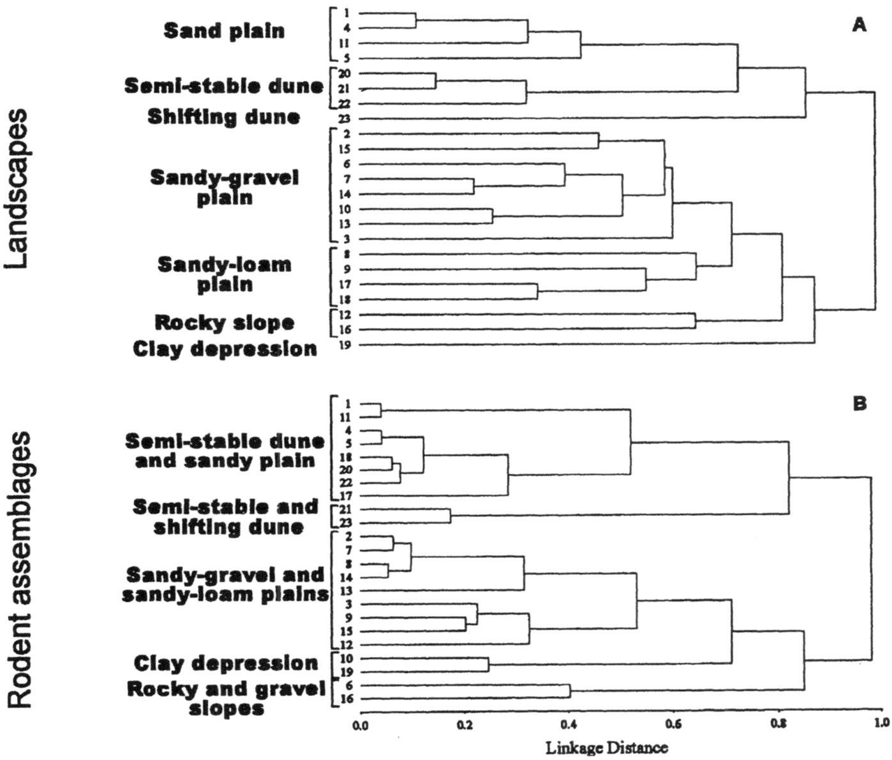

Fig. 7.3. Dendrogram showing the results of the cluster-analysis of sample plots in the southern Kyzylkum Desert, based on their similarity in environmental parameters (A) and rodent species composition (B). Numerals indicate the numbers of sample plots

desert rodent species (our data, see below). This conclusion is also in agreement with the absence of stable spatio-temporal species combinations found there (Brown and Kurzius 1987, 1989).

The above-mentioned argument supposes that rodents respond to a complex of environmental factors that may differ from those perceived physiognomically. Furthermore, different rodent species can react to the environment in different ways. Some species respond sharply to a small number of factors, with a positive reaction only in a narrow range along a gradient of each factor. The coincidence of these narrow ranges of several factors in a real environment is relatively rare, so a species with this type of reaction acts as a habitat specialist. The opposite case occurs when there is a relatively weak response to a large number of factors with a positive reaction to a broad range of each of them. These factor combinations are widely represented in reality, so it is difficult to distinguish a landscape unit that is optimal for such

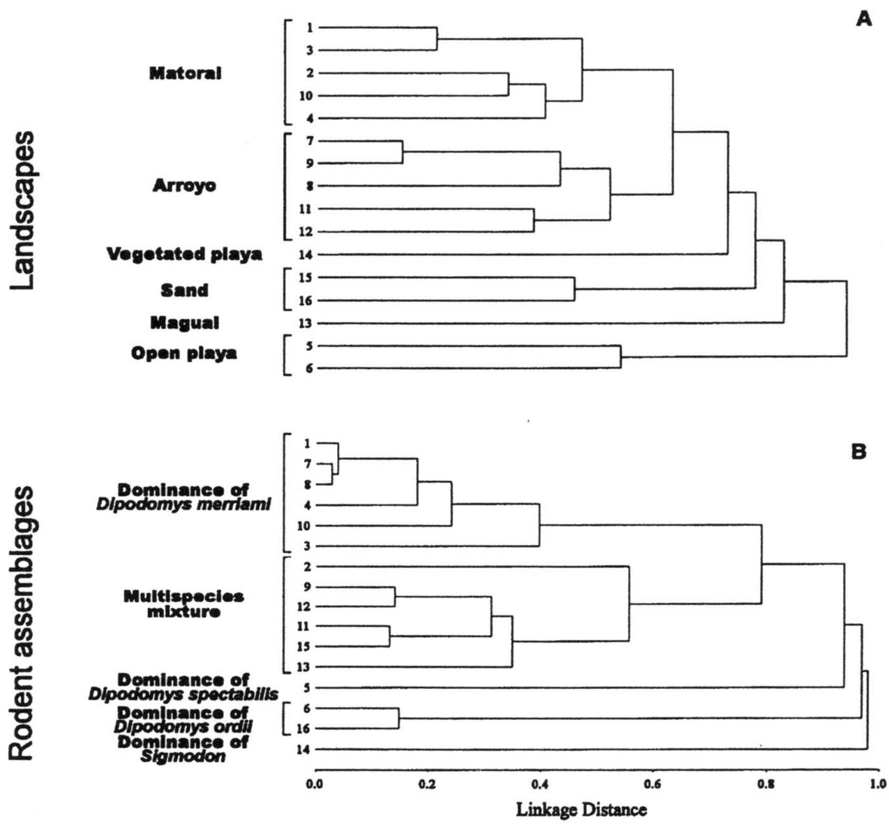

Fig. 7.4. Dendrogram showing the results of the cluster-analysis of sample plots in the northern Chihuahuan Desert, based on their similarity in environmental parameters (A) and rodent species composition (B). Numerals indicate the numbers of sample plots

a species. Such species, therefore, act as habitat generalists. There can also be an intermediate case when a species demonstrates a moderate response both in the number of significant factors and in the intensity of its reactions (determined by the range of positive response along each factor gradient). In this case, the necessary environmental conditions for a given species can be found mainly in a certain part of its geographic range (its optimum of the range), whereas in other parts of the geographic range these conditions are rare (i.e. at the ecological periphery of the range). Consequently, the species behaves as a habitat generalist in the optimum part of its geographic range and as a habitat specialist on the periphery.

Examples of all three types of response of a species to environmental structure can be found. The first is exemplified by the bush-tailed jird *Sekee-*

tamys calurus which only inhabits rocky mountains and outcrops along all its relatively small geographic range (Harrison and Bates 1991). An example of the second type of response is to be seen in Sundevall's jird *Meriones crassus*. This species occupies a huge geographic range and inhabits almost all the landscape units within it (Krasnov et al. 1996b). The brush-footed jerboa *Dipus sagitta* represents an example of the third type of response. It occurs in practically all the landscape units in the Gobi Desert (the range optimum) but only in sand dunes in the southwestern Kyzylkum Desert (its range periphery; Shenbrot et al. 1995).

To understand which environmental features determine the pattern of distribution of a species it is necessary to measure a large number of the parameters (e.g. soil structure, relief, vegetation abundance and architecture) that are supposed a priori to be important for the species' existence. The effects of habitat (environmental) variables can be evaluated using a stepwise multiple regression of the number of individuals of a species recorded at each given point (a dependent variable) on gradients of environmental factors measured at the same points (independent variables). The set of variables obtained in this way can be used to describe the environmental preferences of the species. Generally, the response of a species to a factor is assumed to be hump-shaped (i.e. has an optimum – Whittaker 1972), whereas a linear trend is considered as a particular case where, for some logistic reason, only some part of the entire gradient has been measured. Thus, it is desirable to use a regression model of $y=a+bx-cx^2$, where y is a dependent variable, x is an independent factor, and a, b, and c are regression coefficients. If both b and c are significant, the relationship is hump-shaped (Abramsky et al. 1985b; Abramsky 1988; Sokal and Rohlf 1995). Thereafter, the original values of densities and habitat variables have to be transformed to an equal-interval scale for both habitat and density variables. Environmental electivity can be measured by a variety of indices (see Lechowicz 1982), one of the simplest of which is Ivlev's index:

$$E_i=(u_i-s_i)/(u_i+s_i),$$

where u_i is an average number of records per sample of interval i of the interval scale of an environmental variable, and s_i is an average number of records per sample in the whole sample pool (Manly et al. 1993).

Our analysis of environmental factors affecting the local distribution of the pygmy gerbil *Gerbillus henleyi* in the desert of the Negev Highlands is an example of the approach described above. We captured animals on 24 1-ha sample grids, with 25 trapping stations each, and measured 19 quantitative parameters of the soil structure, relief and vegetation at each trapping station. Stepwise multiple regression analysis produced the following equation:

$$y=0.47+0.82*RC-0.19*RC^2-0.36*CL-0.08*GR+0.37*V1-0.07*V1^2,$$

where *y* is the number of registrations of *G. henleyi*; *RC* and *CL* are the rock and clay content of the soil, respectively; GR is the abundance of grasses (Poaceae); and V1 is shrub crown volume at the level 0-25 cm. Thus, only four of the 19 factors we measured actually influence the local distribution of the pygmy gerbil. Moreover, the reactions of this species to two of these factors are linear, and species density decreases with an increase in the factors' values. The responses to two other factors are unimodal with maximum gerbil density at moderate values of these factors (rock content and shrub crown volume).

These conclusions are further emphasized by consequent calculations and construction of a habitat-use profile for *G. henleyi* (Fig. 7.5). In this figure, values of Ivlev's electivity index are plotted for each rank of the environmental variables selected earlier. The magnitude of the bars therefore denotes the selectivity (positive or negative) of habitat use. One can easily see that the pygmy gerbil prefers patches with moderate rock and low clay contents of the soil, few grasses and moderate shrub crown volume. It clearly avoids patches of soil with high rock and clay contents, moderate or high abundance of grasses and both very low and very high shrub crown volume.

This approach permits us to enrich the qualitative description of a species' response to environmental structure by quantitative evaluation. The qualitative examples of the three types of rodent response to the environmental structure – demonstrated by *Meriones crassus*, *Sekeetamys calurus* and *Dipus sagitta* (see above) – can be presented quantitatively in terms of electivity along one or other environmental gradient (Figs. 7.6 and 7.7). These habitat-use profiles allow us to visualize those environmental conditions that are preferred by a given species and those which are avoided by it.

Sets of environmental factors determining local distribution are easily revealed and, consequently, habitat-use profiles have easily been constructed for most of the species of the Asian deserts that have been analyzed thus far (the Negev Highlands, Thar, Gobi and Kyzylkum Deserts; Shenbrot and Rogovin 1995; Krasnov et al. 1996a; Shenbrot et al. 1997 and unpubl. data). However, attempts to analyze the species-environment relations of the North American species in the same way failed completely. The correlations between environmental factors and abundance of each species were extremely low or nonexistent ($r2 < 0.13$). Consequently, habitat-use profiles demonstrated those conditions which are avoided, but in most cases did not show any environmental preference (Figs. 7.8 and 7.9). This means that the North American species perceive their environment in a fine-grain mode, responding to subtle changes in each separate factor within each landscape unit. In contrast, most Asian species do so in a coarse-grain mode, responding to a turnover of factor complexes among landscape units. One of the consequences of this difference is that the concept of "microhabitat" has been developed mainly by the American ecologists (Brown 1973; Smigel and Rosenzweig 1974; Brown 1989 etc.) whereas Russian scientists preferred to use the term

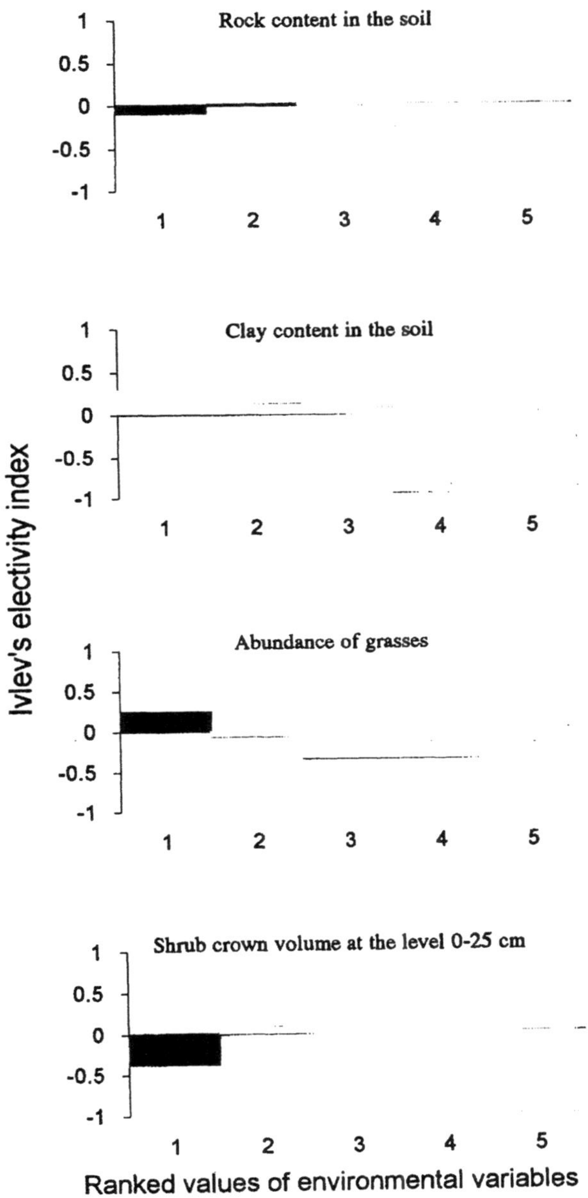

Fig. 7.5. Habitat-use profile of *Gerbillus henleyi* in the Negev Highlands. See text for explanations

"habitat" (="station" in the Russian tradition; e.g. Naumov 1965; Shilov 1977). The result of the fine-grain mode of environmental perception by North American rodents is that many species are able to find suitable conditions in different landscape types and consequently the classification of habitats by

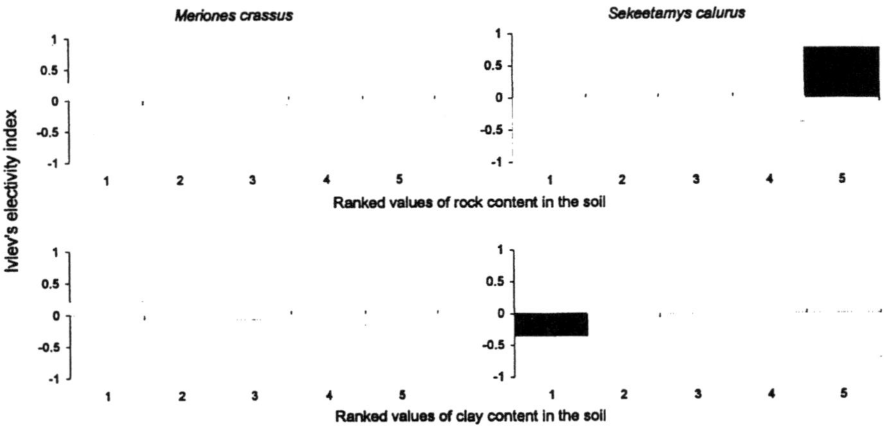

Fig. 7.6. Partial habitat-use profile (soil structure) of *Meriones crassus* and *Sekeetamys calurus* in the Negev Highlands

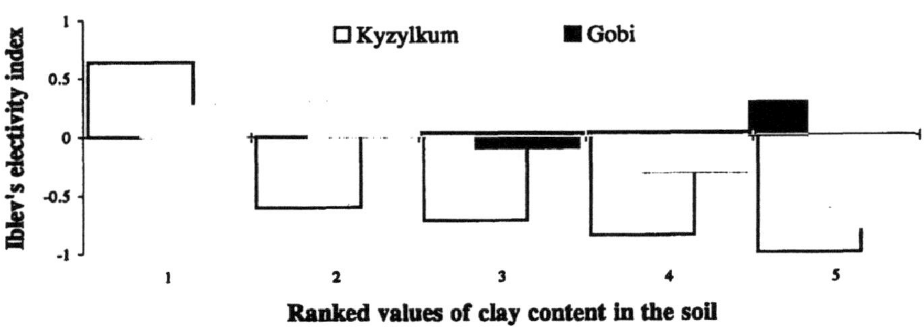

Fig. 7.7. Partial habitat-use profile (clay content of the soil) of *Dipus sagitta* in the Kyzylkum and Gobi Deserts

rodent species composition does not produce an interpretable picture (Fig. 7.10), whereas it does for Asia (Fig. 7.11).

Groups of species with different environmental preferences can be distinguished when rodents respond to complexes of environmental conditions in a coarse-grain mode. For example, five such groups have been described in the central part of the desert of the Negev Highlands (Krasnov et al. 1996a). The first group consisted of rock-dwelling forms – the Egyptian spiny mouse *Acomys cahirinus*, the golden spiny mouse *Acomys russatus* and the bush-tailed jird *Sekeetamys calurus*. These species inhabit rocky areas exclusively throughout their geographical ranges (*A. cahirinus* also occurs occasionally in other types of habitats; Osborn and Helmy 1980, Harrison and Bates 1991).

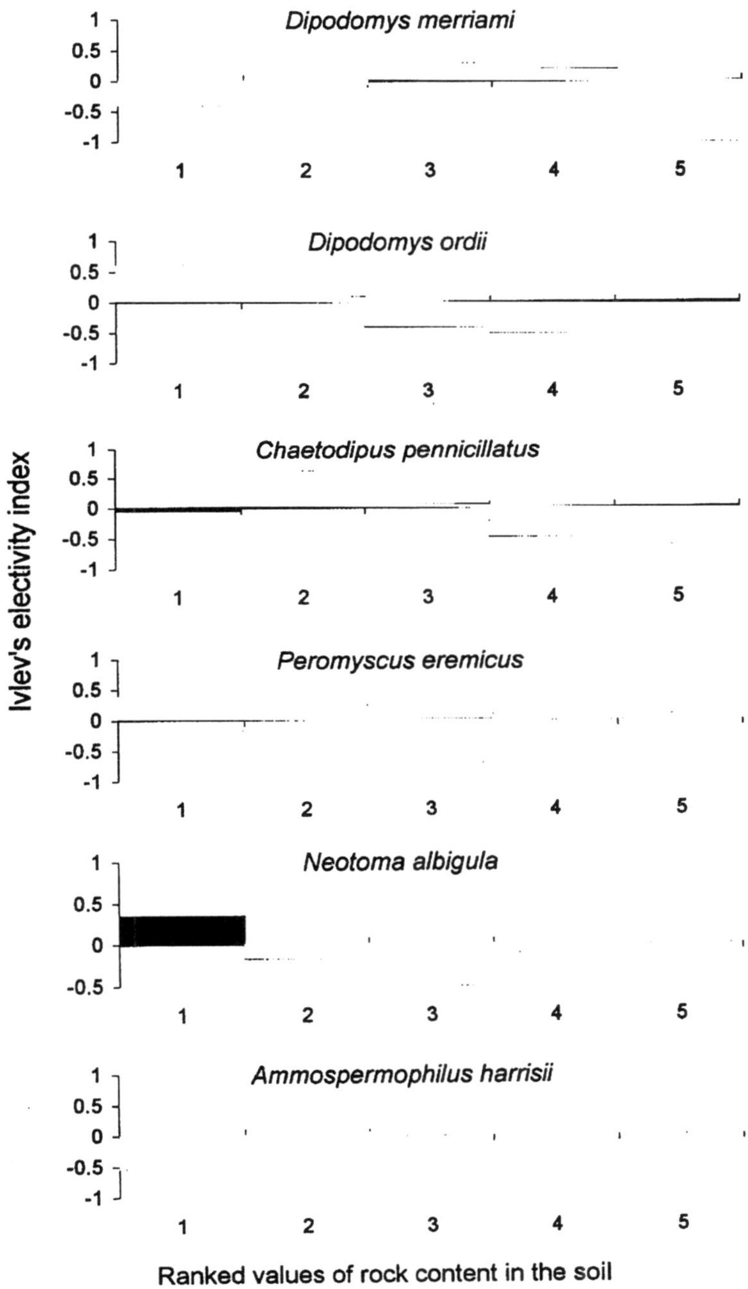

Fig. 7.8. Partial habitat-use profiles (rock content of the soil) of common rodent species in the northern Chihuahuan Desert

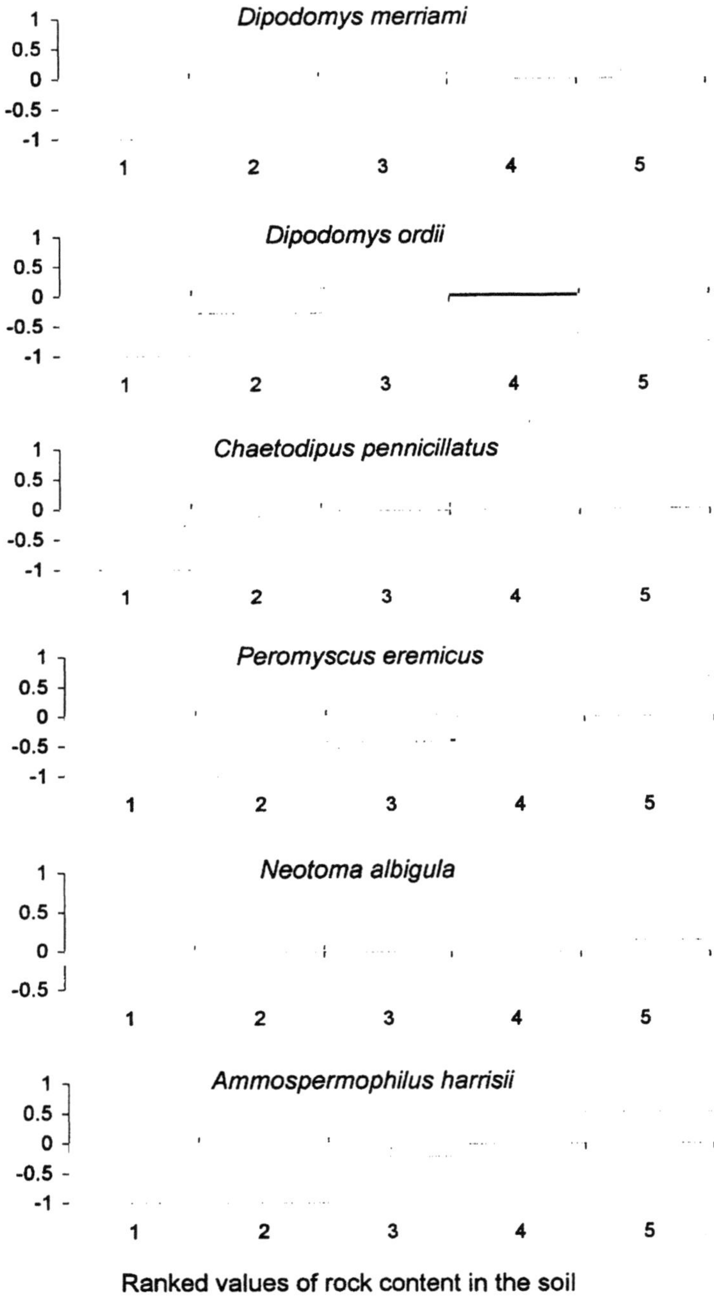

Fig. 7.9. Partial habitat-use profiles (shrub cover) of common rodent species in the northern Chihuahuan Desert

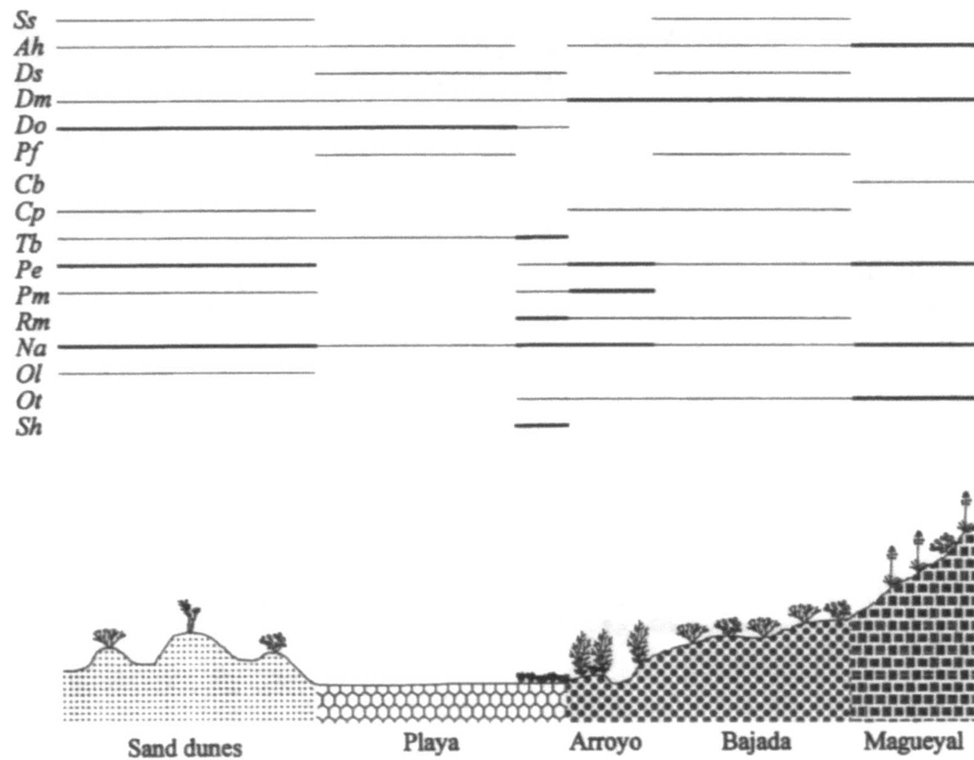

Fig. 7.10. Scheme of rodent species distribution across various habitats in the northern Chihuahuan Desert. Horizontal line breadth corresponds to relative density of the species. Species names are as follows: Ah *Ammospermophilus harrisii*, Ss *Spermophilus spilosoma*, Cb *Chaetodipus baileyi*, Cp *C. penicillatus*, Dm *Dipodomys merriami*, Do *D. ordii*, Ds *D. spectabilis*, Pf *Perognathus flavus*, Na *Neotoma albigula*, Ol *Onychomys leucogaster*, Ot *O. torridus*, Pe *Peromyscus eremicus*, Pm *P. maniculatus*, Rm *Reithrodontomys megalotis*, Sh *Sigmodon hispidus*, Tb *Thomomys bottae*

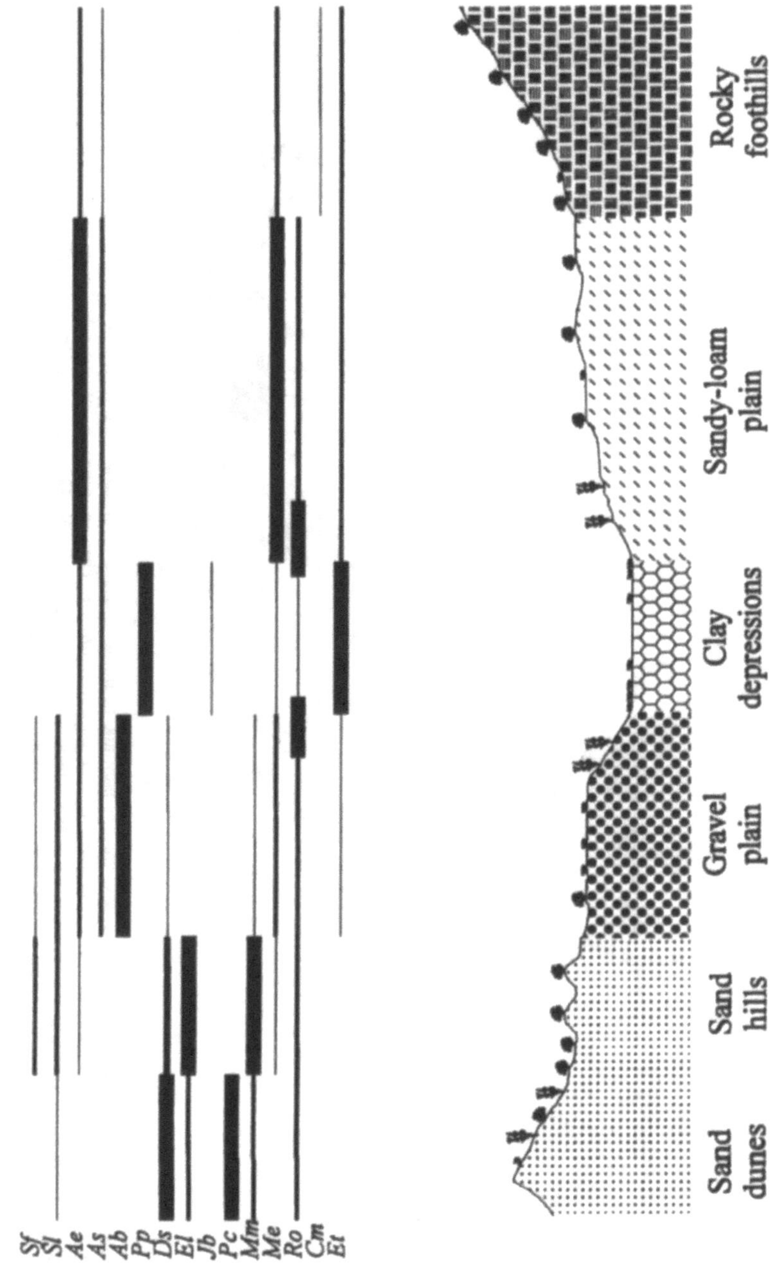

Fig. 7.11. Scheme of rodent species distribution across various habitats in the southern Kyzylkum Desert. Horizontal line breadth corresponds to relative density of the species. Species names are as follows: Sf *Spermophilus fulvus*, Sl *Spermophilopsis leptodactylus*, Ae *Allactaga elater*, As *A. severtzovi*, Ab *Allactodipus bobrinskii*, Pp *Pygeretmus pumilio*, Ds *Dipus sagitta*, El *Eremodipus lichtensteinii*, Jb *Jaculus blanfordi*, Pc *Paradipus ctenodactylus*, Mm *Meriones meridianus*, Me *M. erythrourus*, Ro *Rhombomys opimus*, Cm *Cricetulus migratorius*, Et *Ellobius tankrei*

The second group, the psammophilous or sand-dwelling rodents, is represented only by the Egyptian gerbil *Gerbillus gerbillus*. Depauperation of the local psammophilous fauna is the result of the small size and isolation of sandy areas, whereas the main sand massifs of the Negev maintain a more diverse psammophilous fauna (*Gerbillus gerbillus, G. pyramidum, G. andersoni, Meriones sacramenti*; Abramsky et al. 1985b).

The third group consists of the fat sand rat *Psammomys obesus*, the Asian garden dormouse *Eliomys melanurus* and the house mouse *Mus musculus* which occupy densely vegetated dry riverbeds and slopes. These rodents occur in areas with loess soils and are rarely found on other soil types. The fourth group contains three species, namely the lesser jerboa *Jaculus jaculus*, the pygmy gerbil *Gerbillus henleyi* and the Baluchistan gerbil *Gerbillus nanus*. These inhabit scarcely vegetated patches with low densities of annuals. Previous studies have reported that these three species are considered to be psammophilous (Abramsky 1984, Abramsky et al. 1985a). However, we recorded them on a variety of soils. The last group consists of two species, Sundevall's jird *Meriones crassus* and Wagner's gerbil *Gerbillus dasyurus*, and is characterized by wide distribution and presence in almost all types of habitats (except for *M. crassus* which is not found in rocky sites).

Species with different types of habitat preference have also been distinguished among rodents of the southern Kyzylkum Desert (Shenbrot and Rogovin 1995). These are found to be psammophiles (*Dipus sagitta, Eremodipus lichtensteinii, Paradipus ctenodactylus, Meriones meridianus*), the inhabitants of clay depressions (*Pygeretmus pumilio, Jaculus blanfordi*) and open gravel plains (*Allactodipus bobrinskii*), and species that occupy almost all landscape types (*Allactaga severtzovi, Meriones erythrourus, Rhombomys opimus, Ellobius tankrei*).

The distribution of rodents according to their types of habitat specialization in the Gobi Desert is characterized by the presence of psammophiles (*Salpingotus crassicauda, Salpingotus kozlovi*), petrophiles or rocky-dwellers (*Alticola barakschin, Cricetulus longicaudatus*), the inhabitants of gravel plains (*Cardiocranius paradoxus, Allactaga bullata, A. balicunica, Stylodipus andrewsii*) and habitat opportunists (*Euchoreutes naso, Allactaga sibirica, Dipus sagitta, Meriones meridianus, Rhombomys opimus, Phodopus roborovskii*; Rogovin and Shenbrot 1995). The Thar Desert contains psammophiles (*Gerbillus gleadowi*), petrophiles (*Cremnomys cutchicus, Mus saxicola*), occupants of open gravel plains (*Millardia gleadowi*) and habitat generalists (*Tatera indica, Gerbillus nanus, Meriones hurrianae, Mus booduga*; Rogovin et al. 1994).

Among South African desert rodents, sand-dwellers (e.g. *Bathyergus janetta, Gerbillurus tytonis*), rock-dwellers (e.g. *Petromus typicus, Petromyscus collinus, Graphiurus platyops*), inhabitants of open plains (e.g. *Gerbillurus vallinus, Desmodillus auricularis*), and species widely distributed across several landscape types (e.g. *Tatera brantsii, Malacothrix typica, Zelotomys woosnami*) can be distinguished (De Graaf 1981). Strict occupants of sand

(*Pseudomys desertor, Notomys fuscus*), rocks (both *Zyzomys* species) and open plains (*Pseudomys australus, Notomys cervinus*) occur in the Australian desert fauna (Strahan 1983).

In contrast to the Old World deserts, it is difficult to distinguish habitat specialists in the North American deserts. For example, our observations have shown that rodents limited to one or other landscape type in the northern Chihuahuan Desert appeared to be either locally rare species that can be more widespread when their densities increase (e.g. *Chaetodipus hispidus*), or non-desert species that only inhabit mesic intrazonal patches (e.g. *Peromyscus maniculatus*). However, strong preference for a particular habitat type has been reported in some North American desert rodents. For example, *Dipodomys deserti, Microdipodops megacephalus* and *Microdipodops pallidus* are considered to be strict psammophiles (Genoways and Brown 1993). Habitat preferences of South American desert rodents are poorly known. Nevertheless, the presence of the sand-dwelling *Eligmodontia moreni* (Mares et al. 1981), the rock-dwelling *Lagidium viscacia* (Rowlands 1974), *Octomys mimax* (Mares 1973) and *Abrocoma benneti* (Redford and Eisenberg 1992), and the habitat opportunists *Calomys callosus* and *Graomys griseoflavus* (Mares et al. 1981) should be noted.

Our main conclusion from the above observations is that types of habitat specialization are very similar in different deserts. This is understandable. Indeed, all deserts have similar landscape structures (see Chap. 2). If there are, for example, sand dunes in a particular desert, the occurrence of species specialized to life in sand dunes should be expected there. Consequently, the presence of psammophilous species in almost all the world's deserts is due to the presence of sands in these deserts. However, reality is somewhat more complex. Sand patches in the Chihuhuan Desert are well developed, but there are no strictly psammophilous rodents there. Sand is inhabited mainly by Ord's kangaroo rat *Dipodomys ordii* and the desert pocket mouse *Chaetodipus penicillatus*. Both these species also occupy other landscape types. Another example is afforded by the absence of petrophilous species in the main regions of the deserts of Middle Asia in spite of the availability of rocky slopes with numerous niches and crevices. Rock and cliffs there are inhabited by the gray hamster *Cricetulus migratorius* which is regarded as a habitat generalist in other parts of its geographic range. The only location in which two strict rock-dwellers (the mouselike hamster *Calomyscus bailwardi* and the Persian jird *Meriones persicus*) are recorded is southwestern Turkmenistan. The list of exceptions is long. Clay depressions (takyrs) in the Turan deserts are the specific habitats of jerboas *Pygeretmus pumilio* and *Jaculus blanfordi* while, in the Gobi Desert ,takyrs do not have specific inhabitants. The absence of psammophilous species in the Chihuahuan desert, of petrophilous species in the Turan deserts, and of sclerophilous (=takyr-dwelling) species in the Gobi Desert is probably caused by some historical processes. On the other hand, these exclusions bring us back to the question of "what is a habitat?" Is a habitat determined merely by environmental variation or should it be considered in conjunction with the species that recognize it as such? Are there sandy habitats in the desert of Negev Highlands? The answer is surely positive. There are sands and there are species that perceive sand as something specific. Are there rocky habitats in most of the Turan deserts ? The answer is negative. There are no rocky habitats in the Turan deserts although there are rocks. This is because rodents do not recognize the specificity of a rocky landscape type as a place for life. Thus, the application of the term "habitat" to a pattern of responses of a set of species to a set of landscape units seems to be preferable. According to Rosenzweig (1992, 1995), habitat diversity depends on species diversity or "species diversity is a horse; habitat diversity is a cart" (Rosenzweig 1995, p. 176). Nevertheless, landscape types are frequently very distinct and there is at least one species that recognizes one as a specific habitat, so the two approaches coincide.

The existence of species that are adapted to a more or less narrow range of landscape variation permits us to classify these species by types of habitat specialization. Four main types of habitat specialization can be distinguished in deserts. These are (1) adaptations to a sandy horizontal substrate, namely psammophily; (2) adaptations to a hard horizontal substrate (takyrs) namely sclerophily; (3) adaptations to a horizontal substrate with alternating soft- and hard-soil patches (open gravel plains); and (4) adaptations to a hard vertical substrate (rocks and cliffs), namely petrophily. Each of these types has specific morphological characters that evolved as an adaptive response to specific landscape features (see below).

Types of habitat specialization emerge independently in different phylogenetic lineages and/or desert regions. As a result, representatives of only two to three types of habitat specialization are usually found in each local fauna. Cases in which all types of habitat specialization occur are rare and seem to be exceptions rather than the rule (Fig. 7.12). The most widespread type of habitat specialization is psammophily. Sand specialists occur in the faunas of most provinces in all the continents. The reason for the absence of specialized sand-dwellers in regions where sands are rare or absent (the Syrian, South Pakistan, Tzaidam and Kalahari provinces) is clear and does not need any additional explanation. It should be remembered, that the large massifs of non-stabilized and semi-stabilized sand dunes of the southern part in the Kalahari Desert belong faunistically to the Namib province (Chap. 3), whereas the main landscapes of the Kalahari province sensu stricto are alternating densely vegetated sandy plains, clay depressions and rocky outcrops. Taking into account that psammophily is actually an adaptation to a life on non-stabilized substrates (given that to inhabit stabilized sandy plains does not demand any specific adaptations), we can conclude that there are no conditions for the evolutionary development of psammophily in the Kalahari province.

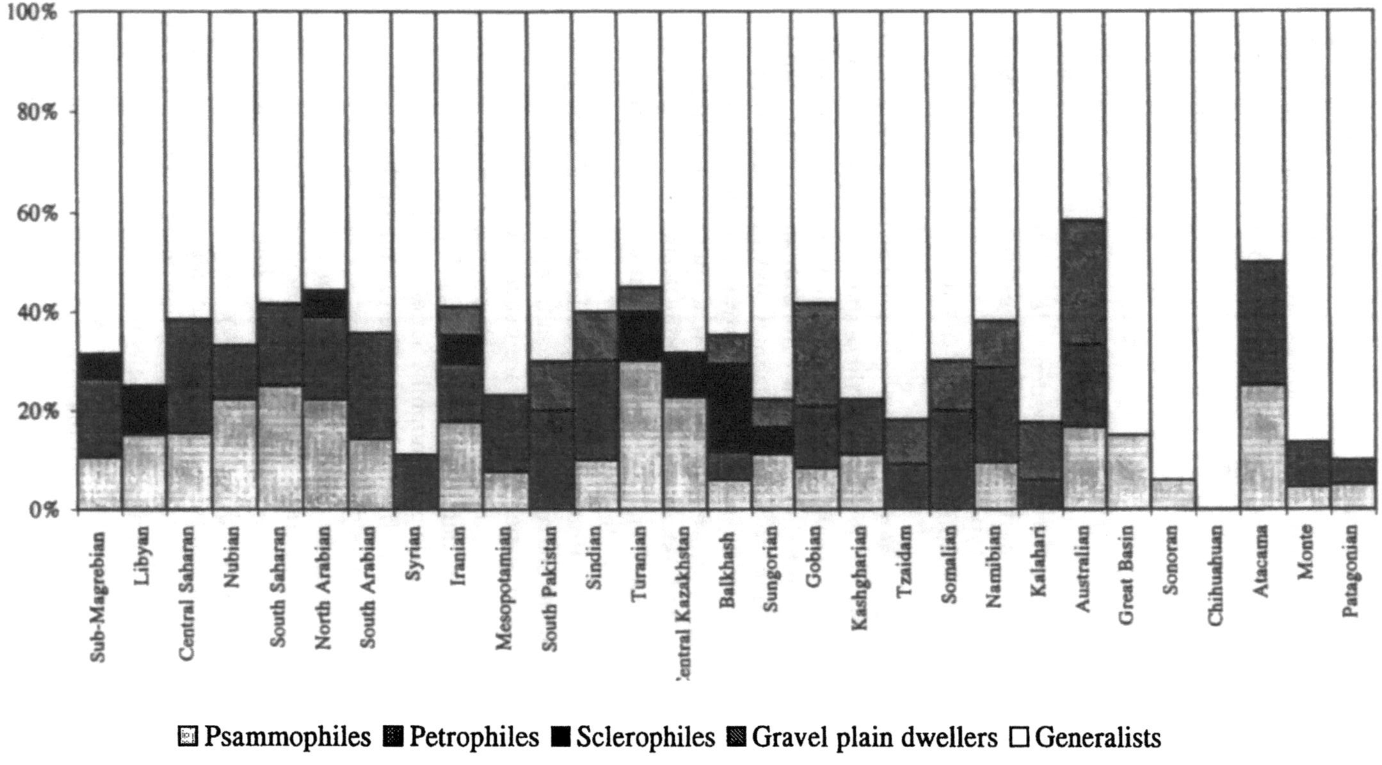

Fig. 7.12. Distribution of rodents with different types of habitat specialization in different desert provinces. Species inhabiting more than one landscape type were assigned to the generalist category

Strict psammophiles are also absent from two regions in which sandy landscapes are well developed (the Somalian and Chihuahuan provinces). The case of the Somalian fauna is somewhat unclear on account of the scarcity of data on rodent biology. In the Chihuahuan province, however, the absence of a psammophilous component can be explained on ecological/ historical grounds. For example, sands are widely represented, but only by small patches in lake valleys (playas). This small-scale mosaic does not permit the persistence of isolated populations, and thus, may prevent the evolution of specialized species. As a result, sandy patches are occupied by generalists.

Psammophily evolved independently in several rodent taxa, but is most frequent among jerboas and gerbils. Three of five subfamilies of jerboas have psammophilous species (Cardiocraniinae, Dipodinae and Paradipodinae). Moreover, psammophily in these subfamilies is manifested on a generic level. Psammophilous (*Salpingotus*, *Dipus*, *Eremodipus*, *Paradipus*) and non-psammophilous (*Cardiocranius*, *Stylodipus*, *Jaculus*) genera can be distinguished. Psammophily occurs in three gerbil genera. It is most common in *Gerbillus* (eight out of 23 desert species), less so in *Meriones* (two out of 11 desert species) and *Gerbillurus* (one out of four desert species). In addition, specialized sand-dwellers can be found among the Heteromyidae (*Microdipodops*, *Dipodomys deserti*), Murinae (*Notomys fuscus*), Sigmodontinae (*Phyllotis gerbillus*, *Eligmodontia moreni*) and Bathyergidae (*Bathyergus janetta*).

Petrophily is less widespread among desert rodents than is psammophily. Petrophilous species are absent from the faunas of those provinces in which there are no rocky environmental features (the Lybian and Central Kazakhstan provinces). There are also provinces where rocky landscapes are present whilst specialized desert rock-dwellers are absent. In most of these cases (three North American provinces and the Sungorian province) the absence of specialized desert petrophiles is caused by the occupation of rocky habitats by non-desert species which have penetrated them from high mountains nearby. For example, rocks and cliffs in Sungoria are inhabited by voles *Alticola argentatus* and *Alticola macrotis* and the daurian pika *Ochotona daurica*, all of which are characteristic of non-desert high mountain habitats. The latter species belongs to the order Lagomorpha but is ecologically similar to rodents. Of particular interest is the Turanian province, where two petrophilous species occur but only in the extreme southwest (see above), whereas most of the rock massifs are not inhabited by specialized species. It is difficult to explain this phenomenon in any other way than by referring to some unknown historical reason.

Petrophilous species can be found among almost all rodent lineages. Some taxa consist exclusively of rock specialists. These are Ctenodactylidae, Chinchillidae, Abrocomidae and Petromuridae (the last taxon is monotypic). Other taxa either have petrophilous genera (Sciuridae: *Atlantoxerus*; Gerbillinae: *Sekeetamys*; Murinae: *Acomys*, *Cremnomys*, *Zyzomys*; Cricetinae: *Calomyscus*; Arvicolinae: *Alticola*; Petromyscinae: *Petromyscus*; Octodontidae: *Octomys*) or petrophilous species within prevalently non-petrophilous genera (Sciuridae: *Xerus princeps*, *Spermophilus alashanicus*; Gerbillinae: *Meriones persicus*; Murinae: *Mus saxicola*; Cricetinae: *Cricetulus longicaudatus*; Myoxidae: *Graphyurus ocularis*).

Rodents which exclusively prefer open gravel plains occur in the Old World deserts only. Gravel plain specialists are absent in the Saharo-Arabian subrealm, and from those Old World desert regions in which hammada-like landscapes are widely distributed. In the Turano-Gobian subrealm, however, gravel plain specialists are represented by jerboas of the genera *Cardiocranius*, *Allactodipus*, *Allactaga* and *Stylodipus* (but not by all species in the last two genera). There are two specialized gravel plain inhabitants in the Irano-Sindian subrealm, namely Hotson's jerboa *Allactaga hotsoni* in Iran and the soft-furred rat *Millardia gleadowi* in the Thar Desert and southern Pakistan. In Somalia and South Africa, all the gravel plain specialists are gerbils (monotypic *Ammodillus* and *Microdillus* in Somalia and *Gerbillurus vallinus* and *Desmodillus auricularis* in the southern Africa). The two Australian representatives showing this type of habitat specialization belong to the murine genera *Pseudomys* (*P. australis*, *P. chapmani*) and *Notomys* (*N. cervinus*).

Sclerophilous specialists can only be found in some of the Palaearctic provinces. They are jerboas belonging to three genera, *Pygeretmus*, *Allactaga* and *Jaculus*. In the last two, sclerophiles belong only to the subgenera *Scarturus* and *Haltomys*, respectively.

The absence of both sclerophiles and gravel plain specialists from the American deserts requires some explanation. It seems probable that the differences in the perception of the environment between the North American and the Asian species (see above) could be the main cause. As to South America, the scarcity of ecological data leaves this question open, although the absence of bipeds there could well be an explanation for the absence of species which are able to distinguish subtle variation of the substrate.

7.3 Morphological Correlates of Habitat Specialization

The spatial heterogeneity of the desert environment is expressed in the mosaic distribution of patches that provide different conditions for sheltering and burrowing (Rosenzweig 1973; Thompson 1982a,b; Price et al. 1984; Kotler and Brown 1988; Kotler et al. 1991, 1992; Shenbrot et al. 1995). In addition, it provides different compositions and types of aggregation of food (Reichman 1981; Reichman and Price 1993). Different types of substrate affect locomotion and foraging (Price and Waser 1985; Price and Longland 1989; Shenbrot et al. 1995). The environmental variation among landscape units results in optimal behavior being selected for in each spatial (environmental) unit. Both the optimization of individual behavioral tactics and the natural selection toward optimal behavioral strategies seem to be primary steps in the evolution of a morphological innovation. In turn, morphological novelties appear to be permanently tested and corrected by the environment through behavior. Both the directional changes in a system and its equilibrium state are based on a close interrelationship between the contours of positive and negative feedback (Fig. 7.13). For example, imagine that the climate becomes more arid and the environment more patchy and poor in feeding resources. How can a hypothetical quadrupedal rodent species persist? Energetically, the cheapest decision seems to be a change in foraging behavior. At first, this can be achieved by either more intensive (increase in foraging time) or more extensive (increase in foraging territory) use of space, including open patches, with no significant change in diet. However, the risk of predation increases with the extension of either the duration or the area of foraging activity. Structural change in the morphology of the locomotory system of the prey towards increased effectiveness in escaping predators, is the simplest way to compensate for an increased risk of predation. Consequently, individuals that are able to accelerate more rapidly in the initial stage of locomotion should be those that are selected. Observations on the home range ecology and foraging behavior of jerboas and kangaroo rats that differ in their level of morphological adaptation for bipedal locomotion fit this hypothetical scenario (Rogovin, in press). Jerboas, especially fast running species, have large home ranges, whereas the home ranges of kangaroo rats of a similar size are smaller. Moreover, the size of the home range among jerboa species is positively correlated with body mass, whereas this is not the case with kangaroo rats. The latter tend to use their home ranges much more intensively, with respect to the utilization of feeding resource than do jerboas. They also seem to be more dependent on shelters. This is especially true of species with large body size (*Dipodomys spectabilis*, *D. nelsoni*) which spend more than 90% of their time in their burrows. Territoriality (in terms of home range defence) is not characteristic of jerboas, but there are examples of well pronounced territoriality among kangaroo rats (Jones 1993; Randall 1993).

A considerable amount of evidence suggesting a close relationship between morphology and adaptive behavior in rodents has been accumulated (i.e. Dice 1947; Webster 1962; Eisenberg 1963, 1967; Webster and Webster 1971; Smirin and Orlov 1975; Wilson et al. 1976; Dewsbury et al. 1980; Randall 1981, 1994). Results of unpublished studies by Rogovin and Surov of the individual behavior of different desert rodents in a series of so called open-field tests (in clean, unfamiliar cage in which an animal has been placed to study its exploration behavior) showed a positive linear correlation ($r^2 = 0.436$; $p < 0.05$) between the frequency of vertical postures adopted when exploring the new cage and the position of the rodent's eyes (i.e. the frontal distance between the eyes, which determines a sector of the field of vision). Vertical postures have been most frequently demonstrated in species that inhabit sparcely

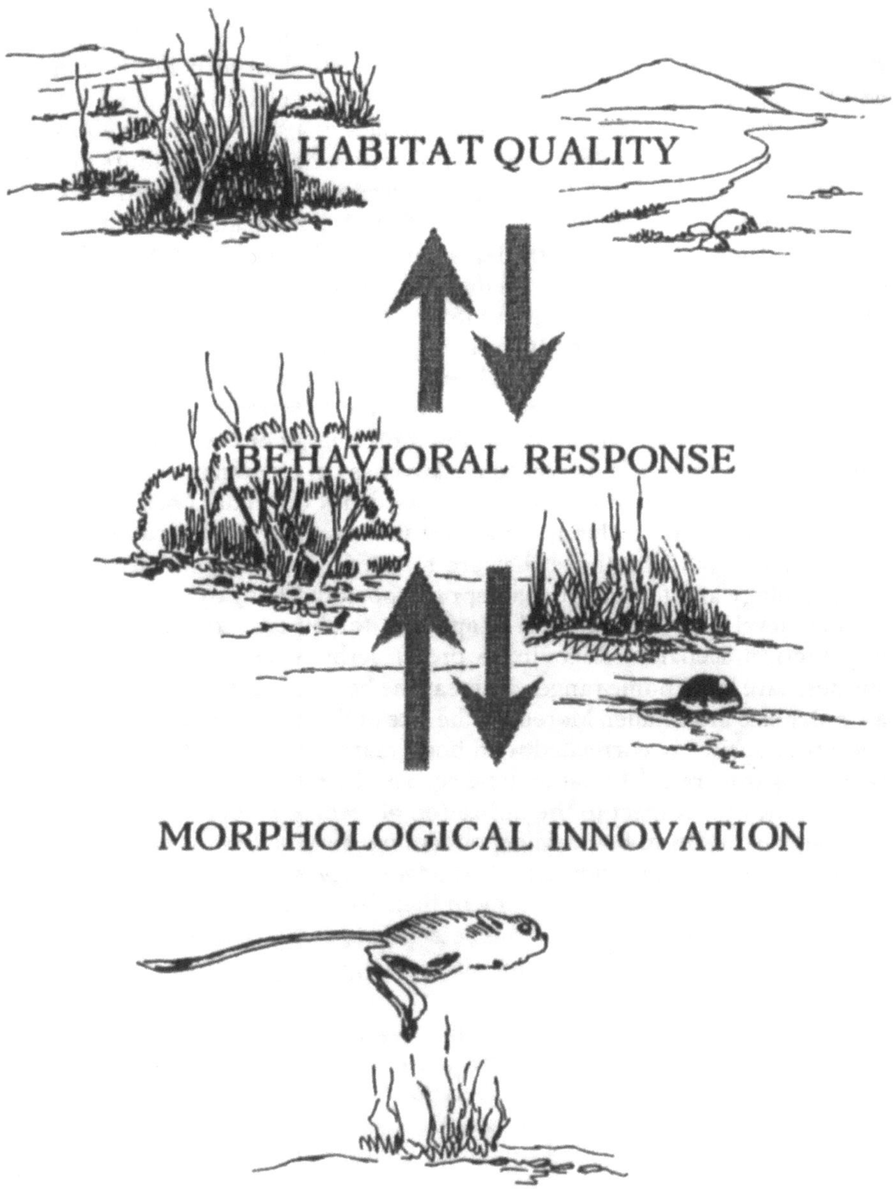

Fig. 7.13. Schematic relationship between habitat quality, individual behavior and morphological innovations. See text for explanations

vegetated or open areas, have relatively large or medium sized bodies and shelter in deep burrows or escape predatory attacks by fast bipedal running

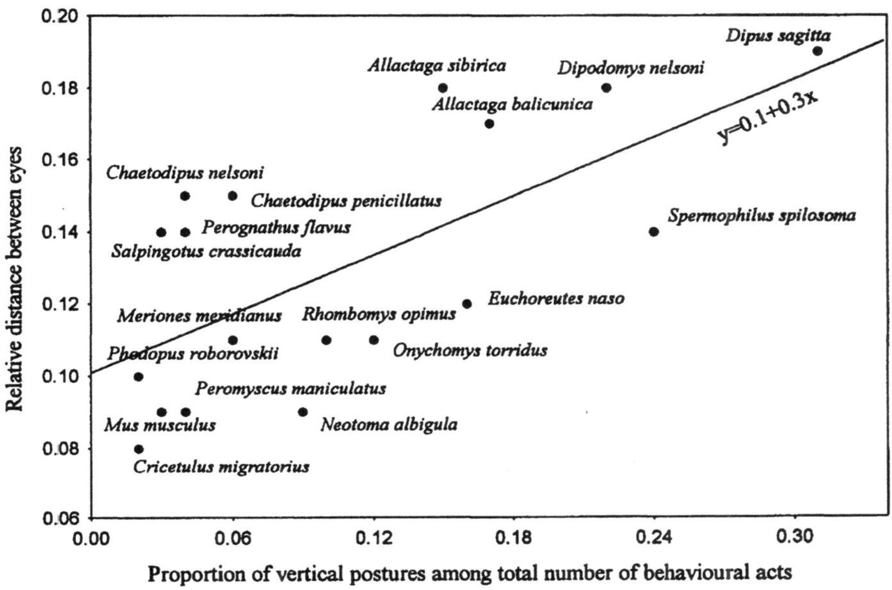

Fig. 7.14. Position of eyes (mean distance between eyes corrected per 1 cm of head-and-body length) versus proportion of vertical postures among total number of behavioural acts registered per 10 min when an animal explored unfamiliar territory. Five males and five females of each species were tested. The unfamiliar territory was a clean cage (1 x 1m) with fresh sand as a substrate

(Fig. 7.14). These species also have a greater frontal distance between the eyes, suggesting that a wide field of vision is an important feature for inhabitants of open areas (see also Smirin and Orlov 1973). In contrast, small species that are tied to natural shelters such as shrubs or patches of dense grass have no chance of identifying predators at a distance. Because they usually do not depart far from their shelters, the main problem for them is to choose the correct direction of escape from the approaching source of danger. The eyes of these species are close to one other on the front of the head. In consequence, their sight is oriented forwards.

The external morphological characters of the rodents of "open" habitats have traditionally been considered to be the most useful characters to use in the classification of rodent life forms (e.g. Kashkarov 1938; Formozov 1956; Mares 1980). It is not surprising, therefore, that most of these characters reflect adaptations to specific types of habitat. Habitat partitioning appears to be the main way in which ecological resources are shared by coexisting species of terrestrial vertebrates (Schoener 1974). Consequent habitat specialization does not usually require deep functional (physiological or biochemical) transformation, but only some structural rebuilding (changes in the morphology) of an organism. Such innovations can evolve in accordance with the above-mentioned scheme through the selection of adaptive behavioral strategies (Dolnik 1986).

Which morphological features are correlated with habitat preferences in desert rodents? In fact, many of the external morphological characters of a particular species reflect its adaptation to the specific features of its preferred habitat. This has been partly discussed in Chapters 5 and 6. For example, the external morphology of the hind feet, and especially of their distal parts, reflects the properties of the predominant type of substrate used. This relationship is developed most highly in bipedal rodents. The reason is that the mechanism of bipedal jumping demands exact conformity between the properties of the substrate surface and the morphology of the supporting part of the hind foot (Ognev 1948; Fokin 1978). Kangaroo rats which are not highly specialized for bipedal running possess rather "generalist" hind feet (Berman 1985) and do not distinguish between substrate types as finely as do most jerboa species. The only exception is the kangaroo rat (*Dipodomys deserti*) which has well developed fur brushes on its toes. This is a feature typical of all sand-specialized rodents (both bipedal and quadrupedal). Among fast-running jerboas of the subfamilies Allactaginae, Dipodinae and Paradipodinae, species inhabiting hard clay soils have large callous pads at the ends of their toes and a prominent conic callus on the underside of the foot near the base of the toes. The toe pads are subdivided into several portions, thereby providing better cohesion with the hard soil surface (Fig. 7.15). Jerboas that inhabit gravel substrates (*Allactodipus bobrinskii*, *Allactaga balikunica*, *Euchoreutes naso* etc.), not only have well-developed callous pads and a conic callus but also the proximal parts of their toes have brushes. These characters prevent an unstable ground surface from slowing them up. In addition, the shape of the distal part of the hind foot in jerboas reflects the diversity of habitats that each species occupies (Fig. 7.15).

Morphological features of the hind feet reflecting adaptations to locomotion on a sandy substrate are expressed to varying degrees in different species. They are positively correlated with the running speed typical of each particular species. In other words, the faster an animal runs, the better are its morphological adaptations to sand locomotion expressed. In quadrupeds, psammophilous species of the genus *Gerbillus* have hind feet with the same proportions as do other species of the genus, but the toes and soles of their feet are more densely hairy than are those of their non-psammophilous congenerics. In psammophilous jerboas, the third (central) toe of the hind foot is much shorter than that in non-psammophilous species. Consequently, all three toes of the hind feet in psammophiles are the same length and their points of support lie in one line. Furthermore, psammophilous jerboas also exhibit variation in the structure of the footbrush. Primitive bipeds (*Salpingotus*) have a dense brush of long and moderately soft hairs underneath the soles and toes of their hind feet. Species with well-developed bipedal loco-

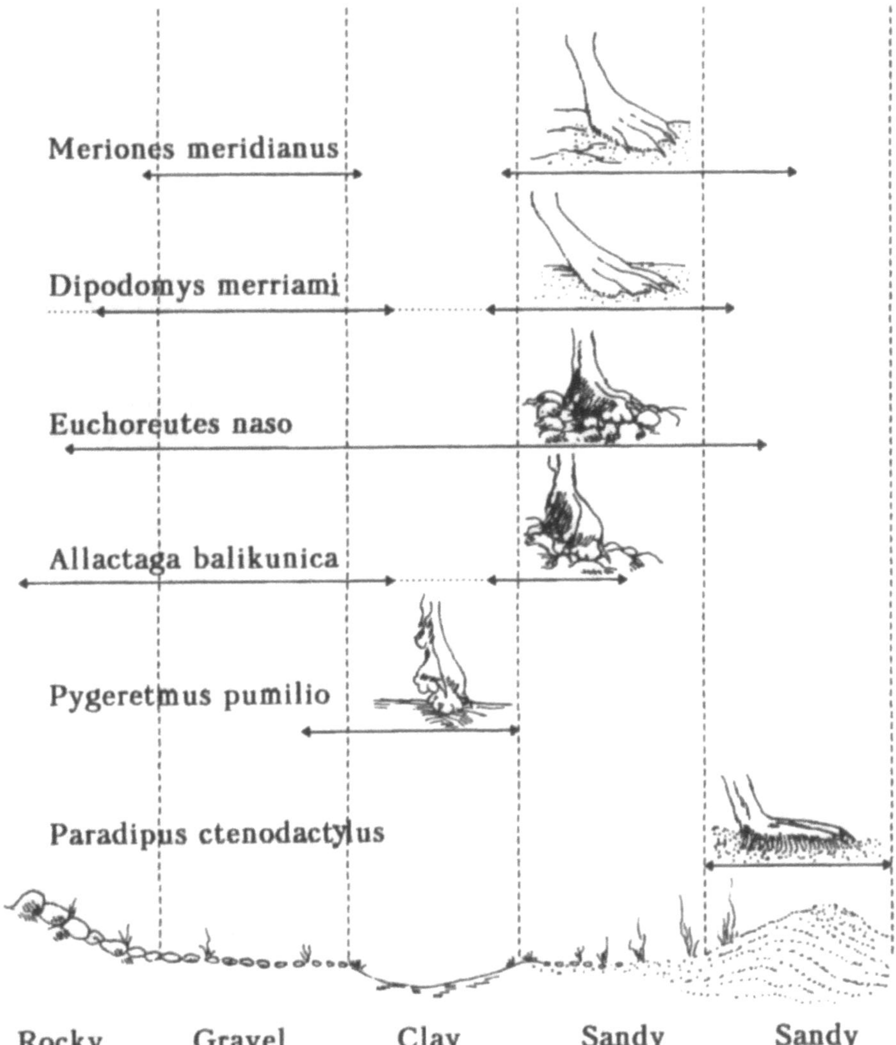

Fig. 7.15. External morphology of distal parts of the hind foot of some desert rodents and schematic ranges of the habitats they occupy

motion whose running speed is moderate (*Dipus*, *Eremodipus*) possess a footbrush consisting of homogenous and relatively short wiry hairs. The most complicated footbrush is found in the high-speed runner *Paradipus ctenodactylus*. It consists of two types of hairs, namely short wire bristles and long soft hairs. Each toe bears two rows of the bristles. Both rows of the central toe are located on its lower surface and their bristles are directed downward. The first row of bristles of the lateral toes is located on the external side of the toe and the bristles are directed downward. In contrast, the second row is located on the internal side and its bristles are directed forward and internally. The long, soft hairs of the brush are located between the rows of bristles and are directed downward.

The shape and size of toes and claws is another group of characters that reflects the habitat preferences in desert rodents. However, it is necessary to point out that these characters reflect two different aspects of habitat use, namely digging and locomotion. As far as toes and claws are concerned, digging adaptations are reflected in the morphology of the forelimbs only, whereas locomotory adaptations are reflected mainly in the morphology of the hind feet.

There are two ways of digging, namely using the forelimbs or using the incisor teeth. Each rodent species can use both these techniques, but the predominant use of one or other depends on the mechanical structure of the substrate. Psammophilous species of the three-toed jerboas (Dipodinae) have relatively long toes and claws: they dig into soft soils mainly with the forelimbs. This behavior seems to be conservative because other species of the subfamily which occupy habitats with hard clay soil (*Stylodipus telum, Jaculus blanfordi*) also use their forelegs when digging. These species also use their teeth for digging. Their incisors are more massive than those of the psammophilous Dipodinae, although they have a similar shape (Zubtzova 1980). Nevertheless, the two species need specific conditions for burrowing within their habitats. *Stylodipus telum* prefers to use the holes of other rodents, especially those of *Rhombomys opimus* (Lobachev et al. 1976). When occupying areas of hard clay, *Jaculus blanfordi* does not dig in the extremely hard flat surface of the takyr, but uses sandy mounds around the shrub bases in which to begin burrow excavating (Babaev and Ataev 1962; Shenbrot et al. 1995). Jerboas of the subfamily Allactaginae that inhabit hard soils dig predominantly with their incisors. These are at obtuse angles of the axis of the jaw (Gambarian et al. 1978; Zubtzova 1978, 1980). Another example of a specific manner of digging, dependent on the substrate type is represented by the African mole-rats (Bathyergidae). In psammophilous *Bathyergus* the foreclaws are the main digging tool whereas, in other Bathyergidae that occur in a variety of substrates, the incisors are used to a greater extent than are the claws (Nowak 1991).

The locomotory adaptations that are reflected in the shape of claws are clearly exemplified by the five-toed jerboas. The claws on the three central toes of their hind feet are reduced to short and wide triangular plates. These plates are directed upward and, thus, do not retard fast running. A specific shape of the claws is also characteristic of rodents that use vertical surfaces (e.g. rocks or tree trunks). These have relatively small claws that are curved and very sharp. For example, the rock-dwelling *Acomys* have much more arched and pointed claws than the non-specialized *Mus musculus* (Fig. 7.16).

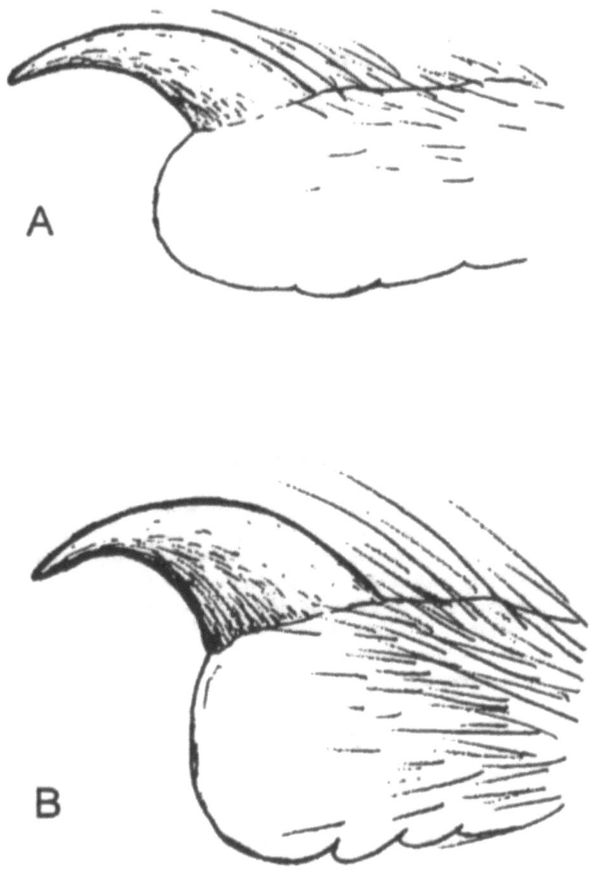

Fig. 7.16. Claws of *Mus musculus* (A) and *Acomys russatus* (B)

Desert rodents that build aboveground nests (*Selevinia*, *Neotoma*, *Peromyscus*) also have relatively small claws. The highly developed manipulative ability of their forepaws may be partly connected with intricate nest-building activity and the ability to climb shrubs.

Another group of morphological features that reflects the habitat preferences of desert rodents concerns characteristics of their analyzer systems. We already discussed the role of the position of the eyes, and here we will pay more attention to the auditory organs. It has been noted repeatedly in literature that many desert rodents have highly developed auditory bullae (Vinogradov 1937; Webster 1962; Webster and Webster 1971; Lay 1972; Klimova 1975; Pavlinov 1988). The complex middle ear cavities play a important role in detecting low frequency sounds produced either by terrestrial predators sneaking up to their prey or by hunting owls (Webster 1962; Webster and Strother 1972; Webster and Plassman 1992). The low humidity of the desert

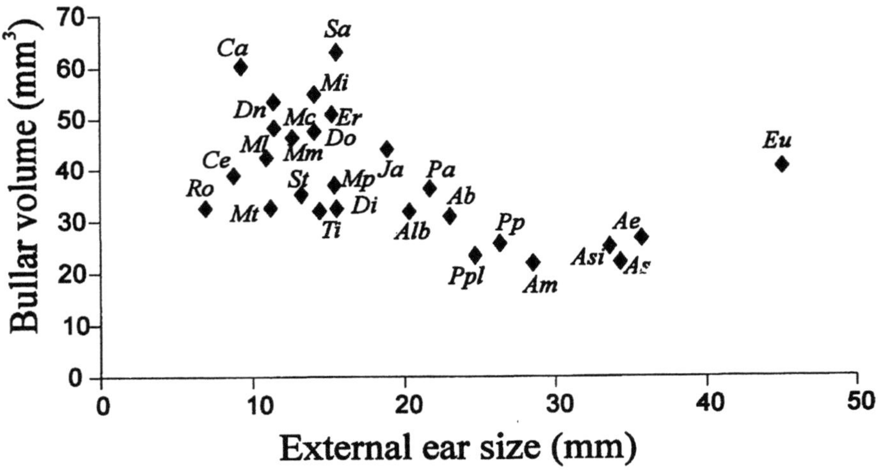

Fig. 7.17. Relationship between bulla size and external ear size in some desert rodents. Species names are as follows: Ce *Chaetodipus nelsoni*, Do *Dipodomys ordii*, Dn *D. nelsoni*, Mi *Microdipodops pallidus*, Ca *Cardiocranius paradoxus*, Sa *Salpingotus crassicauda*, Eu *Euchoreutes naso*, St *Stylodipus andrewsii*, Di *Dipus sagitta*, Ja *Jaculus blanfordi*, Er *Eremodipus lichtensteinii*, Pa *Paradipus ctenodactilus*, Ae *Allactaga elater*, Am *A. major*, As *A. severtzovi*, Asi *A. sibirica*, Ab *A. bullata*, Alb *Allactodipus bobrinskii*, Pp *Pygeretmus pumilio*, Ppl *P. platyurus*, Ti *Tatera indica*, Mt *Meriones tamariscinus*, Mp *M. persicus*, Ml *M. lybicus*, Mc *M. crassus*, Mm *M. meridianus*, Ro *Rhombomys opimus*

air prevents the suppression of sounds of this type. Different desert soils have different characteristics that influence the transmission of low frequency sounds. In desert habitats with hard clay soil, the air-borne component of the sounds produced by the movements of predators is more pronounced than is the seismic component. In sandy habitats, the seismic component is transmitted and detected over a greater distance than it is in hard soil habitats. This may explain why the volume of the tympanic bullae of psammophilous species is usually larger than that of sclerophilous rodents (I.Y. Pavlinov and K.A. Rogovin, unpubl. data; Fig. 7.17).

The main conclusion from the above discussion is that the degree of expression of morphological adaptations to a particular habitat type (i.e. habitat specialization) depends both on the specificity of this habitat type (e.g. non-stabilized sands are more specific than stabilized sands) as well as on the specificity of the pattern of habitat use (e.g. slow walking on a sand surface demands fewer specific features than does fast running).

7.4 Habitat Specialists and Habitat Generalists in Deserts

7.4.1 Definitions

The terms "habitat specialist" and "habitat generalist" are common in the ecological literature. These terms are easily understandable intuitively although it is difficult to clarify them unequivocally. No species exists that uses the entire spectrum of environmental variation. Rather, each species uses some portion of the environmental hypervolume. Furthermore, species differ in the range of environmental variation that they use. These differences can be presented either in terms of the number of habitats occupied or of the range of environmental conditions preferred. This environmental range or number of habitats is nothing other than the spatial component of the niche breadth of a given species. Consequently, the measurement of this component of the niche breadth can be based either on qualitative environmental units (using diversity indices) or on the estimation of variance of a species' position along the environmental gradient (using descriptive univariate or multivariate statistics).

Rosenzweig (1981) has noted that the terms "specialist" and "generalist" are used in the literature in two senses. Ecologists usually attribute these terms to what a species *does*, whereas evolutionists understand it to mean what a species *can do*. To distinguish between these two uses, Rosenzweig (1981, 1987b) has proposed the use of the terms "specialist" and "generalist" for the description of evolutionary trends and patterns and the terms "selector" and "opportunist" for the description of behavioral and ecological patterns. However, in the context of the present volume we prefer to use the initial terms, bearing in mind both their evolutionary and ecological components.

7.4.2 Comparative Aspect

Are specialists and generalists absolute or relative categories? In other words, can we assign a species to one of these categories without considering the community in which it exists ? Analyzing the distribution of a species among qualitatively different habitats, it can be stated that a species occupying one or a few similar habitats is a specialist, whereas a species using a broad range of disparate habitats is a generalist (e.g. McPeek 1996). However, as emphasized above, habitat structure is itself determined by the interactions of the local species assemblage with environmental diversity rather than merely with landscape structure.

On the other hand, when species distribution is analyzed along continuous environmental gradients, the niche breadth of each separate species

Table 7.1. Comparison of different methods of estimation of spatial niche breadth in rodent community of the Negev Highlands. Each species' rank, in ascending order of niche breadth, is in parentheses

Species	Estimation of niche breadth			
	H[a]	HD[b]	RA[c]	PC[d]
Jaculus jaculus	4	3.33 (10)	53.3 (10)	1.33 (7)
Gerbillus dasyurus	6	2.90 (8)	45.1 (9)	1.65 (10)
G. henleyi	4	3.24 (9)	34.8 (8)	1.63 (9)
G. gerbillus	2	1.02 (2)	7.3 (1)	1.09 (3)
Sekeetamys calurus	1	1.00 (1)	8.1 (2)	0.78 (1)
Meriones crassus	6	4.17 (11)	56.5 (11)	1.92 (11)
Psammomys obesus	3	1.11 (3)	9.9 (7)	1.32 (6)
Mus musculus	3	1.15 (4)	9.0 (4)	1.19 (5)
Acomys cahirinus	4	1.72 (7)	8.5 (3)	1.03 (2)
A. russatus	2	1.55 (6)	9.4 (5)	1.50 (8)
Eliomys melanurus	2	1.37 (5)	9.7 (6)	1.13 (4)

[a]Number of occupied habitats.
[b]Diversity of habitat occupation, calculated using Hill's N_2 index (Hill 1973).
[c]Measure of habitat breadth with special correction to eliminate the density-dependent effect, calculated using the method of Rosenzweig and Abramsky (1985).
[d]Niche breadth calculated as the average deviation of observation points from the species centroid in the multidimensional space of environmental variables, transformed by factor analysis for the orthogonalization of the axes.

can be measured without taking the patterns of distribution of other species into account. However, it is clear that neither species without any variation in their distribution (absolute specialists) nor species equally distributed along the entire gradient (absolute generalists) actually exist. Consequently, estimates of niche breadth produce values that cannot be interpreted in terms of specialization or generalization without comparison either with other co-occurring species, or with the same species in other localities. For example, the jerboa *Pygeretmus pumilio* exclusively inhabits clay soils in the Middle Asian deserts whereas it occupies a variety of substrates in the Gobi Desert (Shenbrot et al. 1995). Calculations of the substrate component of niche breadth results in a value of 1.42 for the Middle Asian deserts and 2.15 for the Gobi desert. These values are meaningful only in comparison with one another. Thus, each species should be considered as a specialist or a generalist only in this comparative aspect. Habitat specialists and habitat generalists are concepts that are meaningful only in their relationships with one another.

Moreover, the specialist-generalist dualism is an ideal logical concept. In reality, there is a continuum of the levels of relative specialization and generalization. In other words, the species of each community can, according to their niche breadths, be arranged in a continuum; it is up to the interpreter to divide them into categories of degree of specialization. Table 7.1 illustrates this continuum for a rodent assemblage of the Negev Highlands. Different methods for the estimation of niche breadth produce slightly different results. Nevertheless, the results of all methods coincide in the case of species that occupy marginal positions (extremely narrow or extremely wide niches). All methods indicate that *Sekeetamys calurus* and *Gerbillus gerbillus* are habitat specialists, whereas *Meriones crassus* is a habitat generalist.

7.4.3 Types of Specialization

The reasons why a species prefers a particular habitat (or a set of habitats) and avoids other habitats may be different. Above all, a species is usually adapted to the physical characteristics of its preferred habitat. The types of adaptive responses of species to specific habitats have been discussed above. Another reason for preference for a specific habitat may be the consequence of specialization to one or other type of food resource found there. The fat sand rat *Psammomys obesus* in the deserts of Israel demonstrates distinct habitat selectivity which differs in different regions. In the Negev Highlands it occurs mainly in loess hills, whereas in the Arava Valley it can only be found in regions of sands. The reason for this pattern of habitat distribution is that the species is a strict folivore and needs large amounts of succulent green vegetation all the year round. The sites that satisfy its foraging requirements in the Negev Highlands are patches of *Atriplex halimus* and *Anabasis articulata* on the flood plains of wadis among the loess hills, while in the Arava Valley it forages on dense stands of *Haloxylon persicum* that grow on sand. This example can be considered as locally-restricted habitat specialization. Finally, the third group of reasons is related to patterns of resource use. These reasons, for example, are responsible for the differences in habitat requirements between bipedal and small quadrupedal species that are determined by their patterns of foraging activity. Bipeds are able to cover long distances between resource patches and avoid predation by sprinting from open patches to shelter. However, the dense vegetation in the lower layer decreases their speed of locomotion. Consequently, bipedal species either select open habitats or a mosaic of open and vegetated patches. In contrast, small nocturnal quadrupeds prefer either densely vegetated habitats or the vegetated patches within a mosaic.

7.4.4 Types of Generalization

As in the case of specialists, the reasons why a species displays "generalist" features can be various. McPeek (1996) argued that "generalists" are not uniform but rather include several categories. For example, a generalist may be a polymorphic species in which various specialist genotypes for different environmental conditions are maintained, and each of them can be manifested in a specific situation (Dempster 1955; Hoekstra et al. 1985; Hedrick 1986; De

Meeûs et al. 1993). Assuming that this statement is correct, we have nevertheless not found any evidence in the literature that illustrates it in the case of rodents. Another type of generalist is a phenotypically plastic species in which a genotype expresses a particular phenotype in one set of ecological conditions and another phenotype in another (Levins 1968; Van Tienderen 1991; Moran 1992). This type of generalist is exemplified by the house mouse *Mus musculus*. The spatial structure of populations of this species has been shown to be determined mainly by the level of individual aggression and varies between habitats (Krasnov and Khokhlova 1989, 1994). Generalists can also be species in which the phenotypic distribution represents a compromise between the optimal phenotypes for the different environmental conditions experienced (MacArthur 1972; Felsenstein 1979). This is sometimes known as a "jack-of-all-trades" phenotype. Again, we have not found any rodent examples. McPeek (1996) noted that all three types of generalists exist because different phenotypes (or genotypes) confer maximum fitness in different environments. Contrarily, he suggested an additional (fourth) type of generalist that might exist because its position in food webs presents no trade-off in fitness among two communities (or habitats). This implies that McPeek's statement is based on the assumption that the same niche exists in several communities. This accords with the Eltonian niche concept.

7.4.5 Specialist/Generalist Relationships

The ecological consequences of specialization and generalization, as well as the patterns of coexistence of specialists and generalists within a community, have long attracted the attention of community ecologists. Why do species differ in their level of specialization of habitat use (e.g. MacArthur and Pianka 1966, MacArthur 1972)? McPeek (1996), using food web models, explained this among-species difference by trade-offs in the abilities of species to perform in various ecological interactions. Trade-offs occur when the ability of a species to perform in one interaction decreases its ability to perform in another interaction (Tilman 1987). Examples of such trade-offs include the conflicting requirements for avoiding different predators (e.g. Kotler et al. 1992; Werner and McPeek 1994), and the trade-off between acquiring resources and avoiding predators (e.g. Kotler et al. 1991). Some species may be limited by trade-offs to occupy a single habitat, while others experience no trade-off among habitats and may therefore be able to generalize their habitat use. Furthermore, specialists perceive the environment in a fine-grained way, whereas the generalists do so in a coarse-grained way. The same change in the environment can be perceived by specialists as a major interruption that causes some habitats to be hospitable and others to be inhospitable while a generalist perceives it merely as a minor alteration that would not influence its persistence in either habitat.

The difference between habitat specialists and generalists in the pattern of their spatial distribution is clear by definition. But what about their abundance? Two alternative opinions have been expressed about this as follows:

1. A specialist species is well adapted to a certain habitat type where it attains high abundance; other habitats are unsuitable. A generalist species is adapted to several types of habitat but the degree of adaptation is moderate and the species occurs at moderate or low densities in each habitat. In other words, the generalist is a "jack-of-all-trades and a master of none" (MacArthur and Connell 1966; Rosenzweig 1979a), an aphorism that from time to time appears in the discussions on specialization and generalization. This opinion is widely accepted and serves as a basis for both evolutionary and ecological theories (Lively 1986; Rosenzweig 1987b; Van Tienderen 1991, 1997; Moran 1992).
2. A specialist species is usually not so well adapted to the few places in which it occurs that it attains high densities there. Generalists typically outnumber specialists in all habitats, so that "jack-of-all-trades appears also to be master of all" (Brown 1995). It should be noted, however, that there are two provisos to this point of view, namely (a) specialists are species with narrow habitat distributions and restricted geographic ranges, while generalists are species with broad habitat distributions and wide geographic ranges; (b) specialist/generalist relationships should be compared between relative taxa and/or within the same guild.

Proponents of both points of view cite a number of arguments (both theoretical and empirical) supporting their opposing concepts (see references above). Here we shall look at real communities and try to understand to which opinion desert rodents conform (Fig. 7.18). Direct census data cannot be used for the evaluation of each species' abundance in a habitat because different habitats can support different numbers of rodents according to differences in the availability of the resources they provide. Moreover, there are huge size differences among rodent species. It is sometimes difficult to compare their abundance because the biomass of food they consume is a function of both numbers and size. To avoid these problems, we assumed that the carrying capacity of a given habitat is correlated with the overall biomass of rodents occurring within it. So, we calculated the proportional biomass of a given species in a given habitat and used this parameter as an index of species abundance instead of the actual number of individuals present. Habitat niche breadth for each species was calculated as the diversity of habitats occupied by the species using an index analogous to Hill's N2 index (the reciprocal of Simpson's index) for species diversity (Hill 1973).

Rodents in real communities demonstrate different relationships between the degree of their specialization and mean abundance (Fig. 7.18). There is a clear trend of mean abundance decreasing with an increasing degree of spe-

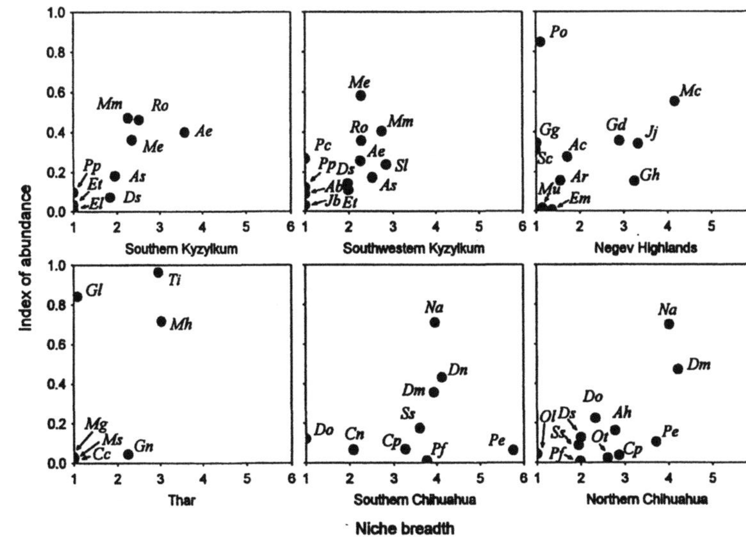

Fig. 7.18. Relationship between relative abundance and habitat niche breadth of rodents in six desert communities. Relative abundance was calculated as the proportional biomass of a given species in a given habitat, habitat niche breadth was calculated as a diversity of habitats occupied by a species using an index analogous to Hill's N2 index for species diversity (Hill 1973). See text for explanation.Species names are as follows: El *Eremodipus lichtensteinii*, Ds *Dipus sagitta*, As *Allactaga severtzovi*, Ae *A. elater*, Pp *Pygeretmus pumilio*, Mm *Meriones meridianus*, Me *M. erythrourus*, Ro *Rhombomys opimus*, Et *Ellobius tancrei* (southern Kyzylkum); Sl *Spermophilopsis leptodactylus*, Ds *Dipus sagitta*, As *Allactaga severtzovi*, Ae *A. elater*, Pp *Pygeretmus pumilio*, Jb *Jaculus blanfordi*, Ab *Allactodipus bobrinskii*, Pc *Paradipus ctenodactylus*, Mm *Meriones meridianus*, Me *M. erythrourus*, Ro *Rhombomys opimus*, Et *Ellobius tancrei* (southwestern Kyzylkum); Jj *Jaculus jaculus*, Mu *Mus musculus*, Ac *Acomys cahirinus*, Ar *A. russatus*, Gd *Gerbillus dasyurus*, Gg *G. gerbillus*, Gh *G. henleyi*, Mc *Meriones crassus*, Sc *Sekeetamys calurus*, Po *Psammomys obesus*, Em *Eliomys melanurus* (Negev Highlands); Mg *Millardia gleadowi*, Ms *Mus saxicola*, Cc *Cremnomys cutchicus*, Gl *Gerbillus gleadowi*, Gn *G. nanus*, Mh *Meriones hurrianae*, Ti *Tatera indica* (Thar); Ss *Spermophilus spilosoma*, Cp *Chaetodipus penicillatus*, Cn *C. nelsoni*, Dm *Dipodomys merriami*, Dn *D. nelsoni*, Do *D. ordii*, Pf *Perognathus flavus*, Pe *Peromyscus eremicus*, Na *Neotoma albigula* (southern Chihuahua); Ah *Ammospermophilus harrisii*, Ss *Spermophilus spilosoma*, Cp *Chaetodipus penicillatus*, Dm *Dipodomys merriami*, Do *D. ordii*, Ds *D. spectabilis*, Pf *Perognathus flavus*, Pe *Peromyscus eremicus*, Na *Neotoma albigula*, Ol *Onychomys leucogaster*, Ot *O. torridus* (northern Chihuahua)

cialization among species of the communities of the southern and southwestern Kyzylkum Desert and the southern Chihuahuan Desert. This is not so in the case of rodents of the Negev Highlands, the Thar Desert and the northern Chihuahuan Desert. Some generalist species attain high abundance and outnumber all or at least most of the specialists in their community (e.g. *Allactaga elater* in the southern Kyzylkum Desert, *Meriones crassus*, *Tatera indica*, *Neotoma albigula*, *Dipodomys merriami*). Thus they are in accordance with the second opinion listed above; we will refer to them as B-type generalists (after Brown 1995). There are also generalists that occur in low abundance (*Gerbillus henleyi*, *Spemophilopsis leptodactylus*, *Peromyscus eremicus*) although not very much lower than that of most of the specialists in their communities. Nevertheless, they accord with the first opinion and we will refer to them as M/R-type generalists (after MacArthur and Connell 1966; Rosenzweig 1979a). The same classification can be applied to specialist species. Some specialists (B-type) are less abundant than most of the generalists of their communities (*Pygeretmus pumilio*, *Eremodipus lichtensteinii*, *Ellobius tancrei* in the southern Kyzylkum; *Pygeretmus pumilio*, *Allactodipus bobrinskii*, *Jaculus blanfordi* in the southwestern Kyzylkum; *Eliomys melanurus*, *Mus musculus* in the Negev Highlands; *Millardia gleadowi*, *Mus saxicola*, *Cremnomys cutchicus* in the Thar Desert; *Dypodomys ordii* in the southern Chihuahuan Desert and *Onychomys leucogaster* in the northern Chihuahuan Desert). M/R-type specialists occur in certain communities only (*Psammomys obesus* in the Negev Highlands and *Gerbillus gleadowi* in the Thar Desert). In addition, both types of specialists and generalists often occur within the same community. Furthermore, Fig. 7.18 shows that B-type relations are more often observed than M/R-type relations, so the famous adage can be rephrased as the "jack-of-all-trades appears to be master of (at least) some of them".

Strictly speaking, the abundance of specialists and generalists should be compared within the same habitat. In such comparisons, the observed abundance (in terms of proportional biomass) of species, all else being equal, can reflect the degree of their relative adaptation to this habitat. Furthermore, we do not need to assume that there is a correlation between overall biomass and environmental carrying capacity. We calculated the relative abundance of species for each pair of specialists and generalists within the same habitat (that of the specialist) in six communities of desert rodents (see above) and plotted them against each other (Fig. 7.19). The abundance of the generalists was significantly higher than that of the specialists in five cases, significantly lower in two cases and not significantly different in nine other cases. Again, we can state that there are different types of specialist/generalist relationships, but B-type relationships occur more often that M/R-type relationships.

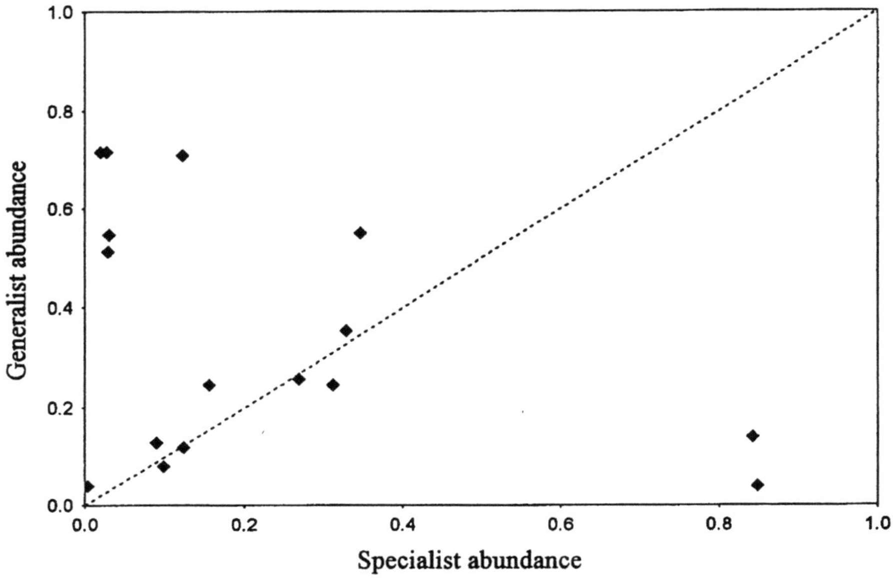

Fig. 7.19. Relative abundance of specialist/generalist pairs of species within the same habitat for six communities of desert rodents. See Fig. 7.18 for the list of communities. The dashed line denotes equal abundance

7.4.6 Spatial Variation of the Degree of Specialization

As mentioned above, some species behave as specialists in one part of their geographic range and to different degrees as generalists in other parts of their range. This means that the realized habitat niche of such species is geographically variable. Sometimes, a species can be regarded as a habitat generalist in the main part of its geographic range whereas it demonstrates habitat specialization at the marginal parts. Classic examples are afforded by the gray hamster *Cricetulus migratorius* and the brush-footed jerboa *Dipus sagitta* (see above). Such cases can be defined as geographically restricted specialization. In other instances, the places in which a species behaves as a specialist and the places where it displays the features of a generalist both lie in the main part of its geographic range. For example, the midday jird *Meriones meridianus* is a more or less strict sand-dweller in the Turan deserts where it coexists with *Meriones erythrourus*, while it is a habitat generalist in the deserts of Central Asia where it is the only representative of its genus. One of the reasons for the exclusion of this species from certain habitats in part of its range may be interspecific competition (Shenbrot and Rogovin 1995), although direct evidence of such competition is absent. Such cases can be, therefore, considered as ecologically restricted specialization. No clear border occurs between these two types of specialization however, because the mechanisms are the same in both cases, namely either physical absence of certain landscape units, or the presence of a successful competitor in a certain part of the geographic range (in the main part of the range or at the edge). Another example is provided by a shift of habitat use determined by geographic variation in the physical properties of the habitat type. For example, the mole-vole *Ellobius talpinus* and the jerboa *Stylodipus telum* shift to strict psammophily in the northwestern part of their geographic ranges because in this region they can find an optimal soil temperature regime only in sand (Naumov 1965). In general, geographic variation in the degree of specialization or generalization can be determined either by geographic variation of habitats (abiotic, variation in physical properties, and/or biotic, presence or absence of competitors) or geographic variation of a species (genotypic, e.g. Hedrick 1986, or phenotypic variation, e.g. Moran 1992) or both.

7.4.7 Specialization and Geographic Range

Brown (1984, 1995) proposed a niche-based model explaining the negative correlation between the degree of specialization and geographic distribution. According to this model, a specialist is a species that can endure restricted abiotic conditions, is able to use few types of resource and is highly tolerant of a very limited set of competitors, predators, parasites and diseases. In contrast, a generalist species tolerates a wide range of physical conditions, uses a broad range of resources and survives in the presence of many natural enemies. The same attributes that enable a species to occur in either many or a few habitats cause it respectively to have either a broad or narrow geographic range.

We plotted the niche breadth of rodents from the above-mentioned six desert communities against the size of the species' geographic ranges. The niche breadth calculation we used has been described above. The geographic range of a species was estimated as the number of 2° by 2° quadrats occupied by it (see Chap. 4). The predicted pattern of a positive correlation between habitat niche breadth and the size of geographic range is clearly revealed in the Asian deserts but is absent from those of North America (Fig. 7.20). Considering the main taxonomic groups of desert rodents, this pattern is well expressed among jerboas, somewhat less so among gerbils and jirds, and is not found among heteromyids (Fig. 7.21). It should be noted that unbiased estimation of the niche breadth of a species should be based on the parameters of its fundamental niche. However, this is impossible in practice. The nearest to fundamental niche estimation can be obtained in conditions optimal for the species. Actually, in our examples there are certain species among each taxon for which niche breadth is underestimated because it was measu-

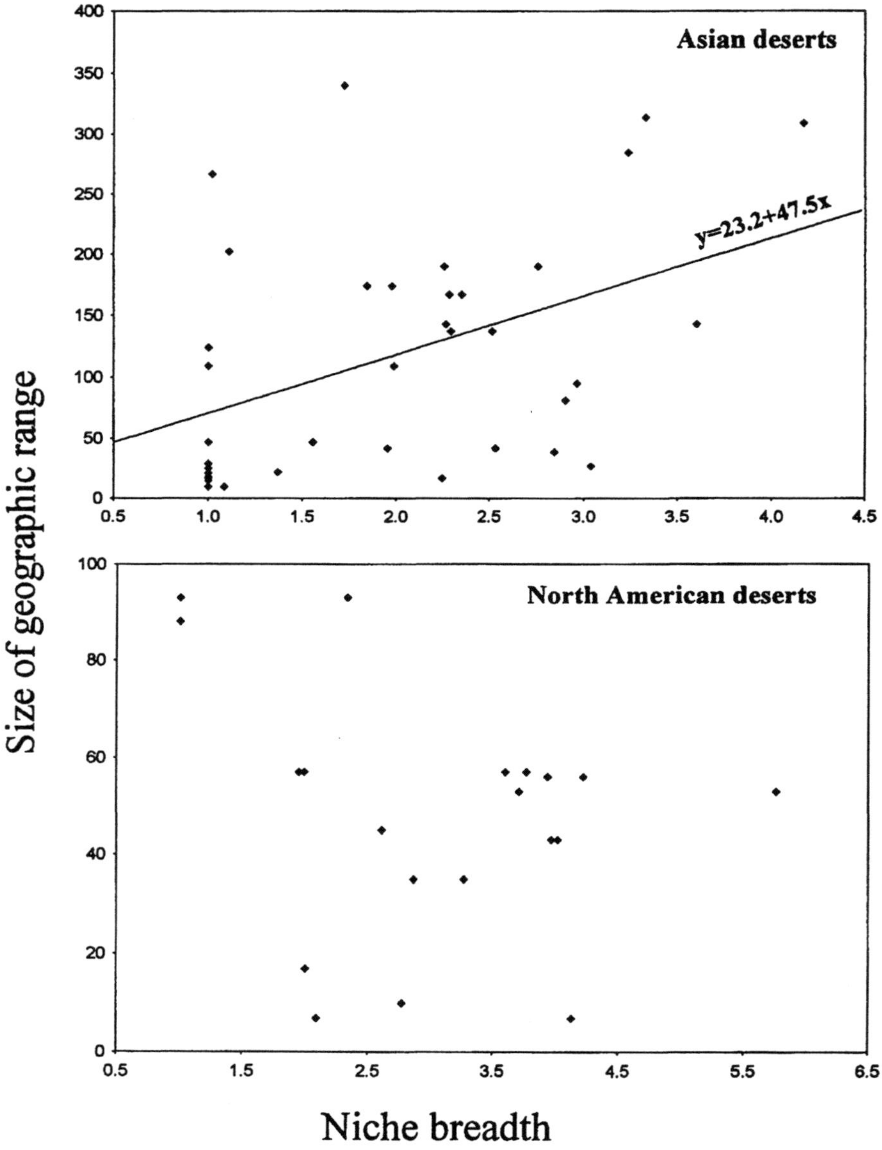

Fig. 7.20. Relationship between habitat niche breadth and size of geographic range in rodents of four Asian and two North American desert communities

red outside the optimum (geographic or ecological) of the range. Species with underestimated values of niche breadth include *Pygeretmus pumilio* and

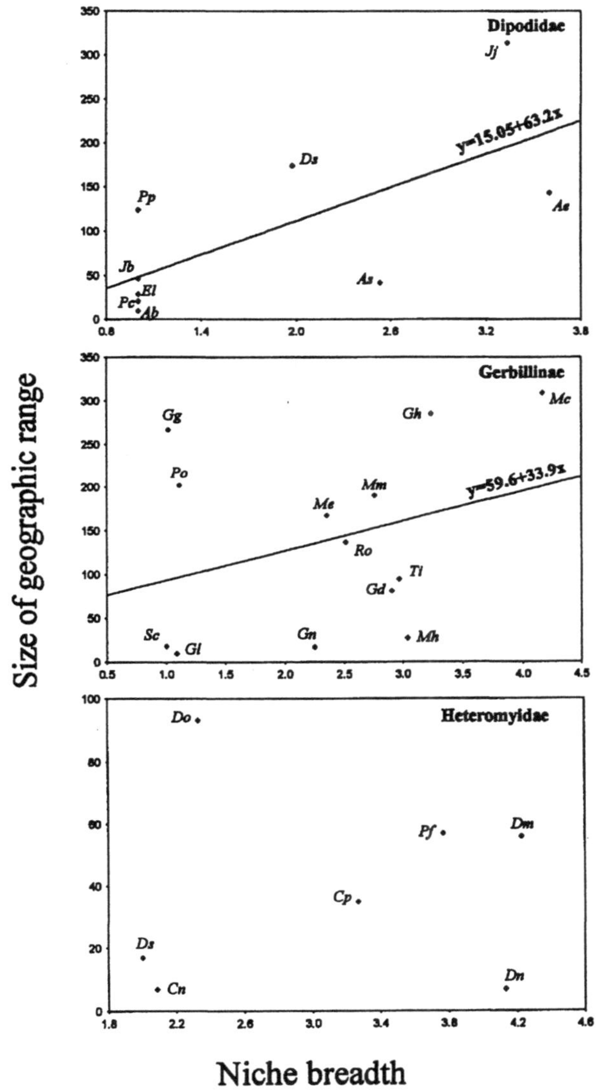

Fig. 7.21. Relationship between habitat niche breadth and size of geographic range in Dipodidae, Gerbillinae and Heteromyidae. Dipodid species names are as follows: Ab *Allactodipus bobrinskii*, Ae *Allactaga elater*, As *A. severtzovi*, Ds *Dipus sagitta*, El *Eremodipus lichtensteinii*, Jb *Jaculus blanfordi*, Jj *J. jaculus*, Pc *Paradipus ctenodactylus*, Pp *Pygeretmus pumilio*. Gerbilline names are as follows: Gd *Gerbillus dasyurus*, Gg *G. gerbillus*, Gh *G. henleyi*, Gl *G. gleadowi*, Gn *G. nanus indus* (geographic range was calculated for the subspecies), Mc *Meriones crassus*, Me *M. erythrourus*, Mh *M. hurrianae*, Mm *M. meridianus*, Po *Psammomys obesus*, Ro *Rhombomys opimus*, Sc *Sekeetamys calurus*. Heteromyid names are as follows: Cn *Chaetodipus nelsoni*, Cp *C. penicillatus*, Dm *Dipodomys merriami*, Dn *D. nelsoni*, Do *D. ordii*, Ds *D. spectabilis*, Pf *Perognathus flavus*

Dipus sagitta among the jerboas, *Gerbillus gerbillus* and *Psammomys obesus* among gerbillids, and *Dipodomys ordii* among heteromyids. Niche breadth of the two jerboa species was estimated in the southern or southwestern Kyzylkum Desert, near to the southwestern border of their geographic ranges, although optimal conditions for *Pygeretmus pumilio* and *Dipus sagitta* are in the central Kazakhstan and in the Gobi Desert, respectively (Shenbrot et al. 1995). The two gerbillid species were studied in the central Negev Desert which is at the northeastern edge of their ranges, whereas their optima are considered to be in North Africa (Le Berre 1990; Harrison and Bates 1991). Estimation of niche breadth of *Dipodomys ordii* was made in the Chihuahuan Desert, although its optimum seems to be in the center of its geographic range at the border between the Sonoran and Great Basin deserts (Hall 1981). Correction of the cases mentioned above would shift the positions of these species to the right side of the graphs. This would make the pattern for the Dipodidae and Gerbillinae more pronounced, but the absence of a pattern for the Heteromyidae would remain.

The sharp contrast between the Asian and American rodent faunas in the presence of the macroecological pattern describing the dependence of the size of geographic range on niche breadth, suggests the presence of some fundamental differences in community organization and/or biogeographic history in these regions. It is obvious that such a pattern can be found only if the geographic distribution of each species is determined mainly by the limits of its tolerance to environmental conditions. In addition, a relatively long history of species-environment interaction (a long period of occurrence of a particular species in a particular environment in a particular geographic region) is necessary for such a pattern to become established. The long history of species-environment interaction will result in an equilibrium between demands of a species and the environmental resources available. Recent environmental changes, as well as competitive interactions can disturb this pattern. This is true of the North American deserts which underwent drastic fluctuations during the Quarternary (see above). Moreover, there is some evidence that the North American communities of desert rodents are strongly affected by interspecific competition, whereas this does not play an important role in Asian communities (Kelt et al., in press).

The same reasons, although in a softer form, can be used to explain differences in the degree of expression of the pattern between Dipodidae and Gerbillinae in the Asian deserts. The evolutionary history of jerboas from the late Miocene was played out in Asia (Shenbrot et al. 1995), while gerbils and jirds evolved mainly in Africa and penetrated into Asia only during the late Pliocene (Pavlinov et al. 1990). In addition, there is no evidence of interspecific competition among jerboas, while interspecific competition among gerbillines has been demonstrated both directly (Ziv et al. 1993; Chabovsky and Popov 1994) and indirectly (Yom-Tov 1991). It is possible, that this explanation can be applied to outliers from this trend. For example, the geographic distribution of the Indian desert jird *Meriones hurrianae* may be limited by competition with the ecologically similar fat sand rat *Psammomys obesus* and the great gerbil *Rhombomys opimus*, although the causes of this non-conformation to the trend may be more specific features of the natural history of the species.

7.5 Habitat Selection as a Mechanism of Species Distribution

7.5.1 The Theory of Habitat Selection

The observed patterns of habitat distribution of rodent species can be explained by the modern theory of habitat selection (Rosenzweig 1974, 1981, 1987b, 1991, 1995). This theory is based on the model of the so-called ideal free distribution of individuals among habitats (Fretwell and Lucas 1970; Fretwell 1972). This model assumes that individuals are equally *ideal* in their ability to select the habitat where their fitness would be the greatest for any given time and are *free* to move into any habitat without constraint. These assumptions seem to be unrealistic at first glance, but thorough observations and experiments have demonstrated the fact that the distributions of some animals fit the ideal free model rather well (Harper 1982; Pierotti 1982; Milinski 1984; Cassini and Foeger 1995; but see Tregenza et al. 1996). The direct outcome of the model of ideal free distribution is that density-dependent habitat selection occurs with an increase in the number of occupied habitats with increasing population density. Imagine that there are several habitats that differ in their suitability for a species. At equal densities, the greater the quality of a habitat, the greater will be the fitness of the individuals occupying it. At low density, individuals select the best habitat only. However, within this habitat, fitness declines with the increasing density due to intraspecific competition. When fitness drops below a certain threshold, the best habitat has reduced value to each individual, and some individuals move to less suitable (lower quality) habitat to maximize their individual fitness. The result is the equalization of fitness among the habitats. Following increases in population density, this process expands to more and more habitats.

The higher the density, the more habitats are occupied by a species (Fig. 7.22). This prediction is in agreement with numerous field observations, and any experienced field zoologist can probably recall some examples from his or her own field studies. Examples from among desert rodents can be found in Rosenzweig and Abramsky (1985). However, agreement between the prediction and the empirical data is not sufficient to validate the theory. It should also be demonstrated that the basic assumption of the theory (ideal free distribution) is fulfilled. The adequacy of the observed distribution of individuals to the ideal free mode should also be proved. It has frequently been shown that animals are not ideal or free in their distribution. For example, territoriality inherent to almost all rodent species violates the "free" component of the assumption. The distribution of individuals is described better by an ideal despotic model (Fretwell 1972), when socially dominant individuals occupy the highest quality habitat and prevent further density-dependent resource depletion through agonistic behavior. Subordinate individuals are

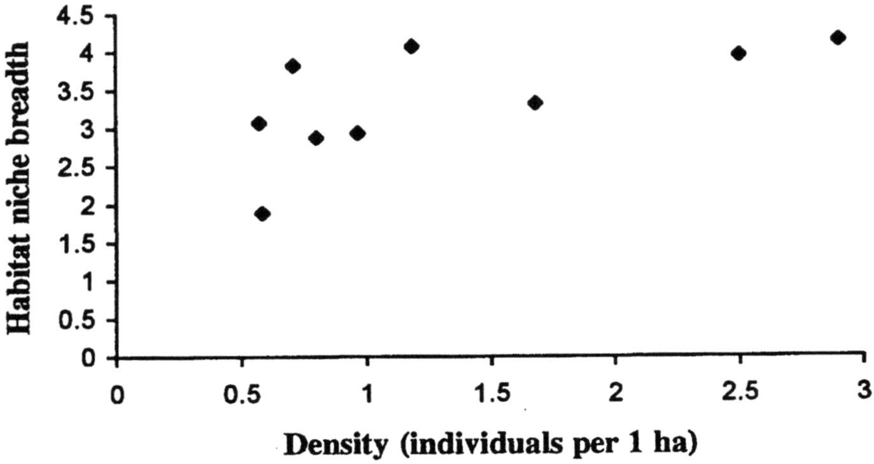

Fig. 7.22. Changes in habitat niche breadth (the diversity of occupied habitats) with density in *Meriones crassus* in the Negev Highlands

forced to enter lower-quality habitats, incurring a cost in terms of lower fitness than could be achieved in a prime habitat. This type of distribution has been shown for the white-footed mouse *Peromyscus leucopus* by Halama and Dueser (1994). Moreover, in strict specialists, the number of occupied habitats does not change with density fluctuations, i.e. habitat selection is density-independent (Abramsky et al. 1985b; Rosenzweig and Abramsky 1985; Shenbrot and Rogovin 1995). Furthermore, some species demonstrate habitat shifts with density changes. For example, the pygmy gerbil *Gerbillus henleyi* in the Negev Highlands prefers sand dunes and clearly avoids large wadis at high density, whereas at low density it avoids sand dunes completely and prefers complexes of narrow wadis and loess hills (Fig. 7.23).

7.5.2 The Isodar Theory

A more sophisticated version of the theory of habitat selection permits us to solve the above-mentioned problems, This is the isodar theory proposed and developed by Morris (1987, 1988, 1990, 1996). This theory explains the mechanisms of both density-dependent and density-independent habitat selection and predicts habitat shifts (switched preference at crossover regulation). Consider the situation in which individuals of a species inhabit two adjacent habitats, and we have replicated estimations of species densities in both habitats for two different time periods (with low and high density levels). The theory assumes that habitats can differ either qualitatively (by efficiency of

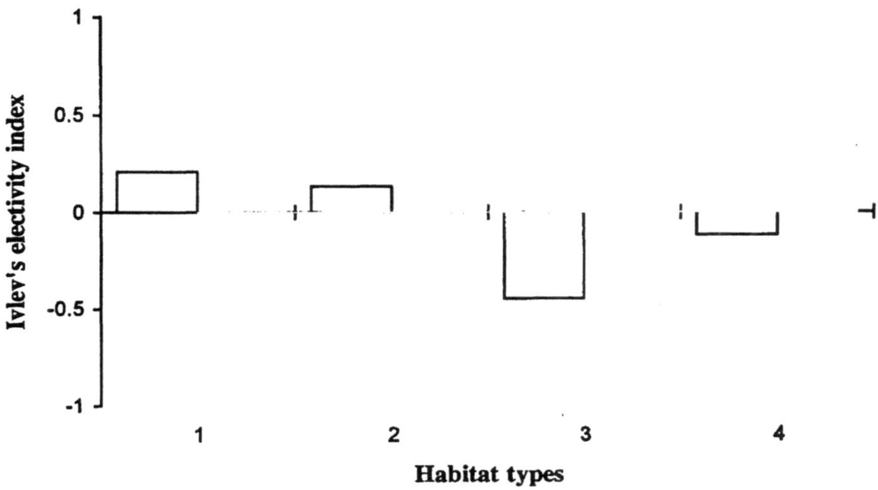

Fig. 7.23. Changes in summer habitat distribution of *Gerbillus henleyi* in the Negev Highlands under high (white columns) and low (black columns) density. Values of Ivlev's electivity index (Manly et al. 1993) reflect habitat preferences. Habitat types are as follows: 1 sand dunes, 2 gravel plains, 3 broad dry riverbeds, 4 complex of narrow dry riverbeds and loess hills

resource acquisition), or quantitatively (by the amount of the resource), or both. If individuals select habitats so as to maximize their reproductive success or survival, the density in each habitat will reflect its relative suitability. Plotting density estimates in one habitat against another habitat will produce an isodar sensu Morris (1987). Each point on this line corresponds to equilibrium densities in two different habitats at equal fitness. The line traces population changes among time periods, responding to growth or declines or immigrations, but maintaining the equal fitness of an ideal free distribution (Morris 1987, 1988). The resultant isodar will have a non-zero intercept when habitat selection is density-dependent and a zero intercept if habitat selection is density-independent. The slope of an isodar indicates the presence or absence of a habitat shift. If the slope is equal to 1, the habitat shift is absent. A slope >1 indicates the amplification of selectivity of the best habitat with increasing density, whereas a slope <1 indicates a shift in selectivity from one habitat at low density to another habitat at high density. Morris (1994) demonstrated that the isodar theory is applicable not only under conditions of ideal free distribution, but also under conditions of ideal despotic or preemptive distribution (see below). The isodar theory has been successfully applied to the estimation of differences in pattern of habitat selection between two psammophilous gerbils (*Gerbillus pyramidum* and *Gerbillus andersoni*) in the Negev Desert (Ovadia and Abramsky 1995). Both species appeared to be density-dependent habitat selectors. Another example is pre-

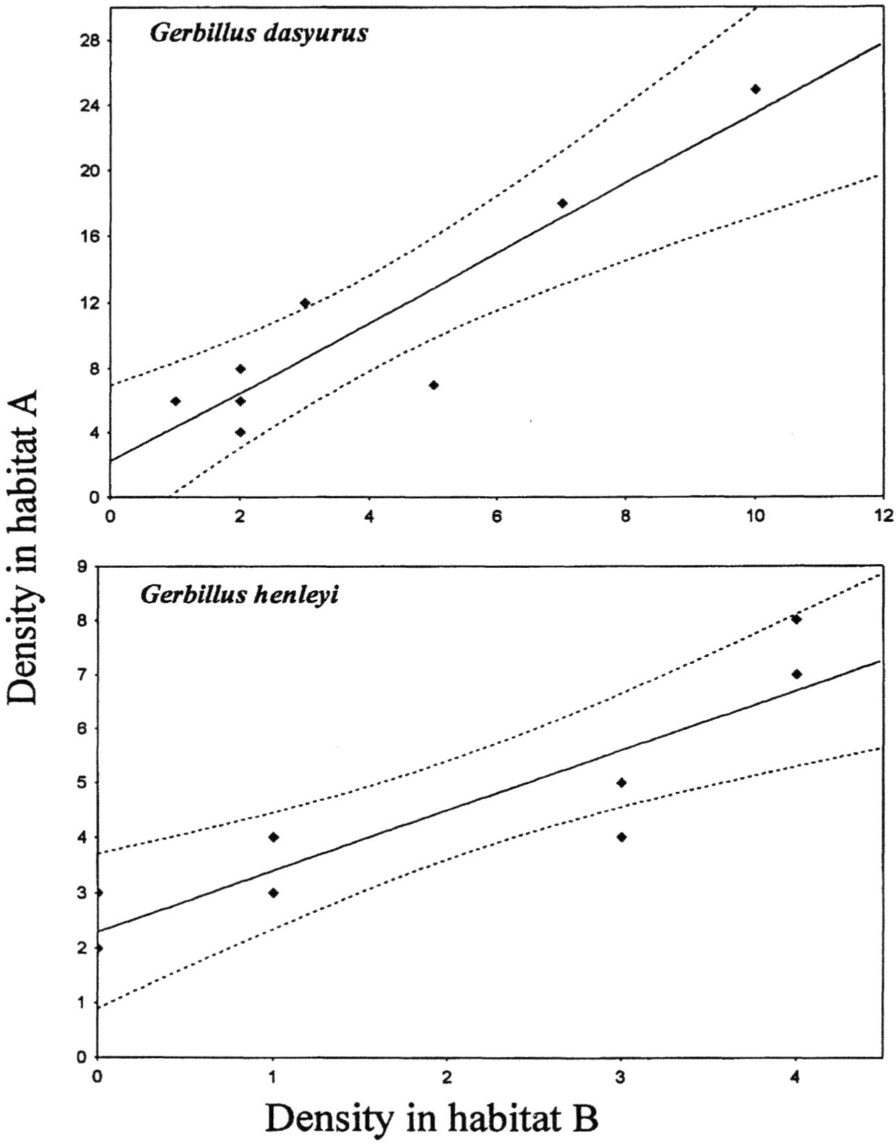

Fig. 7.24. Isodars for two rodent species of the Negev Highlands. Dashed lines are 95% confidence bands

sented in Fig. 7.24. Analysis of the isodars shows that *Gerbillus dasyurus* demonstrates density-independent habitat selection, whereas *Gerbillus henleyi* demonstrates density-dependent habitat selection.

7.5.3
Habitat Selection along an Environmental Gradient

The isodar theory considers only a two-habitat system. However, many species occur in more than two habitats. Furthermore, under some natural conditions it is difficult to determine the border between two habitats because there is a gradual transition between them. In a broader sense, the distinction between discrete versus continual habitat transitions is a problem of spatial scale. For example, a mixture of two habitat types with an increase of the first type toward one end of a transect and an increase of the second type toward the opposite end will, on a small scale, be represented as a gradient. On the other hand, a situation that appears on the small scale as a sharp ecotone between two habitat patches, will look like a gradient on the large scale. Thus, analysis of habitat selection along gradients can be an important addition to the analysis of habitat selection among discrete habitat patches.

We have proposed a set of models of density-dependent habitat selection along an environmental gradient (Shenbrot and Krasnov, in press) that uses an approach similar to Morris' isodar method. It was assumed that the values of the qualitative and/or quantitative parameters of habitats are normally distributed along a gradient. Density estimates along the gradient in one time period, plotted against those in another time period, will produce a line, which we called a "paraisodar". Each point on this line corresponds to the equilibrium densities of two different time periods with different overall population sizes, but maintaining constant fitness ratios between habitats due to density adjustments. Depending on the parameters of the normal distribution of resource abundance and habitat quality along a gradient, the models predict density-independent or density-dependent habitat distribution with or without density-dependent habitat shifts. We tested these models using data on desert rodents of the central Negev Desert. Four species (*Jaculus jaculus, Gerbillus dasyurus, G. gerbillus* and *Mus musculus*) were found to be density-independent habitat selectors, four (*Meriones crassus, Psammomys obesus, Acomys cahirinus* and *A. russatus*) appeared to be density-dependent habitat selectors and one (*Gerbillus henleyi*) demonstrated density-independent habitat selection in winter and density-dependent habitat selection in summer. Examples of paraisodars are presented in Fig. 7.25.

7.5.4
Sources and Sinks

Most ecologists use density as an indicator of habitat suitability for a species. However, habitats may differ, not only in their ability to sustain a certain level of a species' density, providing it with a certain amount of a resource and conditions for its acquisition, but also in their roles in population dynamics. There is evidence that populations of the same species in different habitats

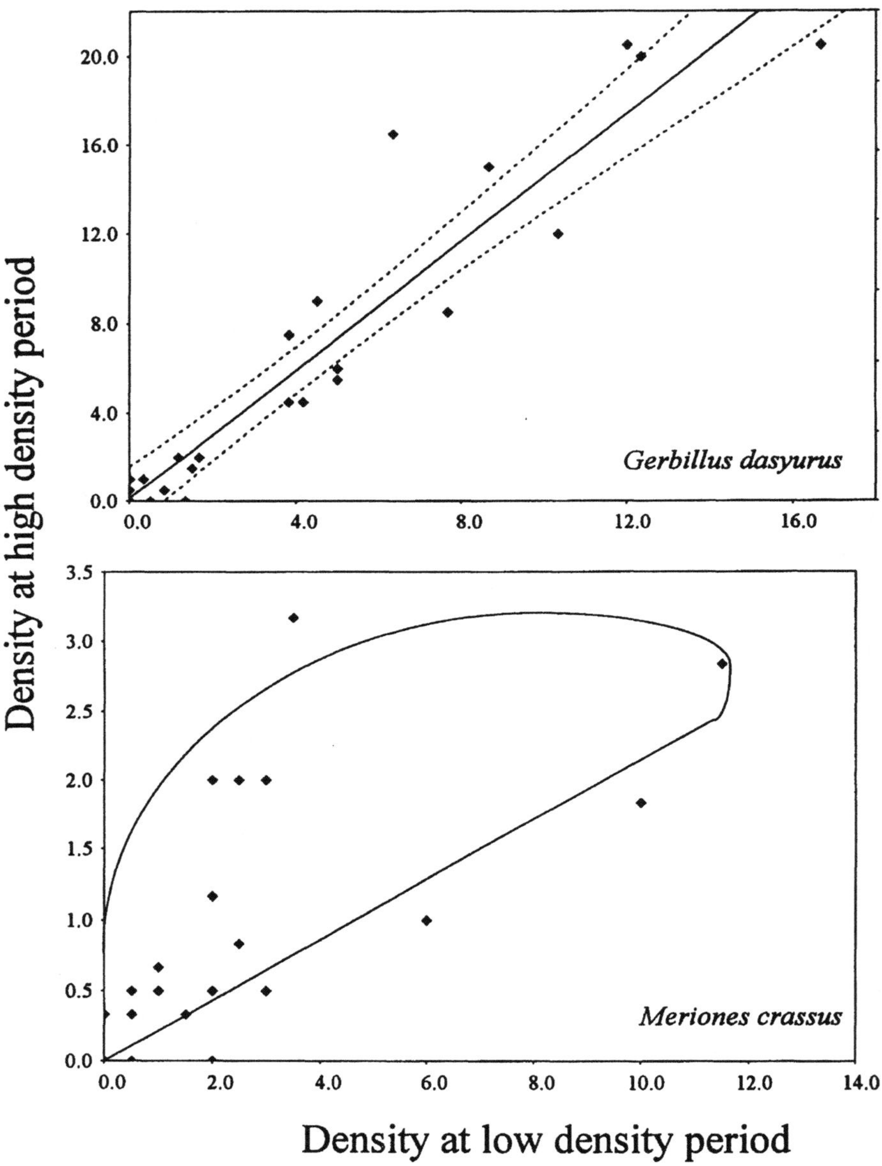

Fig. 7.25. Paraisodars for two rodent species of the Negev Highlands. The black line for *Gerbillus dasyurus* is the linear regression line. The black curve for *Meriones crassus* is the theoretical line for the model describing simultaneous change in habitat quality and resource abundance along a gradient with distinct maxima (Shenbrot and Krasnov, in press)

may differ in their reproductive success as well as in mortality level. These differences result in a different demographic balance among habitats. For example, Grant (1975) demonstrated that the survival of adults and the reproductive success of the meadow vole *Microtus pennsylvanicus* is lower in forest habitats than in grassland habitats. Breeding activity and overwinter survival of the deer mouse *Peromyscus maniculatus* vary among serial stages of coniferous forest (Van Horne 1981). Populations of the harvest mouse *Rheithrodontomys megalotis* demonstrated sharp among-habitat differences in demography (Skupski 1995).

The first ideas about the differing functional roles of different habitat types in density dynamics were proposed as early as in the middle of this century by Naumov (1937, 1948, 1965). He distinguished "survival stations" and "dispersal stations". Animal populations persist in "survival stations", in which they maintain high and stable densities. In contrast, "dispersal stations" are only occupied by animals during increasing and peak phases of their population dynamics, and density is highly variable within them. Dispersal is always unidirectional from "survival" to "dispersal" stations. This model was used by Naumov and most Russian ecologists to explain both the seasonal and the long-term spatial population dynamics of rodents (Bashenina 1962; Panteleev 1968; Naumov et al. 1972; Naumov 1975; Ivanter 1975).

Analogous ideas in terms of dispersal patterns and population demography were formulated independently in the 1970s-1980s (Lidicker 1975; Van Horne 1983), and developed in late 1980s and 1990s (Holt 1985, 1996; Pulliam 1988; Pulliam and Danielson 1991; Hanski 1991; Morris 1991; Kawecki 1995; Gilpin and Hanski 1997 etc.). Among-habitat differences in reproductive success and/or adult survival cause the among-habitat variation in fitness. This conflicts with the model of ideal free distribution and requires other theoretical models to explain the observed patterns. To avoid this contradiction, the model of ideal pre-emptive distribution was proposed by Pulliam (1988) and Pulliam and Danielson (1991). This distribution supposes that each habitat contains a fixed number of qualitatively different individual breeding sites. The assortment of breeding sites of different quality varies among habitats. First, individuals select the better sites that occur in the better habitat. The poorer habitat is used only when it contains the next-best breeding sites etc.

Some proportion of individuals regularly occurs in less favourable "sink" habitats, where reproduction is inadequate to compensate for local mortality. These populations are maintained in such habitats only under continuous immigration from a more favourable "source" habitat nearby. Sometimes, the reproductive surplus in a "source" habitat is high and the reproductive deficit in a "sink" habitat is low. In such cases, only a small number of individuals reproduce in the "source" habitat, whereas the population in "sink" habitats can have a relatively high density – although this latter population would vanish without immigration (Pulliam 1988). There may be also "pseudosinks" which support a population without immigration, but immigration increases population density above the local equilibrium (Holt 1985; Watkinson and Sutherland 1995).

We deliberately put aside the theoretical problems connected with the evolutionary consequences of "source-sink" dynamics, namely the evolution of the ecological niche, evolutionary stable strategies (ESSs), models of species coexistence and so on (see Morris 1991; Holt and Gaines 1992; Holt 1996; Loreau and DeAngelis 1997 etc.). The "source-sink" concept is interesting in the present context as a possible explanation of the observed patterns of habitat distribution in desert rodents. Unfortunately, in spite of a large body of theoretical publications, empirical data are extremely scarce.

The feral house mouse *Mus muculus* in the northeastern Caspian Desert may exemplify the "source-sink" mechanism of habitat distribution (Krasnov 1984; Khokhlova and Krasnov 1986; Krasnov and Khokhlova 1989, 1994). The "source" habitats are those along the shores of salt lakes where the density of mice is stable all year round. After the spring and summer peaks of reproduction in the "source" habitat, young individuals were observed to disperse to densely vegetated patches far away from the shore. The reproductive rate in these "sink" habitats was found to be extremely low and most recruited individuals were recorded as immigrants from the "sources". Reproduction in the "source" habitats terminated in autumn, and mice in the "sink" habitats became extinct during winter.

The habitat shifts in the pygmy gerbil *Gerbillus henleyi* described above can also be explained in terms of "source-sink" habitat dynamics. It seems that open gravel plains with relatively low but stable densities can be considered as the "source" habitat, whereas sand dunes with periodically high but extremely variable density (with occasional extinction) are clearly "sinks".

7.5.5 Behavioral Mechanisms of Habitat Selection

As mentioned above, the theory of habitat selection presupposes quantitative and qualitative differences between habitats. Quantitative differences between habitats are differences in the amount of resources available, whereas qualitative differences assume differences in the conditions of resource acquisition (physical properties of the habitat determining efficiency of resource search and manipulation, predation risk, the presence of competitors and so on). Theoretically, both these differences are introduced as parameters in the equations of population dynamics (see Morris 1988, 1990). Estimation of these parameters is not simple in practice. Direct measurement of resource abundance is possible, although very difficult. For example, standing crop and the soil bank of seeds of certain sizes must be measured for a granivorous species; biomass and production of certain green parts of certain plant species should be measured for a folivore, and so on. However, direct estimation of habitat quality is impossible. A solution to this problem was

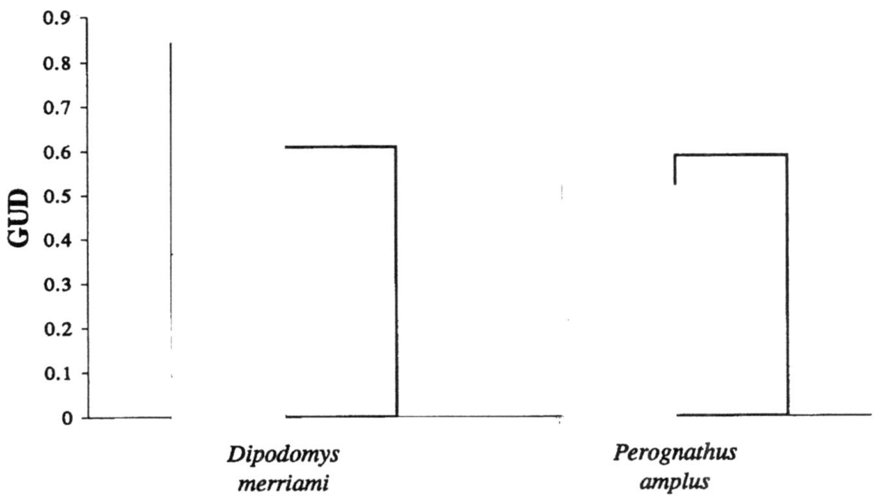

Fig. 7.26. Giving-up densities of *Dipodomys merriami* and *Perognathus amplus* in open (white columns) and under-bush (black columns) patches (based on the data of Brown 1989)

found to lie in the application of the concepts of behavioral ecology, particularly optimal foraging theory (e.g. MacArthur and Pianka 1966; Stephens and Krebs 1986). The main point of this theory is that an individual optimizes its foraging behavior so that the benefit/cost ratios are be maximal. As a forager exploits a patch, the amount of resource within it continually declines. After the resource declines below a certain threshold, the cost of foraging is too high to be worthwhile. At this point, the forager should leave a patch and either move to another or change its activity type (Charnov 1976). Based on these assumptions, the method of measurement of giving-up densities (GUD) has been proposed to estimated habitat quality (Brown 1988, 1989). The core of this method is the measurement of the amount of food that remains in the controlled exploited patch (seed tray) after the forager leaves it (see details in Valone and Brown 1989; Mitchell and Brown 1990; Mitchell et al. 1990; Kotler and Brown 1990; Brown et al. 1994 etc.). The lower the GUD (the amount of food remaining), the higher the habitat quality from the point of view of a species that exploits it.

The application of the GUD method allows one to demonstrate that the same habitat/microhabitat can be of different attractiveness for different species and that a species can evaluate the quality of a habitat. For example, studies in the Sonoran Desert (Brown 1989) showed that *Perognathus amplus* did not distinguish the qualitative differences between open and under-bush patches, whereas *Dipodomys merriami* did. In addition, the open patches were more satisfactory than the under-bush patches from the point of view of *Dipodomys merriami* (Fig. 7.26).

Another example is provided by the habitat distribution of the psammophilous gerbil, *Gerbillus andersoni allenbyi* in the Negev desert. This species occurs mainly in stabilized and semi-stabilized sand dunes and avoids unstabilized sand dunes (Abramsky et al. 1985a). However, this pattern of habitat distribution is the result of competition with *Gerbillus pyramidum*. In the absence of the latter, *Gerbillus andersoni allenbyi* uses all sand habitats equally (Abramsky et al. 1990). The measurements of GUD support this finding, demonstrating that all sandy habitats are qualitatively similar as far as *Gerbillus andersoni allenbyi* is concerned (Brown et al. 1994).

It should be noted that the method described has clear limitations. It takes only qualitative among-habitat differences into account, whereas the evaluation of quantitative differences demands other techniques. The interpretation of the results obtained by this method should, therefore, be cautious. This method of investigation has been well developed for granivorous animals. To date, it is difficult to foresee how it will work when applied to species with other modes of feeding, especially those with mixed diets.

It can be concluded that in some cases optimal foraging decisions explain, at least partially, the observed patterns of habitat distribution. It should be remembered, however, that the optimal behavior is only one of the mechanisms that, both separately and combined, affect habitat distribution. We do not yet know the relative importance of optimal behavior in this complex of mechanisms.

CHAPTER 8

Spatial Organization of Desert Rodent Communities

8.1 Definitions

A general definition of community is "any assemblage of populations of living organisms in a prescribed area or habitat" (Krebs 1994, p. 431). This definition is too broad. A random species collection could also be considered to be a community according to this definition. However, a community should be a system (e.g. a set of interacting elements sensu Von Bertalanffy 1968) to be a subject for scientific analysis. This understanding of the term "community" was developed by the "evolutionary ecology" school founded by Hutchinson (1957) and MacArthur (1958). According to this, a community is a system of co-occurring interacting populations of different species of one taxon of indeterminate size that constitutes one or several contiguous trophic levels (but see Drake 1990). Competition was originally considered to be the main type of interaction structuring a community, according to the concept of limiting similarity (MacArthur and Levins 1967; MacArthur 1970; May and MacArthur 1972). This concept has been strongly criticized (Rosenzweig 1974; Roughgarden 1976; Turelli 1978) and is no longer a dominant paradigm in ecology. Nevertheless, many theories that are attempting to replace it (e.g. Roughgarden 1976; Rosenzweig 1979a,b, 1987a,b, 1991, 1995; Rosenzweig and Abramsky 1986; Pimm and Rosenzweig 1981, but see Wiens 1989) still rely on interspecific competition as a main motive force.

The most important problems of community ecology are to determine the mechanisms that enable species to coexist and to analyze community structure. According to the general system theory, a structure is determined as the relation between elements of a system (Von Bertalanffy 1968). In the case of ecological communities, these relations can be either functional (competition, mutualism, etc.) or topological (distribution of species niches in resource space, distribution of species in the space of morphological characters, etc.) or both. A structured community is one in which the species composing the community differ more in their morphology, behaviour and environmental preferences than expected by chance (see Brown 1995).

The concept of limiting similarity (MacArthur and Levins 1967; May and MacArthur 1972) predicts that the niches of coexisting species that use one resource dimension are evenly distributed along this resource gradient. This distribution has been called "overdispersion of niches" (Schoener 1974). Furthermore, the limiting similarity concept can serve as a theoretical base for Hutchinson's empirical rule (Hutchinson 1959). This rule states that coexisting congenerics have a constant magnitude of morphological character difference. Hutchinson suggested that a ratio of 1:1.3 in the body sizes of small:large species is necessary to maintain coexistence. If body size (or the size of certain body parts) is considered to be an indicator of the mode of resource use (e.g. niche position), Hutchinson's rule is indeed a direct reflection of limiting similarity. Supporting evidence has been found among many animal taxa (see Giller 1984 for review), including granivorous desert rodents (Brown 1973, 1975). Cody (1974) has demonstrated that the overdispersion of niches also occurs in the case of multidimensional resource space.

There are good reasons to view the spatial organization of a community as a basis for research into the backgrounds of community structuring (Schoener 1974, 1986). In many cases, spatial displacement or habitat selection is an important mechanism facilitating the coexistence of sympatric species (Schoener 1974). The importance of spatial segregation has been demonstrated many times among desert rodents (Brown and Lieberman 1973; Meserve 1976, 1981a; Nel 1978; Stamp and Ohmart 1978 etc.).

8.2 The Qualitative Approach to the Analysis of Community Composition

By definition, a community is composed of a set of species and each species is characterized by its own unique manner of resource use. These patterns of resource use (niches) are different among different species. The qualitative approach to analyzing the structure of a community assumes the existence of functional groups of species that differ qualitatively in their manner of resource use. Supposedly, species of the same group are similar in their manner of resource use and compete with one another, whereas species belonging to different groups differ sharply in this parameter and seldom interact. These groups can be distinguished both directly by the ecological-behavioral features of the species as well as indirectly by morphological indicators. The qualitative approach to the analysis of community structure is based on the idea that a community is composed of that combination of species (from the available species pool) which maximizes resource use with simultaneous minimization of competition. These combinations are non-random and only some of them really exist. This concept was elucidated by Diamond (1975) as "assembly rules". These rules are related to intra-guild patterns only (e.g. a group of ecologically similar species, see below). The main point is that the co-occurrence of species is non-random. Some species pairs co-occur more frequently than would be expected by chance (permissible combinations), whereas other combinations are more rare or even non-existent (forbidden combinations). Furthermore, the pattern of species co-occurrence depends on the abundance and diversity of the resource base. A combination that is forbidden in poor environments can become permissible in rich environments. This pattern has been explained mainly by interspecific competition (Diamond 1975). This original concept initiated a stormy discussion (i.e. Connor and Simberloff 1979; Simberloff and Boecklen 1981; Diamond and Gilpin 1982; Gilpin and Diamond 1984) which has only recently died away (Brown 1995).

Further development of Diamond's ideas is seen in the assembly rules of M'Closkey (1978, 1985) and Fox (1987, 1989 etc.) and the nested subsets of Patterson (Patterson and Atmar 1986; Patterson 1987, 1990 etc.). M'Closkey (1978) measured the mean niche separation for possible combinations (assemblies) from a pool of four heteromyid granivores. There were 11 possible species combinations, but only four have ever been observed – those with minimum niche separation. The main conclusion of this study was that the co-occurring rodents were not a random subset of the available species pool. Niche separation has been shown to increase with species richness, thus supporting the hypothesis of May and MacArthur (1972). The assemblages observed were considered to be saturated because limited resource space may be best utilized by a certain species set which leaves fewer resources unused. Non-observed assemblages were regarded as under-saturated because not all available resources were used (M'Closkey 1978). Fox (1987, 1989), studying communities of small mammals in southeastern Australia, demonstrated that there was a clear tendency for genera (or other taxonomically-related groups of species with similar diets) to be represented as equally as possible in local assemblages. His assembly rule states that "there is a much higher probability that each species entering a community will be drawn from a different genus ... until each group is represented, before the rule repeats" (Fox 1987, p. 201). Fox and Brown (1993) evaluated the possible application of this rule to North American desert rodents. They showed that the rule can be applied to different functional groups of rodents (primarily granivorous bipedal heteromyids, quadrupedal heteromyids and quadrupedal non-heteromyids), and to assemblages at different spatial scales (the Great Basin biogeographic province and the North American desert realm). They also suggested that interspecific competition can be a mechanism underlying this rule. Competition influences the probability that species in different functional groups differ sufficiently in resource use to be able to coexist. The application of the rule cannot be evaluated for rodent communities in other deserts of the world. The reason for this is that the regional pool of species (a source for composing local assemblages) is too small in each functional group. It has been demonstrated by Kelt et al. (1996) that the regional species pool for separate functional groups is large enough to provide numerous species combinations in two cases only, namely granivores in the North American deserts and carnivores in the Australian deserts. The latter case concerns marsupials rather than rodents.

A nested pattern of assemblages is a pattern in which species comprising faunas of lower species richness constitute subsets of the species occurring in successively richer faunas (Patterson and Atmar 1986; Brown 1995; Hecnar and M'Closkey 1997; Wright et al. 1998). This pattern is well known (e.g. Darlington 1957), but attempts to quantify it only began recently (Patterson and Atmar 1986; Patterson 1987; Atmar and Patterson 1993). There are a number of metrics of nestedness (see review in Wright et al. 1998). Most of them are strongly dependent on the size of the input matrix (number of species by number of sites) which makes comparison of nestedness between different data sets difficult. The only metric that is not dependent on matrix size is *T* (matrix "temperature") proposed by Atmar and Patterson (1993). This metric provides a standardized measure of matrix disorder by assessing the deviation of an observed matrix from one of the same rank and fill that is perfectly nested. *T* varies from 0, for a perfectly nested matrix, to 100 for a completely disordered matrix. Thus, nestedness is an estimate of the degree of non-random pattern in the mutual species distribution. However, the value of nestedness will reflect the existing pattern only if the nestedness evaluation is applied to a homogenous set of data. This issue is easily resolved on a geographic scale. Nobody will try to combine data from different archipelagoes or biogeographic units into the same data set. Nevertheless, it is not so simple at an ecological scale. Data combined from a homogenous biogeographic area will reflect the existing pattern only if there are no spatial (habitat) guilds. Otherwise, the pattern will be masked, i.e. the value of nestedness will be lower than the real value (if it is significant at all).

The first attempt at analysis of nestedness in desert rodent communities was undertaken by Patterson and Brown (1991). They used data from 202 sampling sites across the North American deserts and analyzed them in three ways, namely (1) all data combined, (2) separately for two different regions (the Great Basin and Sonoran deserts), and (3) separately for different habitats (desert shrub, desert grassland, sand dunes and shrub-steppe). The results demonstrated nestedness in the entire assemblage of all sites, all Great Basin sites, all Sonoran sites and various habitat groupings of sites within deserts. However, groupings of sites by habitats among deserts did not exhibit nestedness. It was suggested that three conditions may be necessary for the development of nested subset structure: (1) a common biogeographic history, (2) generally similar contemporary environments, and (3) hierarchial organization of niche relationships.

Comparison of the Asian and North American desert rodent communities showed a nested pattern in all the communities studied (Table 8.1), and also that the degree of nestedness in North America is much higher than in Asia

Table 8.1. The degree of nestedness in different desert rodent communities. (Modified from Kelt et al., in press)

Desert	Matrix "temperature" (*T*)
Southern Gobi	26.72
Western Gobi	39.34
Eastern Gobi	38.18
Kyzylkum	36.21
North America	13.84

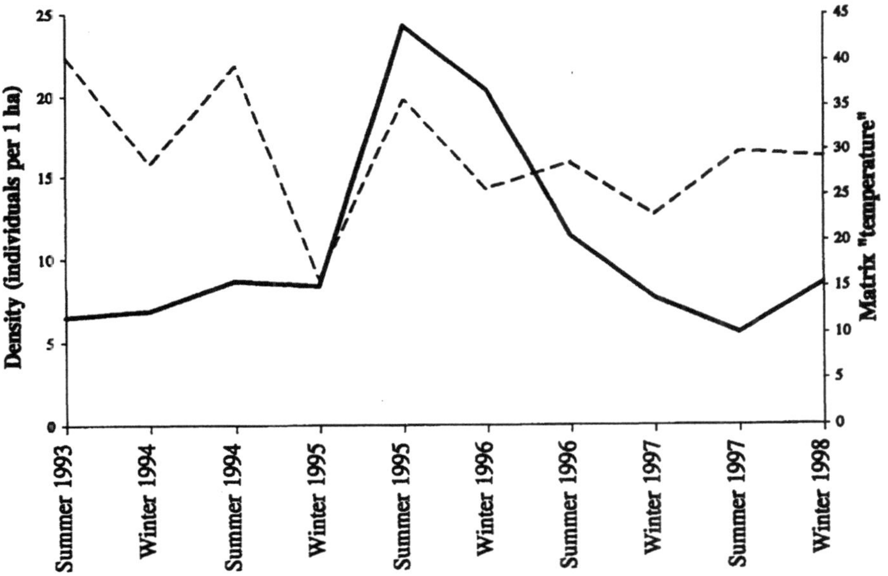

Fig. 8.1. Temporal variation in the degree of nestedness of the rodent community in the Negev Highlands. Bold line Rodent density, dashed line matrix "temperature"

(Kelt et al., in press). This result is not surprising considering the absence of strong habitat specialization in the North American rodents and its presence in Asia (see above) and, consequently, the absence of spatial guilds in North America and the strong expression of these guilds in Asia (see below). Furthermore, the competition that is considered to be the principal dynamic producing nested community structure, is one of the most important factors operating in North American communities, whereas it seems to have a minor role in Asia (Kelt et al., in press).

The degree of nestedness can vary temporally. Figure 8.1 demonstrates temporal dynamics of nested patterns in the rodent community of the central

Negev desert. There were both seasonal and annual fluctuations in the level of nestedness. Nestedness in summer was always lower than that in winter. This reflects clearer expression of community structure in winter (see below). Annual fluctuations in nestedness were not correlated with the density of rodents. These fluctuations are not readily explainable at the present level of our knowledge. The reason may be that nestedness is too general a parameter to reflect several different processes and patterns. Furthermore, these results indicate that the measurement of community nestedness should be based on data collected over several seasons and several years so that averaged value are obtained. Otherwise, the resultant estimation will be of little use.

8.3
Multivariate Description of Spatial Structure

As mentioned above, species live in a characteristic, more or less limited range of habitats; within this range they are most abundant around their environmental optima. Species abundance changes along environmental gradients. Hence, species replacements occur as a function of changes in the environment, or (analogously) with successional time (Peet and Loucks 1977; Pickett 1980; Rusch and Van der Maarel 1992). In other words, each species exists in the space of environmental variables, which may be conceived as the set of axes of its Hutchinsonian niche (see above). The species population, therefore, may be represented as a swarm of points in the coordinate system of *n* axes, whereas relationships of co-occurring species placed in the space of environmental variables represent the spatial structure of a community. Obviously, different environmental factors have different significance for one or other species. Some factors may be of more definitive value for species abundance, others less; yet other factors may play an important role only when linked with one or more additional factors. Such descriptions of community structure reveal the principles of resource partitioning among coexisting species and the factors influencing their distribution and abundance.

Different methods of multivariate statistics provide the possibility of delineating resource space. In recent years progress in the development and application of multivariate methods in ecology has been the result of a change in emphasis in theoretical work on multivariate analysis from the elaboration of formal statistical models towards descriptive techniques for exploring pattern in data sets and providing concise summaries and displays (Gauch 1982; Digby and Kempton 1985). Ordination is one of the most popular sets of methods of multivariate analysis in community ecology because it can provide a compact description of the spatial structure of plant and animal communities. Ordination methods have been used in plant ecology for at least 30 years (see Gauch 1982 and references therein), whereas in animal ecology their application is less common. Nevertheless, a number of animal communities have been analyzed by ordination, from earthworms (Briones et al. 1992) to mammals (Dueser and Shugart 1979; Shenbrot et al. 1994b).

Numerous ordination techniques have been proposed by theoreticians (see Gauch 1982; Pielou 1984; Jongman et al. 1987), but three of them, namely principal component (PCA), correspondence (CA) and discriminant function (canonical variate; DFA) analyses are used most frequently in animal community ecology. It is necessary to note that the meanings of PCA in plant and animal community ecology differ slightly. Plant ecologists use PCA to reduce the dimensionality of species' space (Gauch 1982; Pielou 1984), whereas animal ecologists use PCA in two ways: (1) to reduce the dimensionality of environmental space and subsequently place species in the space obtained, and (2) to calculate weighted average values of environmental variables for each species and to use these values as a data set for PCA (Morrison et al. 1992). All the references to PCA that follow will be used in the last two senses (PCAE and PCAS, respectively).

Each ordination method has its own merits and shortcomings (Jongman et al. 1987; James and McCulloch 1990), and usually one or another is selected a priori by a researcher. As examples of different ordination approaches we can refer to studies on fishes (Meffe and Sheldon 1988), birds (Rotenberry and Wiens 1980) and rodents (Kelt et al. 1994) using PCAE; on birds (Maurer et al. 1981 in Morrison et al. 1992) using PCAS; on fishes (Meffe and Sheldon 1988), birds (Rushton et el. 1994), ungulates (Ben Shahar and Skinner 1988) and rodents (Borghi et al. 1994; Krasnov et al. 1996a) using CA; and on darkling beetles (Krasnov and Shenbrot 1996, 1997), lizards (Shenbrot et al. 1991; Shenbrot and Krasnov 1996), birds (Cody and Walter 1976) and rodents (M'Closkey 1976; Dueser and Shugart 1979; Morris 1979) using DFA.

These ordination techniques differ principally in their approaches to the construction of ecological space. As a result, they produce ecological spaces with different properties. In particular, they are based on different assumptions. PCA relates to a linear response model in which the abundance of a species either increases or decreases with the value of each environmental variable. CA is related to a unimodal response model, in which a species occurs over a limited range of values of each of the latent variables. DFA belongs to linear multivariate methods. In contrast to PCA, it deals with a data set divided into two or more naturally occurring (or a priori defined) groups rather than with one homogenous data set. All three methods assume multivariate normality. Furthermore, DFA also assumes equality of within-group variance-covariance matrices. It is clear, that in ecological studies when groups for DFA are represented by different species, the requirement of homogeneity of variances is not realistic. However, when sample sizes are equal and large, DFA is robust to violation of this assumption (Tabachnik and Fidell 1996). Properties of multivariate methods may be environment-dependent (PCAE), species-dependent (CA) or both environment- and species-dependent (PCAS and DFA). Consequently, parameters of a community estimated in a specific ecological space are method-specific. As different ordination techniques have different powers in representing data for specific research purposes, we have attempted to compare them and estimate their relative effectiveness (Krasnov and Shenbrot, in press). It appears that DFA is the optimal method for describing spatial structure of many animal communities, although not for all taxa. This permitted us to use DFA as the primary method of ordination.

DFA has the following advantages: (1) species densities and environmental variables are considered simultaneously; (2) the subjects of the analysis are complete data sets for each species (not only species means). The latter is important for the estimation of niche breadth based on the dispersion of observation points. The resulting ordination is a model of ecological space, the main property of which is the maximal niche division. The axes of this space are linear combinations of the original environmental variables and may usually be interpreted as composite environmental gradients. This approach to the description of the niche structure of communities was theoretically validated by Green (1974) and has been successfully applied to a number of different animal communities (see references above), although some of the applications of this method deserve criticism (Carnes and Slade 1982; Van Horne and Ford 1982). The result of the application of DFA is the construction of a model of ecological space, which quantifies the segregation of species niches with respect to multiple variables. Mutual location of the niches in the resource space can be described by the distances between niche centers (centroids). This can also be described using measures of similarity of resource utilization (niche overlap). The latter can be estimated either as a cardinality of set crosscut or as the probability of a shared utilization of a resource. A review of the measures of niche overlap can be found in Abrams (1980), Smith and Zaret (1982), Yu and Orloci (1990) etc. If the partition of ecological space among coexisting species is one of the important factors in the structuring of communities, there are good reasons to expect that not only niche placement but also some general characters of the community, such as overall biomass and species diversity, should exhibit meaningful patterns in the model of ecological space produced by DFA.

8.4 Environmental Factors Structuring a Community

To make analyses of the spatial structure of rodent communities from different deserts as comparable as possible, our approach was standardized by measuring almost the same set of environmental variables in a series of 1-ha sites in each study area. The total number of measured or derived variables was from 18 to 24 in different deserts. Three of these described the mechanical composition of the soil, six the abundance and composition of annual plants and geophytes, and the remainder characterized the structure of the

Table 8.2. Description of environmental variables measured for each sample plot in the central Negev desert and included in discriminant function analysis

Variable	Unit
Content of rocks in the soil	%
Content of gravel in the soil	%
Content of clay in the soil	%
Slope	°
Abundance of *Carex*	no m^{-2}
Abundance of Liliacea	no m^{-2}
Abundance of annual Chenopodiacea	no m^{-2}
Abundance of annual forbs	no m^{-2}
Abundance of annual grasses	no m^{-2}
Abundance of annual grasses and forbs	no m^{-2}
Perennial grass cover	%
Cover of aphyllous shrubs	%
Cover of succulent shrubs	%
Cover of microphyllous shrubs	%
Overall shrub cover	%
Perennial plant crown volume at the level 0-25 cm	%
Perennial plant crown volume at the level 25-50 cm	%
Perennial plant crown volume at the level 0.5-1 m	%
Perennial plant crown volume at the level 1-2 m	%

perennial vegetation. The last group of environmental variables comprised cover by perennial grasses, aphyllous, microphyllous and succulent shrubs and the volume of the shrub crowns in four or five vertical layers. Different plants were naturally included in the same category in different deserts. For example, the category of succulents consisted of Chenopodiaceae and Zygophyllaceae in the central Asian deserts, Euphorbiaceae in the Thar Desert, and Chenopodiaceae, Cactaceae and Agavaceae in the Chihuahuan Desert. In addition, parameters of mesorelief were measured. These variables were different in the different deserts, depending on the geomorphology of the region. For example, the angle of slope was measured in upland deserts (the Negev and southern Kyzylkum Deserts), the height of the sand mounds at the base of shrubs was measured in the Kyzylkum and Gobi Deserts and so on. An example list of environmental variables measured is presented in Table 8.2. Sampling sites were distributed across the range of habitat types occurring in each region. Each site was divided into 25 20 x 20 m plots, and all habitat variables were measured in each plot. Two traps were established in each plot at random. Jerboas not entering traps were caught with a net at night, using a searchlight. *Psammomys obesus, Rhombomys opimus, Eolagurus przewalskii* and Asian ground squirrels did not enter the traps either. They were counted in the morning using binoculars. The abundance of subterranean rodents, such as *Ellobius* in Asia and *Thomomys* in North America, was evaluated by counting the groups of fresh mounds. We recorded the

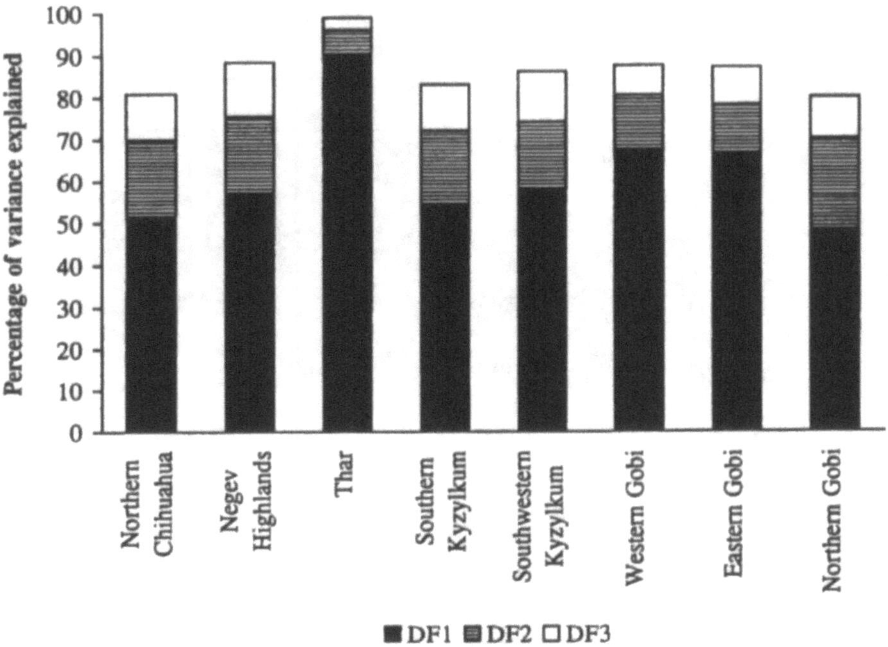

Fig. 8.2. Relative variance of species distribution explained by the three first discriminant axes (DF1, DF2, DF3) in different desert rodent communities

number and species of rodents captured or observed on each plot, over 3 days. For a complete list of the specific habitat variables used and the further details of the methods of analysis, see Rogovin et al. (1991), Shenbrot (1992), Shenbrot et al. (1991, 1994b), Krasnov and Shenbrot (1996, 1997), and Krasnov et al. (1997, 1998).

The first three discriminant axes explained most of the total variation (79-99%) in all regions (Fig. 8.2). The composition of the axes was different for different communities (Fig. 8.3). Among the variables that separated species in ecological space, the characteristics of the soils were the most important in the northern Chihuahuan and Gobi Deserts. In the central Negev and southwestern Kyzylkum Deserts, characteristics of the perennial vegetation contributed most heavily to the discriminant axes. The characteristics of the annual vegetation were most important for the rodent communities of the Thar and southern Kyzylkum Deserts. The relative importance of different environmental features in explaining the distribution of rodent species in different deserts demonstrates that species did not in all cases separate the ecological space described by the discriminant function axes according to the same environmental variables. This difference reflects both the differences in the physical structures of the deserts and the differences in composition of

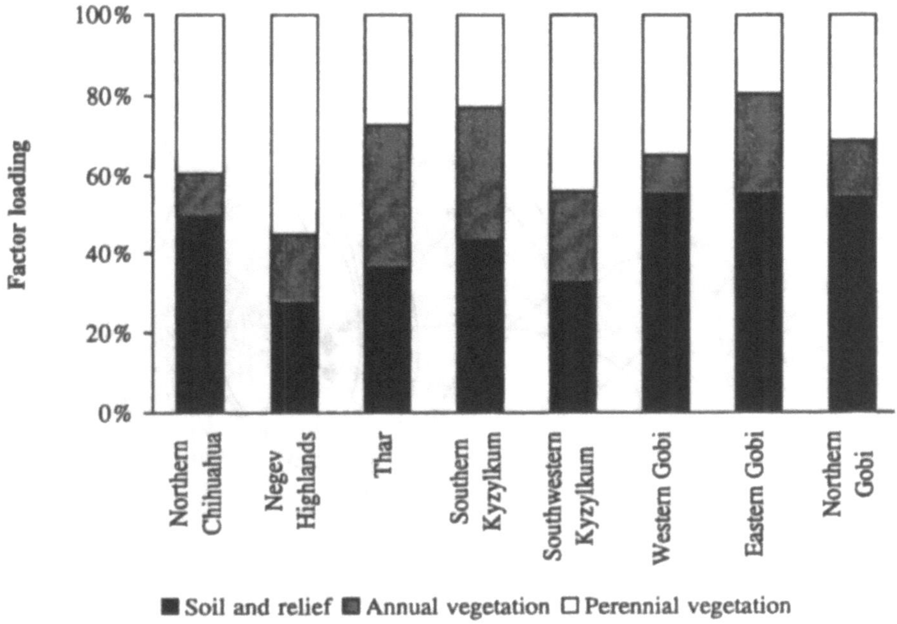

Fig. 8.3. Contribution of soil and relief, annual vegetation, and perennial vegetation variables to the set of three first discriminant axes in different desert rodent communities

rodent communities determined by the presence or absence of species that can display specific reactions to different elements of environmental structure. If a factor or a group of factors is weakly represented in a region (for example, annual vegetation in the Gobi Desert, rocks in the southwestern Kyzylkum Desert, or clay depressions in the central Negev and Thar Deserts), one cannot expect heavy loading (weighting) of this factor on the resultant ecological space. On the other hand, if there are no species that are specially adapted to a narrow interval of a factor gradient (for example, specialized rock-dwelling or sand-dwelling species), there will be no response to the factor (the rock or clay content of the soil in this example) and, therefore, this factor will not contribute to the discriminant axes.

The differences in the relative compositions of the axes suggest that species distributions are associated with somewhat different environmental features in different deserts and, therefore, the niches of species in the ecological space defined by these axes cannot be directly compared among deserts. However, the pattern of niche placement and overlap within each community is still comparable.

The spatial relationships of the species in each community are depicted in Figs. 8.4-8.12. Each figure shows 75%-confidence ellipses representing the

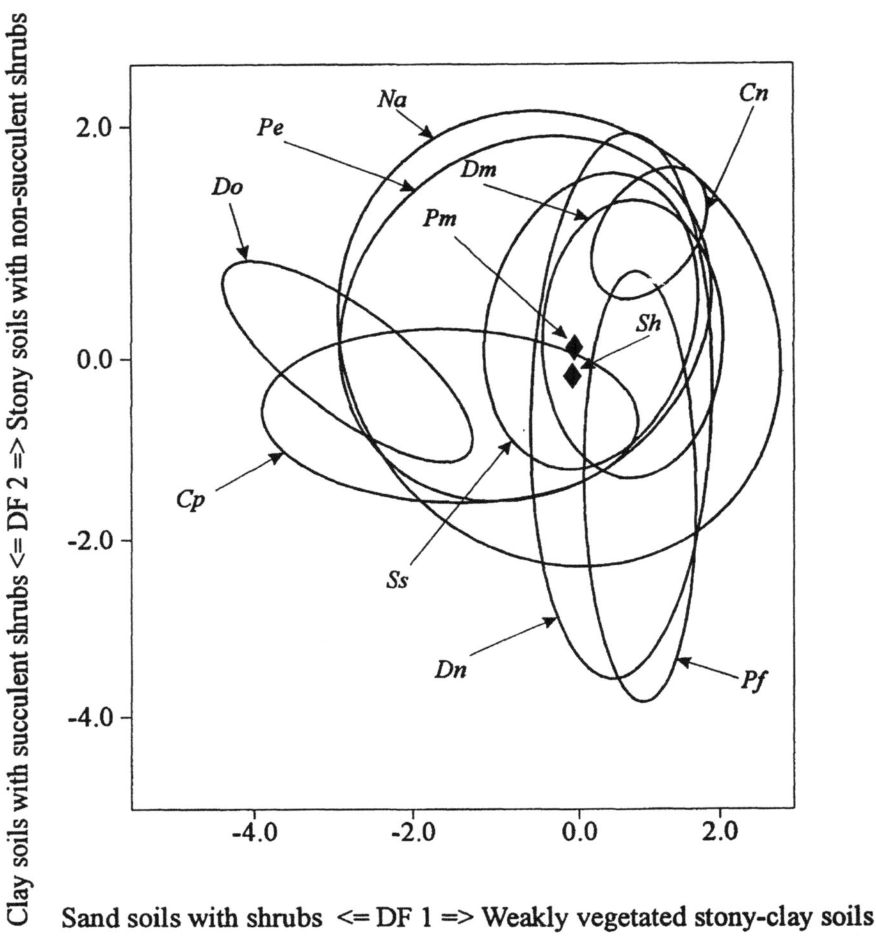

Fig. 8.4. Spatial relationships of niches in the central Chihuahuan rodent community assemblage. Species names are as follows: Ss *Spermophilus spilosoma*, Cn *Chaetodipus nelsoni*, Cp *C. penicillatus*, Dm *Dipodomys merriami*, Dn *D. nelsoni*, Do *D. ordii*, Pf *Perognathus flavus*, Pe *Peromyscus eremicus*, Pm *P. maniculatus*, Na *Neotoma albigula*, Sh *Sigmodon hispidus*

distribution of each species in the space defined by the first two discriminant axes. The figures demonstrate that the species in each community divided the resource space in their own manner, although the responses of species assemblages to environmental gradients were similar. The first discriminant axis correlated with a gradient of soil structure from clay to sand (the Negev, Thar, southern and southwestern Kyzylkum communities), from stony soils to sand (the northern Chihuahuan community), or from stony and clay soils

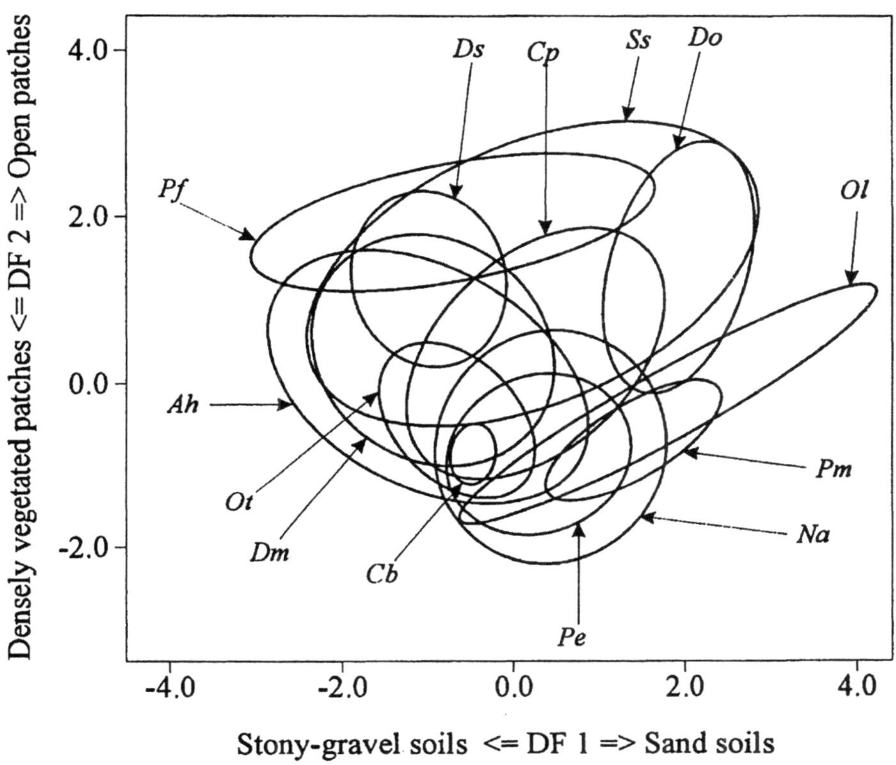

Fig. 8.5. Spatial relationships of niches in northern Chihuahuan rodent community assemblage. Species names are as follows: Ah *Ammospermophilus harrisii*, Ss *Spermophilus spilosoma*, Cb *Chaetodipus bayleyi*, Cp *C. penicillatus*, Dm *Dipodomys merriami*, Do *D. ordii*, Ds *D. spectabilis*, Pf *Perognathus flavus*, Ol *Onychomys leucogaster*, Ot *O. torridus*, Pe *Peromyscus eremicus*, Pm *P. maniculatus*, Na *Neotoma albigula*

to sand (the central Chihuahuan and all Gobian communities). In addition, the loading of the vegetation variables onto the first axis was also considerable, except in the cases of the northern Chihuahuan, southern Kyzylkum, and eastern Gobian rodent communities. Separation along the second axes corresponded mainly with changes in both annual and perennial vegetation, although soil factors also loaded onto these axes in some communities (the central Chihuahuan, Thar, southwestern Kyzylkum, and northern Gobian communities). The only exceptions were found to be the western Gobi and the Negev rodent communities. In these, the second axis either correlated only with the soil gradient or with mesorelief changes from vertical to horizontal surfaces, respectively.

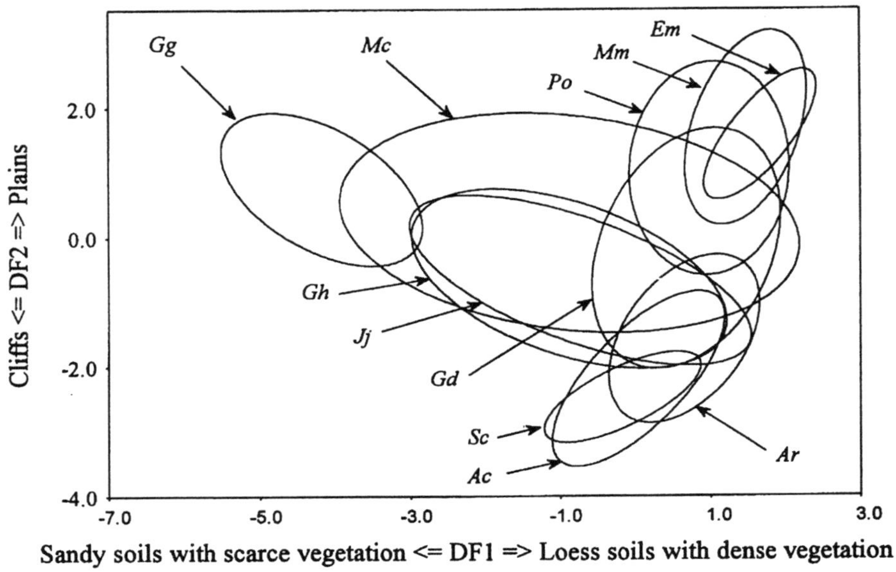

Fig. 8.6. Spatial relationships of niches in the Negev Highlands rodent community assemblage. Species names are as follows: Jj *Jaculus juculus*, Mm *Mus musculus*, Ac *Acomys cahirinus*, Ar *A. russatus*, Gd *Gerbillus dasyurus*, Gg *G. gerbillus*, Gh *G. henleyi*, Mc *Meriones crassus*, Sc *Sekeetamys calurus*, Po *Psammomys obesus*, Em *Eliomys melanurus*

8.5 Guild Structure and Factors Determining It

In most communities of the desert rodents described, the distribution of the projections of species' niches is neither even nor random. Instead, species form some groups within the ecological space. As the ecological space reflects a complex of environmental gradients, these groups can be conceived of as spatial *guilds*. A guild is usually defined as "a group of species that exploits the same class of environmental resources in a similar way" (Root 1967, p. 335). A guild is therefore a unit of ecological classification of species based on their niche parameters as a character set. It should be remembered that space is considered to be a resource in the framework of the concept of ecological niche (see above). On the other hand, a space could be considered not as a resource itself but rather as a condition of resource exploitation. This dualism caused a controversy in the application of the term "guild" to spatial (habitat) species assemblages (Simberloff and Dayan 1991). Nevertheless, application of the term "guild" to spatial groups of species may be accepted.

The occurrence of spatial guilds in a community can be envisaged from the diagram that represents the placement of species in a resource space (centroids or ellipsoids of different probabilities). However, the existence of

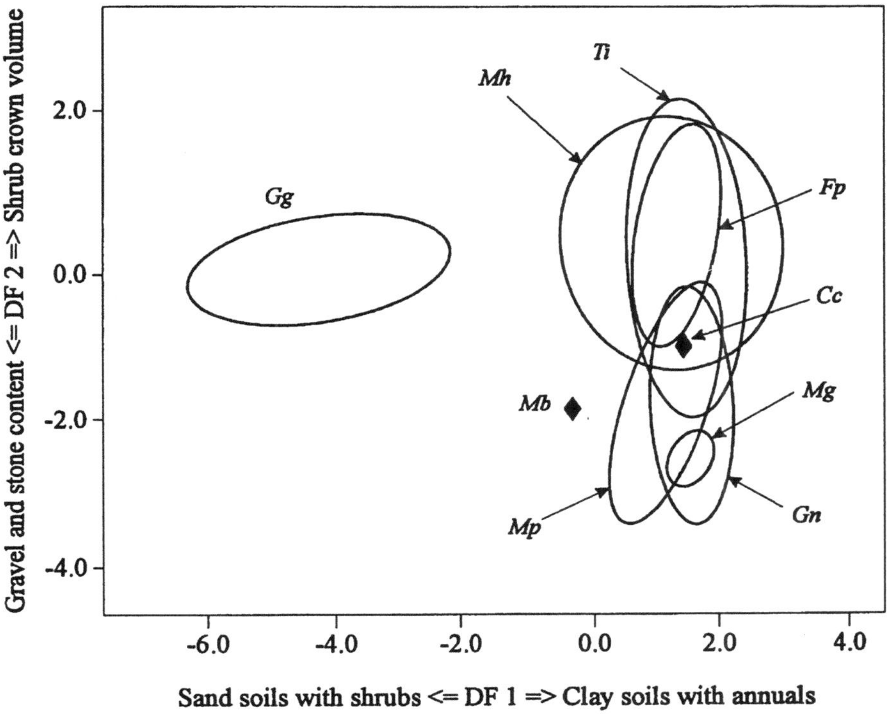

Fig. 8.7. Spatial relationships of niches in the Thar rodent community assemblage. Species names are as follows: Fp *Funambulus pennati*, Mg *Millardia gleadowi*, Mb *Mus booduga*, Mp *M. platythrix*, Cc *Cremnomys cutchicus*, Gg *Gerbillus gleadowi*, Gn *G. nanus*, Mh *Meriones hurrianae*, Ti *Tatera indica*

spatial guilds can sometimes be masked. The causes of this masking vary. They include multidimensionality, large numbers of species, weak expression of structure, and so on. Consequently, some specific technique for revealing spatial guilds is necessary. Such a method is the nearest neighbor analysis proposed by Inger and Colwell (1977) and extended by Winemiller and Pianka (1990). The application of this method consists of rearranging the square matrix of pairwise values of niche overlap by sorting the values of each row in descending order. The resultant rows therefore reflect ranks of closeness. A mean value and standard deviation of overlap can be calculated for each rank of closeness. Plotting the standard deviation of niche overlap against the rank of closeness will produce a curve. This curve will either be continually decreasing (in the case of even or random niche distribution) or be greatest at median ranks of closeness (indicating the unevenness of niche distributions). Until now there have not been either empirical data or models that demonstrate a constant increase in the standard deviations of overlap with

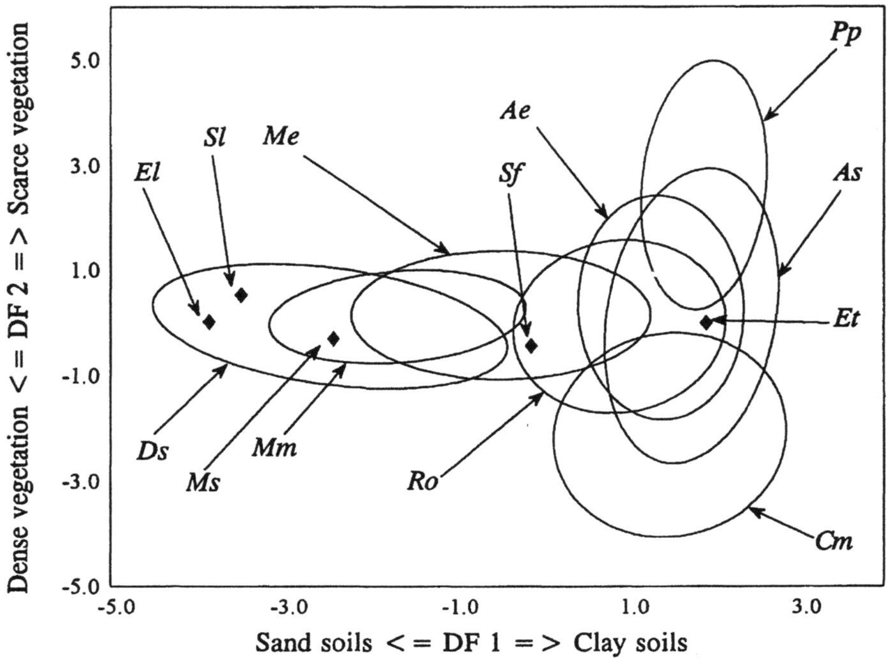

Fig. 8.8. Spatial relationships of niches in the southern Kyzylkum rodent community assemblage. Species names are as follows: Sf *Spermophilus fulvus*, Sl *Spermophilopsis leptodactilus*, Ds *Dipus sagitta*, El *Eremodipus lichtensteinii*, As *Allactaga severtzovi*, Ae *A. elater*, Pp *Pygeretmus pumilio*, Mm *Meriones meridianus*, Me *M. erythrourus*, Ro *Rhombomys opimus*, Cm *Cricetulus migratorius*, Et *Ellobius tankrei*, Ms *Mus musculus*

the rank of nearness. However, the rising part of the unimodal curve has been interpreted as an indicator of species piling up on the same resources. Consequently, the standard deviations were low at similar ranks (Winemiller and Pianka 1990). Figure 8.13 demonstrates the results of nearest neighbor analyses for nine communities of desert rodents. It can be seen that standard deviations of niche overlap plotted against rank of nearness have a secondary peak at 2-5 ranks of nearness in all Asian communities except that of the Thar Desert. At the same time, the curves of the standard deviations of niche overlap decrease gradually with rank of nearness in both North American communities. The shape of these curves indicates that guild subdivision is well expressed in most Asian communities. The opposite is the case for the Chihuahuan and Thar Deserts.

Nearest neighbor analysis allows one to indicate the presence or absence of guilds and the average number of species per guild (the position of a peak of standard deviation curve), but it cannot permit one to judge the number of guilds in a community or the degree of expression of guild structure. In

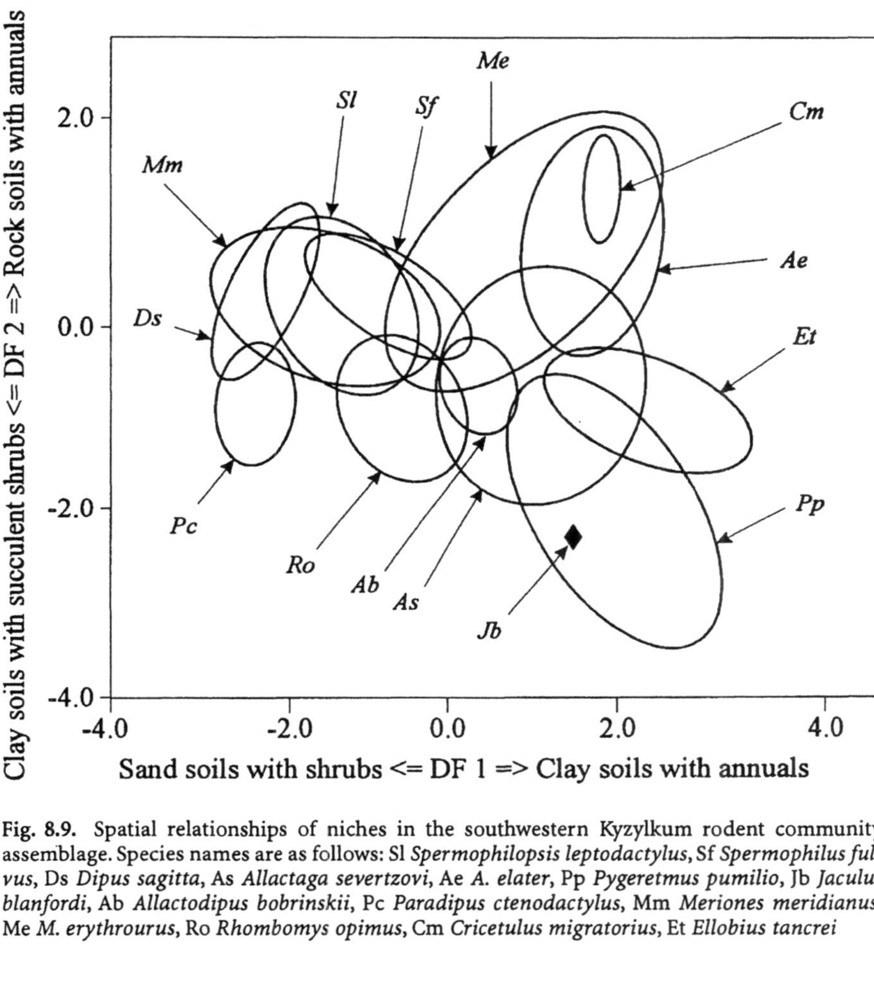

Fig. 8.9. Spatial relationships of niches in the southwestern Kyzylkum rodent community assemblage. Species names are as follows: Sl *Spermophilopsis leptodactylus*, Sf *Spermophilus fulvus*, Ds *Dipus sagitta*, As *Allactaga severtzovi*, Ae *A. elater*, Pp *Pygeretmus pumilio*, Jb *Jaculus blanfordi*, Ab *Allactodipus bobrinskii*, Pc *Paradipus ctenodactylus*, Mm *Meriones meridianus*, Me *M. erythrourus*, Ro *Rhombomys opimus*, Cm *Cricetulus migratorius*, Et *Ellobius tancrei*

addition, this method cannot indicate the occurrence of guild structure if there are two guilds, one of which is composed of only a single species.

Guild structure can, however, be revealed by cluster analysis of a niche overlap matrix. This method has both merits and shortcomings. The results of cluster analysis can identify the number of guilds in a community, and the species composition of each of them. However, the selection of the critical level of distance among species' niches at which to divide their assemblage into guilds is somewhat arbitrary. Nevertheless, the selection of 10-15% of within-group overlap as a critical (minimal) level usually gives the clearest results (Shenbrot et al. 1994b; Krasnov and Shenrot 1996, 1997). Figures 8.14-8.16 represent the results of UPGMA (unweighted pair-group method using

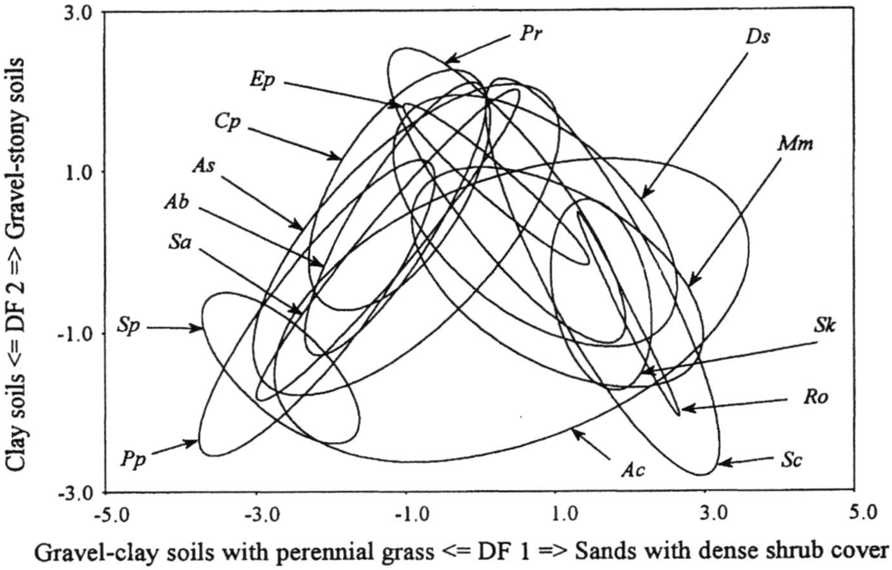

Fig. 8.10. Spatial relationships of niches in the western Gobi rodent community assemblage. Species names are as follows: Sp *Spermophilus pallidicauda*, Cp *Cradiocranius paradoxus*, Sc *Salpingotus crassicauda*, Sk *S. kozlovi*, Ds *Dipus sagitta*, Sa *Stylodipus andrewsii*, Ab *Allactaga bullata*, As *A. sibirica*, Pp *Pygeretmus pumilio*, Mm *Meriones meridianus*, Ro *Rhombomys opimus*, Ac *Allocricetulus curtatus*, Pr *Phodopus roborovskii*, Ep *Eolagurus przewalskii*

arithmetic averages) cluster analyses based on matrices of pairwise niche overlap values for the nine communities of desert rodents considered. These results support the conclusion that guild structure (i.e. the presence of a single spatial cluster) is absent from both Chihuahuan communities as well as from the Thar community. All other Asian communities have well-expressed guild structure, and two to three spatial assemblages of rodent species per community can be revealed.

The degree of niche overlap in spatial resource space reflects the degree of similarity between species in the use of certain environmental gradients. On the other hand, the response of a group of species to environmental gradients determines the habitat structure of an area (see Chap. 7). Consequently, the resultant spatial guilds are no less than habitat guilds (i.e. groups of species occupying one or several similar habitat types). It is clear that the presence of spatial (habitat) guilds assumes the presence of spatial (habitat) specialists. The necessary conditions for the development of a spatial guild in a community are therefore the presence of certain environmental features and of no less than two species which clearly prefer these features.

Spatial guilds are well expressed in all Asian communities and can be easily interpreted in terms of habitat types (compare Figs. 8.7-8.12 and Figs. 8.14-8.16).

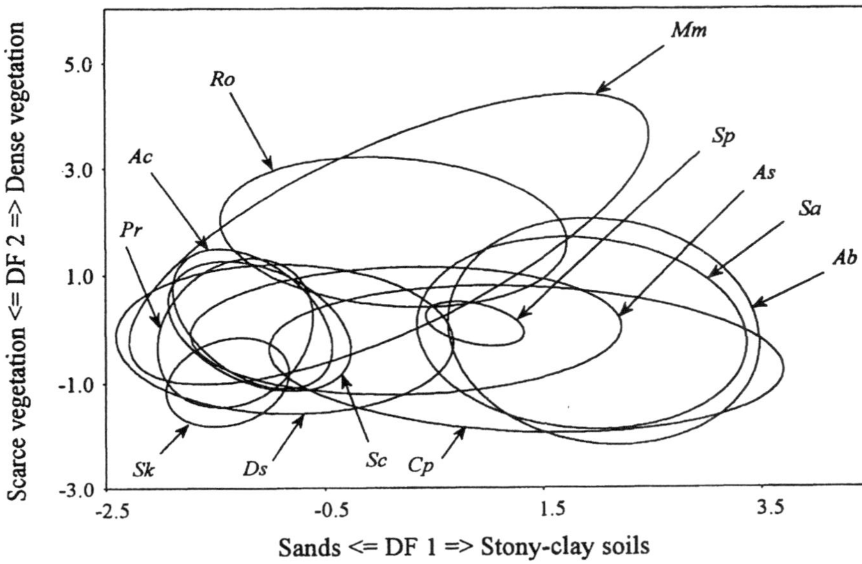

Fig. 8.11. Spatial relationships of niches in the eastern Gobi rodent community assemblage. Species names are as follows: *Sp Spermophilus pallidicauda*, Cp *Cradiocranius paradoxus*, Sc *Salpingotus crassicauda*, Sk *S. kozlovi*, Ds *Dipus sagitta*, Sa *Stylodipus andrewsii*, Ab *Allactaga bullata*, As *A. sibirica*, Mm *Meriones meridianus*, Ro *Rhombomys opimus*, Ac *Allocricetulus curtatus*, Pr *Phodopus roborovskii*

It should be noted, however, that the only guild that is present in almost all the Asian communities is that of psammophiles. Other guilds can be found either in a group of communities which are geographically close, or in a single community only.

Spermophilopsis leptodactylus, *Dipus sagitta*, *Eremodipus lichtensteinii*, *Paradipus ctenodactylus*, *Meriones meridianus*, and *Mus musculus* compose the psammophilous guilds in the rodent communities of the Kyzylkum Desert. The differences in the composition of the psammophilous guild between the two Kyzylkum communities can be caused by the among-year variation in species registrations due to sharp density fluctuations of some species (Shenbrot and Rogovin 1995). Obligate sand-dwellers in the Gobi Desert are represented by two *Salpingotus* species, *Dipus sagitta*, *Meriones meridianus*, *Phodopus roborovskii* and *Allocricetulus curtatus*. In addition, some species behave as psammophiles in one community, whilst they do not inhabit sands in the other. For example, *Allactaga sibirica* is psammophilous only in the eastern part of the Gobi Desert. Contrastingly, *Rhombomys opimus* occupies sand over the most of the Gobi Desert except for the eastern part. The Thar Desert and the Negev Highlands each have a single specialized psammophilous species (*Gerbillus gleadowi* and *Gerbillus gerbillus*, res-

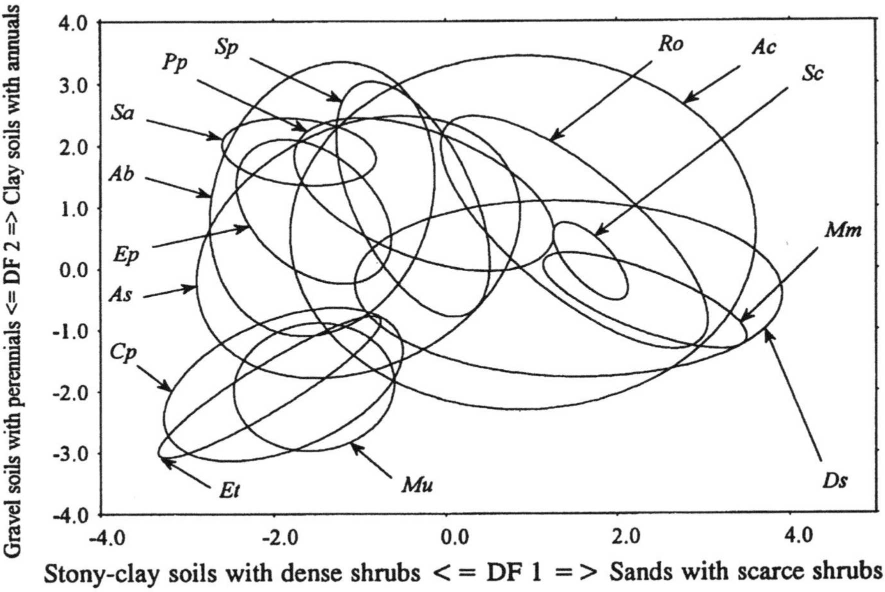

Fig. 8.12. Spatial relationships of niches in the northern Gobi rodent community assemblage. Species names are as follows: Sp *Spermophilus pallidicauda*, Cp *Cradiocranius paradoxus*, Sc *Salpingotus crassicauda*, Ds *Dipus sagitta*, Sa *Stylodipus andrewsii*, Ab *Allactaga bullata*, As *A. sibirica*, Pp *Pygeretmus pumilio*, Mm *Meriones meridianus*, Mu *M. unguiculatus*, Ro *Rhombomys opimus*, Ac *Allocricetulus curtatus*, Ep *Eolagurus przewalskii*, Et *Ellobius tankrei*

pectively). *G. gleadowi* is clearly separated from other rodents in the fauna along DF 1 (Fig. 8.7), while *G. gerbillus* is grouped with species that inhabit sands and sandy-gravel plains (see below).

The guilds of species that occupy loess soils occur among the rodent communities of the western portion of the Great Palaearctic Desert (the Negev Highlands, Thar and Kyzylkum Deserts). These guilds include mainly sciurids and gerbillines. The loess occupants of the Negev Highlands are *Gerbillus dasyurus*, *Psammomys obesus*, *Mus musculus* and *Eliomys melanurus*. Loess habitats of the Thar Desert are inhabited by *Funambulus pennati*, *Tatera indica*, and *Meriones hurrianae*. The composition of the guild in the Kyzylkum Desert is characterized by the presence of *Spermophilus fulvus*, *Meriones erythrourus*, and *Rhombomys opimus*. In addition, *Allactaga elater*, *Cricetulus migratorius*, and *Ellobius tankrei* are included in this guild in the southwestern part of the Kyzylkum Desert.

Species which prefer takyrs and gravel plateau can be grouped into guilds only in the Kyzylkum Desert. The constant component of this guild is represented by *Allactaga severtzovi*, *Allactodipus bobrinskii*, *Pygeretmus pumilio*, and *Jaculus blanfordi*. Interestingly, some facultative loess-dwellers of the

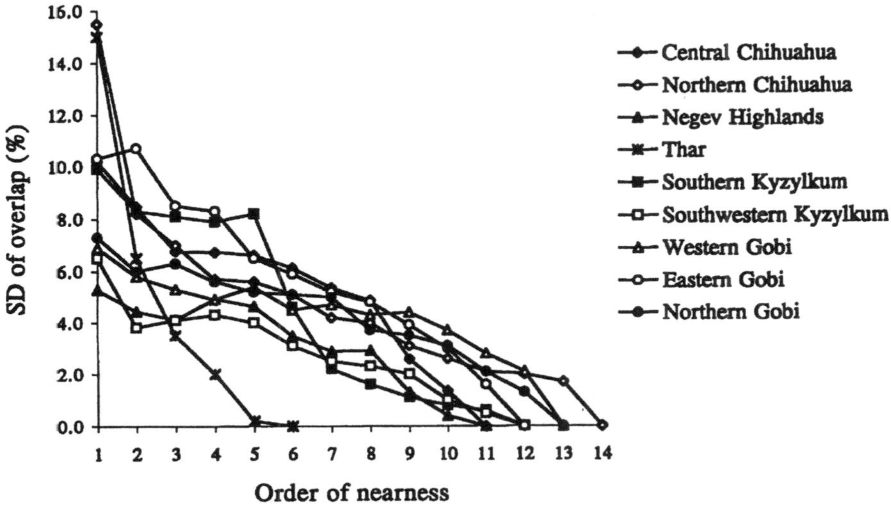

Fig. 8.13. Standard deviations of niche overlaps between first through the nearest neighbors for different communities of desert rodents

southwestern part of the Kyzylkum (*Allactaga elater*, *Cricetulus migratorius*, and *Ellobius tankrei*) become takyr-dwellers in the southern part of this desert.

The guild of species inhabiting sandy gravel and stony ramps is unique to the Gobi Desert. The composition of this guild varies considerably in different parts of the region, but it always includes *Cardiocranius paradoxus* and *Allactaga bullata*. Other species that can be assigned to this guild locally are *Spermophilus pallidicauda*, *Allactaga sibirica*, *Pygeretmus pumilio*, *Stylodipus andrewsii*, and *Eolagurus przewalskii*.

The rodent community of the Negev Highlands has two guilds that have not been found in other Asian deserts. These are a guild of rock-dwellers (*Sekeetamys calurus*, *Acomys cahirinus*, and *A. russatus*) and a guild of species associated with sands and sandy-gravel plains (*Jaculus jaculus*, *Gerbillus gerbillus*, *G. henleyi*, and *Meriones crassus*).

In contrast to the Asian patterns, the rodent fauna of the Chihuahuan Desert cannot be grouped into distinct spatial guilds (Figs. 8.5-8.6). In general, the distributions of species overlap extensively and no distinct clusters are formed (Fig. 8.14).

It can be concluded that communities of desert rodents in different deserts are organized differently in space. In the Asian deserts, species can be clustered into groups based on the spatial overlap of their niches, and these groups are often associated with particular soil types and vegetative structure. This is not the case for the Chihuahuan Desert, and is only true for a single

Central Chihuahua

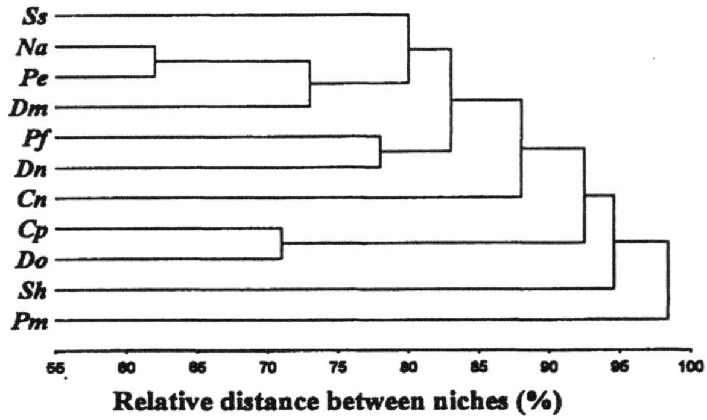

Northern Chihuahua

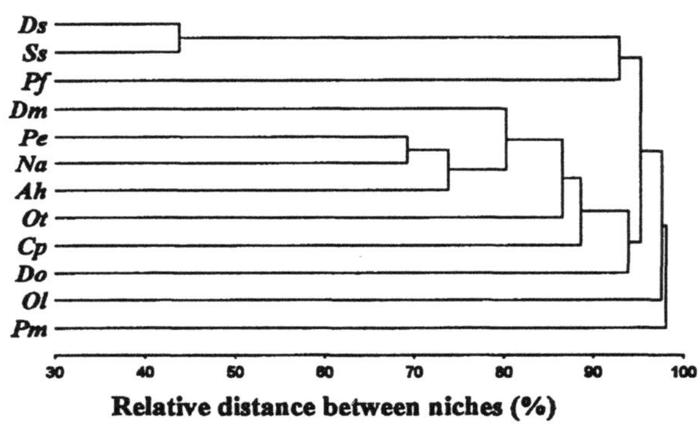

Negev Highlands

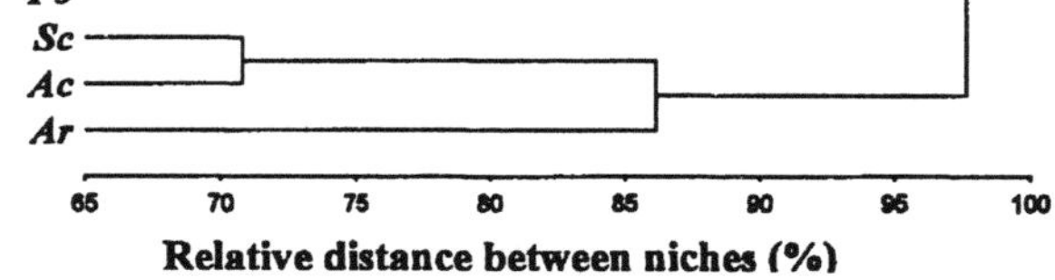

Fig. 8.14. Dendrograms showing niche groupings (spatial guilds) based on the results of UPGMA cluster analyses of spatial niche overlap values for the central Chihuahuan, northern Chihuahuan and Negev Highlands rodent communities. Abbreviations of species names are the same as in Figs. 8.4-8.6

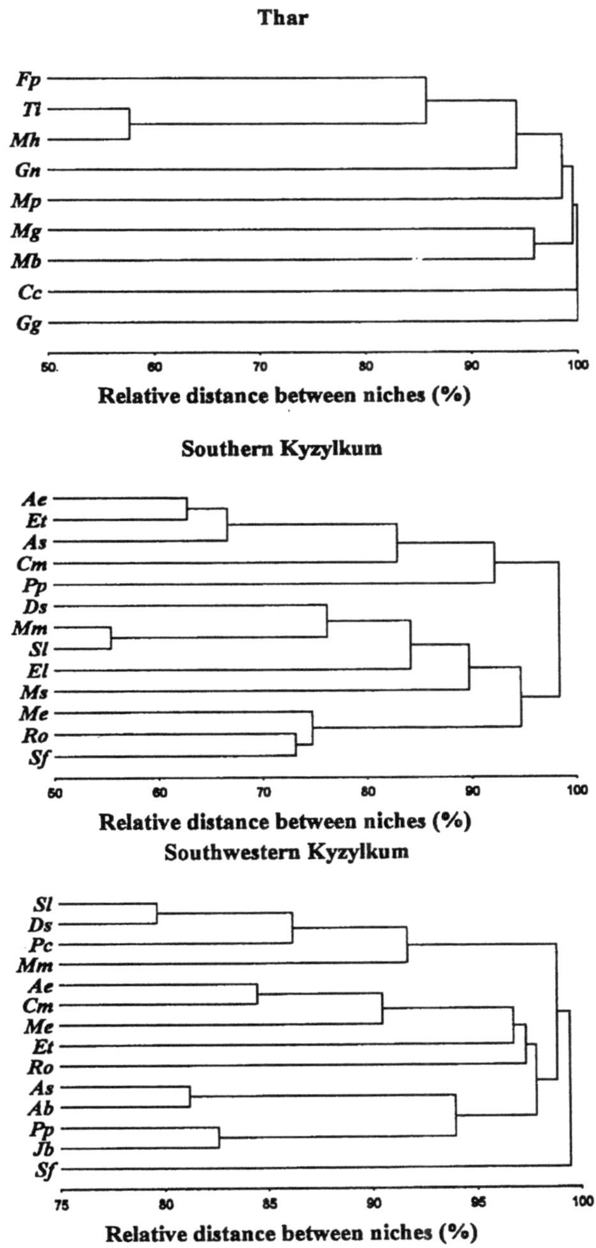

Fig. 8.15. Dendrograms showings niche grouping (spatial guilds) based on the results of UPGMA cluster analyses of spatial niche overlap values for the Thar, southern and southwestern Kyzylkum rodent communities. Abbreviations of species names are the same as in Figs. 8.7-8.9

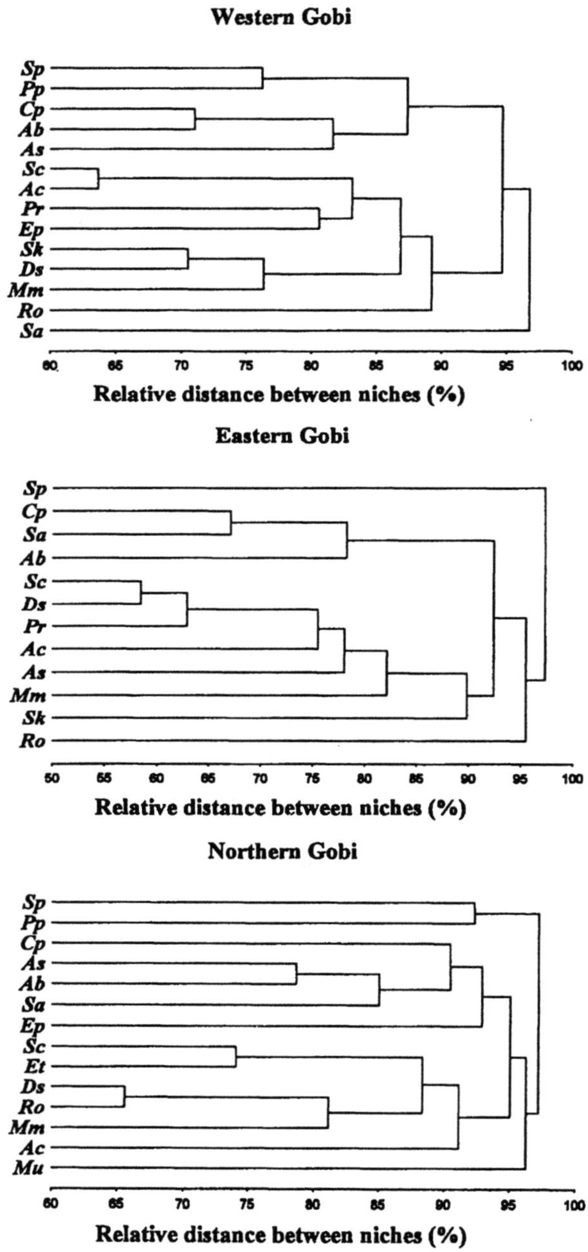

Fig. 8.16. Dendrograms showing niche groupings (spatial guilds) based on the results of UPGMA cluster analyses of spatial niche overlap values for the western, eastern and northern Gobi rodent communities. Abbreviations of species names are the same as in Figs. 8.10-8.12

species in the Thar Desert. This difference implies a more stepwise rate of faunal turnover between habitat types in the Asian deserts.

Many of the differences we have noted between desert rodent communities in Asia and North America can be related to the history of arid land forms and desert habitats on these two continents; the desert landscape in North America is generally younger than that of Asia. In addition, there have been frequent and dramatic contractions and expansions of the desert areas in North America, whereas the major Asian desert landscapes have probably remained relatively extensive since their initial appearance.

The combination of (1) more extensive areas of particular habitat types, such as sand deserts in Asia and northern Africa, (2) greater persistence of desert habitats in Asia than in North America over geological time, and (3) repeated contraction, fragmentation, and re-expansion of desert habitats in North America over geological time, can account for the differences in community organization into spatial guilds. The repeated restriction/expansion, fragmentation, and rejoining of the North American deserts during the Pleistocene, plus a topography with a greater number of barriers, have probably promoted speciation in the desert rodent fauna to a greater extent than in Asian deserts. Thus, the North American rodent genera that are primarily associated with deserts (particularly *Dipodomys*, *Perognathus*, *Chaetodipus* in the family Heteromyidae) tend to be speciose, whereas most genera primarily associated with deserts in Asia contain only one or a few species.

However, this process was not likely to promote greater adaptation to deserts in the North American rodent fauna. In fact, the North American desert rodents in the family Heteromyidae are slightly older than the Asian desert rodent groups. The morpho-ecological types leading to modern kangaroo rats and pocket mice diverged from the ancestral heteromyids in the Oligocene (Hafner and Hafner 1983; Hafner 1993; Wahlert 1993), whereas the origin of jerboas and gerbils is believed to have occurred in the middle Miocene (Pavlinov et al. 1990; Shenbrot et al. 1995). The descent of heteromyids from ancestral savanna forms (see Schmidly et al. 1993), and the repeated occupation of savanna and grassland habitats throughout their evolutionary history, may be one reason why they seem less specialized for deserts in their degree of restriction to desert habitat types. Selection has been more intense, continuous, and uncompromised by trade-offs to those remaining adapted to non-desert habitats in Asia.

The pattern of niche overlap in the Chihuahuan Desert is consistent with the hypothesis that most habitat partitioning occurs at the level of microhabitat use. Many experimental and descriptive studies have shown that species concentrate their activities in different microhabitats (such as under shrubs or in open spaces), and that coexisting species can influence the use of microhabitat by other species (see Reichman and Price 1993; Brown and Harney 1993 for review). A strong difference in foraging microhabitats also occurs in the Asian deserts. Here quadrupedal species are more restricted to areas under cover, and bipedal species are found further from shelter (K.A. Rogovin, unpubl. data). Unfortunately, our data does not allow us to separate macrohabitat from microhabitat partitioning scales so that the patterns of species coexistence in different deserts cannot be compared in this respect. However, even though there is a division into spatial guilds in the Asian desert rodent faunas which is absent in the Chihuahuan Desert, the mean niche overlap between nearest neighbors within each of these groups does not differ. Thus, micro-scale processes of habitat partitioning may still be similar in both these desert regions.

The above-mentioned results demonstrate that every desert rodent community studied has distinguishing characteristics in terms of spatial organization. Does this mean that no generalizations can be made and that every community is unique? On the contrary, the strong similarities noted and our ability to interpret many differences in the light of regional biogeographic histories, implies that readily understandable processes operate in all regions.

8.6 Spatial Patterns of Species Diversity in the Rodent Communities of Different Deserts

Species diversity is one of the most important characteristics of biological communities. Diversity can be considered over a wide spectrum of spatial scales (MacArthur 1965; Whittaker 1975). Two categories of diversity are usually applied to the characterization of communities. These are within-habitat (alpha) and between-habitat (beta) diversities. As has been shown (Chap. 7), the subdivision of a real landscape into habitats is not comparable in different regions. However, diversity characteristics among different communities can be compared if diversity is measured within a continuous resource space for each community.

Because the first two axes of resource space explain, by definition, the maximal proportion of the variance in species composition, measurements of diversity can be restricted to the space formed by these axes. To evaluate the patterns of species diversity along these axes, we calculated probability estimates of abundance for each species at each point of ordination space, based on the assumption of a bivariate normal distribution, together with observational data on maximal species density. As a measure of alpha-diversity, the Shannon index of species diversity ($H = -\Sigma p_i \log p_i$, where p_i is relative abundance of the *i*th species) was used. To evaluate beta-diversity we used species turnover rate (R) between points of ordination space. For determination of R, an approach similar to Whittaker's "half-changes" (Whittaker 1972) was used. As a measure of species turnover rate at point i, we used the equation $R_i = 1/D_{i(0.5)}$, where $D_{i(0.5)}$ is the average distance between point i and surrounding points of 50% similarity in species composition. Similarity was measured as

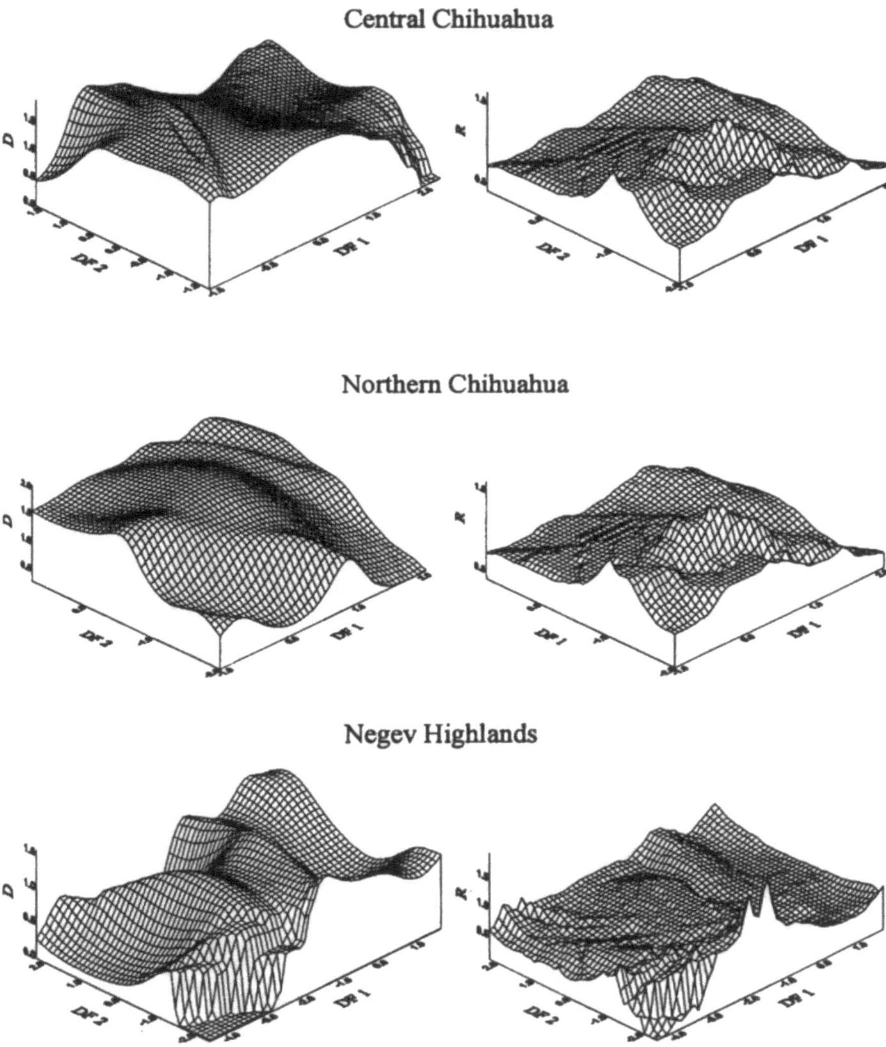

Fig. 8.17. Patterns of alpha-diversity (D) and rate of species turnover (beta-diversity; R) within the space of the two first discriminant axes for the central Chihuahuan, northern Chihuahuan and Negev Highlands rodent communities

$I_s = \min(p_{ik}, p_{jk})$, where p_{ik} is the relative abundance of the kth species at point i and p_{jk} is relative abundance of kth species at point j.

The patterns of diversity change within a resource space in different communities are presented in Figs. 8.17-8.19. The distribution of both parameters is uneven in all communities. The peaks of alpha-diversity corre-

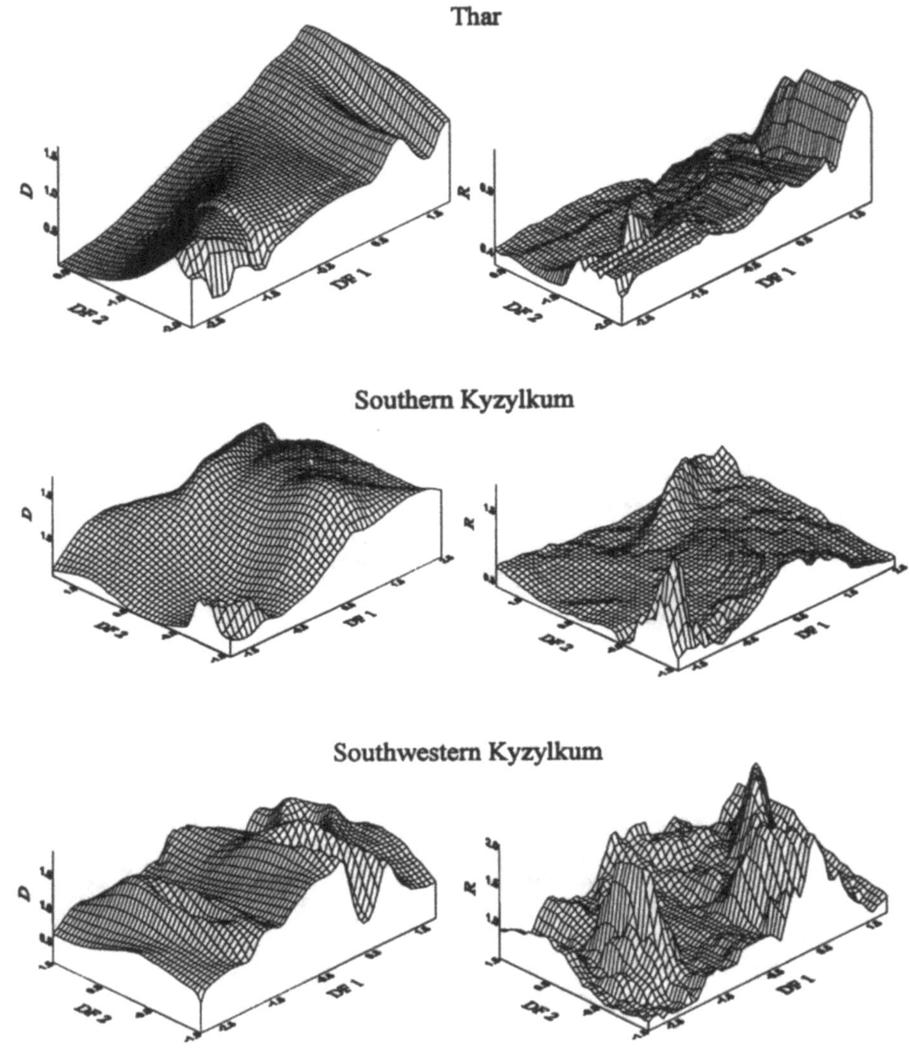

Fig. 8.18. Patterns of alpha-diversity (D) and rate of species turnover (beta-diversity; R) within the space of the two first discriminant axes for the Thar, southern and southwestern Kyzylkum rodent communities

spond either to the most species-rich habitats or to zones of intergradation of adjacent habitats that comprise species from all of them. The depressions of alpha-diversity denote either marginal habitat depauperation or the strong dominance of some species. For example, alpha-diversity of the rodent community of the Negev Highlands has a depression at the positive

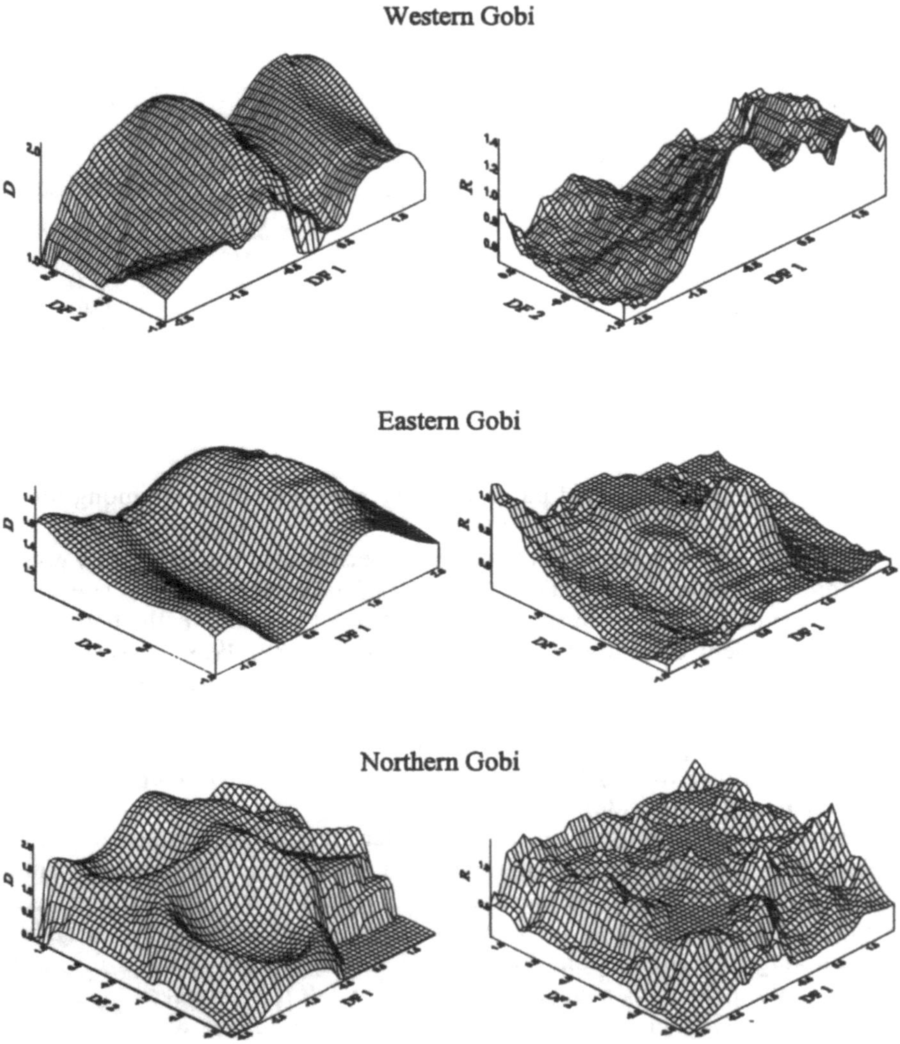

Fig. 8.19. Patterns of alpha-diversity (D) and rate of species turnover (beta-diversity; R) within the space of the two first discriminant axes for the western, eastern and northern Gobi rodent communities

end of DF 1 and zero values of DF 2. This portion of the resource space corresponds to habitats with very high densities of *Psammomys obesus* and *Gerbillus dasyurus* and low abundance of five other species. At the same time, the peaks of alpha-diversity at the positive ends of both axes correspond to a microhabitat having equally high abundance (i.e. low dominance) of four

Table 8.3. Alpha-diversity and rate of species turnover (*R*) in rodent communities in different deserts

Desert	Alpha diversity		*R*	
	Mean	CV	Mean	CV
Central Chihuahua	1.337	27.23	0.720	18.19
Northern Chihuahua	1.444	31.09	0.857	26.95
Negev Highlands	0.770	49.74	0.649	22.80
Thar	0.896	38.06	0.618	30.74
Southern Kyzylkum	1.147	25.98	0.713	30.01
Southwestern Kyzylkum	1.161	23.95	0.983	30.52
Western Gobi	1.579	14.88	0.806	21.34
Eastern Gobi	1.498	13.62	0.659	19.42
Northern Gobi	1.027	54.92	0.789	26.82

species in densely vegetated patches in dry riverbeds flowing among loess hills.

Plateaux and depressions in the rate of species turnover (*R*) signify areas of uniform species composition. In contrast, peaks and ridges correspond with areas of fast change in species composition. For example, the pattern of *R* in the rodent community of the southwestern Kyzylkum Desert has a depression in the central region of DF 1 and ridges on each side of it. The depression denotes the uniformity of the rodent fauna of all the microhabitats of the gravel plateau, whereas the ridges mark the areas of sharp change in species composition between gravel plateaux and either sand dunes (left side) or takyrs (right side).

The levels of both types of diversity differ among different communities (Table 8.3). Alpha-diversity is the lowest in the Negev Highlands and Thar rodent communities and the highest in Chihuahuan and western and eastern Gobi rodent communities. In addition, alpha-diversity is relatively stable within biogeographic regions. The coefficient of variation of alpha-diversity varies without showing a clear geographic trend. It is maximal in the northern Gobi and Negev Highlands communities and minimal in the western and eastern Gobi communities.

The rate of species turnover is lowest in the Thar community and highest in the southwest Kyzylkum community. This parameter is relatively low in the American deserts as well as in the deserts of southern Asia and relatively high in the Turan and central Asian deserts. However, the trend is not clear, and neighboring communities can differ sharply in the level of *R*. The reasons for these differences are either differences in landscape structure (as is the case of the central and northern Chihuahuan Deserts) or the marginal depauperation of some faunas (as is the case of the southern Kyzylkum and eastern Gobi Deserts). *R* is highly variable in the Thar and both Kyzylkum communities, and almost uniform in the central Chihuahuan and eastern Gobi communities. The level of within-community variability in the between-habitat rate of species turnover can serve as a measure of the extent of expression of guild structure. So, it is not surprising that the coefficient of variation of *R* is low in the central Chihuahuan community (see above). It is not clear, however, why the coefficient of variation of *R* is high in the northern Chihuahuan community. The higher variability of *R* in both Kyzylkum communities compared with Gobi communities corresponds to differences in the expression of guild structure between the rodent communities in these regions. A special case is the Thar rodent community. Despite the high level of variability of *R* values, guild structure is not revealed formally there. This may be a result of the real existence of two very distinct spatial guilds, one of which is represented by a single species.

8.7 The Role of Interspecific Competition in Community Structure

Competition holds a central place in community ecology (Southwood 1987). Nevertheless, insufficient data have been collected on competition in natural communities to substantiate this. The direct mode of collecting such data, i.e. by field removal experiments, has revealed a number of examples of pairwise interspecific competitive interactions in natural conditions (see Grant 1972, 1978; Connell 1983; Schoener 1983; Dueser et al. 1989; Gurevitch et al. 1992 for reviews). For example, interspecific competition among desert rodents has been demonstrated by Abramsky and colleagues (Rosenzweig and Abramsky 1986; Abramsky et al. 1991; Ziv et al. 1993; Ovadia and Abramsky 1995) for *Gerbillus andersoni allenbyi* and *Gerbillus pyramidum* in the Negev Desert. In their experiments, *G. a. allenbyi* was found to have no significant effect on *G. pyramidum*, whereas the presence of the latter led to shifts in both microhabitat use and the period of activity of the former. Long-term studies by Brown and colleagues (Munger and Brown 1981; Brown and Munger 1985; Heske et al. 1994) in the northern Chihuahuan Desert demonstrated that the removal of three *Dipodomys* species resulted in an increase in the abundance of small granivorous *Chaetodipus*, *Perognathus*, *Peromyscus*, and *Reithrodontomys*. Short-term removal experiments also revealed a density increase of small quadrupeds (*Perognathus* and *Peromyscus*) after removal of *Dipodomys* (Lemen and Freeman 1983). Other studies demonstrated a microhabitat shift in *Perognathus* and *Microdipodops* after removal of *Dipodomys* (Price 1978; Wondolleck 1978; Harris 1984) and a shift by *Dipodomys* after removal of *Onychomys* (Rebar and Conley 1983). Krasnov (1984) described the pattern of spatial distribution of the house mouse (*Mus musculus*) and the midday jird (*Meriones meridianus*) in habitats of co-occurrence in the northeastern Caspian Lowlands. Mice mainly inhabited patches of dense vegetation from which jirds were absent, and avoided patches occupied by jirds. After jirds had been removed, mice dispersed and populated all possible microhabitats. Furthermore, removal experiments demonstrated that the reproduction of mice was inhibited and their home ranges restricted by the presence of jirds, whereas the number of mice increased and their home ranges expanded when jirds were removed. Laboratory experiments showed that aggressive behavioural interactions were a possible mechanism of spatial segregation between the two species (Krasnov 1983, 1984; Krasnov and Khokhlova 1996). In some cases, removal experiments failed to reveal evidence of interspecific competition between co-occurring species pairs (*Dipodomys ordii – Dipodomys merriami*, Schroeder and Rosenzweig 1975; *Meriones tristrami – Gerbillus andersoni allenbyi*, Abramsky and Sellah 1982). Meserve et al. (1996) found that the results of removal experiments can be both year-specific and season-specific.

Nevertheless, the application of removal experiments is limited by the practical impossibility of performing experimental studies on each pair of species in every habitat. For example, even in the simple case of a community consisting of five species in a single habitat it is necessary to perform 20 experimental treatments with some minimal number of replicates each to estimate a community matrix (i.e. a matrix containing interaction coefficients for each pair of species in a community). It is therefore necessary to develop a technique for the estimation of competition from field census data. The first attempt to measure competition coefficients in the field was made by MacArthur and Levins (1967). The main idea of their approach was that two species compete in proportion to their overlap in resource use. Thus, the degree of niche overlap can be used as a measure of competition. Later it was shown (Abrams 1975) that the results obtained by this technique are distressingly sensitive to assumptions about the shape of the resource utilization curves. As a consequence, the overlap technique is unreliable (Rosenzweig et al. 1984).

Another way to estimate competition from census data based on Lotka-Volterra competition equations was proposed by Schoener (1974). At equilibrium, a competition coefficient can be estimated as the slope of the linear regression of density of one species against the density of the other. Sampling from a variety of different points from an equilibrium population in a homogenous environment should produce a data set from which the value of the competition coefficients can be estimated using linear regression. Hallett and Pimm (1979) demonstrated that the regression method is robust for relatively small deviations from equilibrium.

Crowell and Pimm (1976) expanded this method to take environmental heterogeneity into account. Their technique consisted of two steps: (1) they proposed to correct the data for habitat variation by eliminating the variance which may be due to environmental heterogeneity using stepwise multiple regression analysis; (2) they determined whether the densities of presumed competitors can account for deviations from the habitat regression. The final result of their analysis is expressed in a multiple regression equation describing the relationship between the density of a species versus a set of environmental variables and the density of a competitor; the regression coefficient of the competitor term is the competition coefficient. This method has been used in several studies of small mammal communities (Dueser and Hallett 1980; Hallett 1982; Hallett et al. 1983).

Some modifications of this method were proposed by Rosenzweig et al. (1984). According to these, habitat factors are substituted in the regression analysis by their principal components, or habitat factors are initially used as independent variables to remove variation that is attributable to habitat. The residuals are then regressed against one another as an estimate of interspecific interactions.

Comparison of different modifications of the regression method has revealed quantitative inconsistencies in their results. Moreover, it has been shown that the relative values of coefficients of interaction of two species are determined by the ratio of their average abundances so that rare species always appear to have a stronger influence on common ones than vice versa. It was therefore concluded that this relationship is a statistical artifact (Rosenzweig et al. 1985). These conclusions have been criticized (Pimm 1985; Schoener 1985), and the approach remains in use (Hallett 1991). Recently, Fox and Luo (1996) proposed a procedure for normal standardization of species census data before calculating regression parameters to overcome the artifact mentioned above. They demonstrated that, after the standardization procedure has been applied to the census data, each of the estimated coefficients of interaction matched the observed value well and that the results were consistent with previously published results. This supports the validity of the standardized regression technique.

Calculation of competition coefficients from census data permits us to determine the occurrence of negative interactions between species in a community and to determine the frequency of competitive relationships within communities of desert rodents. In other words, this method allows us to quantify competition as a structuring force in a community. Calculation of competition coefficients for real communities demonstrated that the frequency of competitive interactions varies among different deserts (Table 8.4). The highest frequency of competitive interactions occurs in the northern Palaearctic deserts, whereas in the southern Palaearctic and North American deserts they are relatively less frequent. The extremely high value of competition frequency in sandy habitats of the Negev Desert cannot be considered in this context because the original data did not represent an entire set of the species of a single community but, rather, separate parts of several communities (Rosenzweig et al. 1984).

It should be noted that the determination of competition from census data is estimated under the assumption of an equilibrium state of populations within communities (see above). Almost all of the data presented above

Table 8.4. Occurrence of competitive interactions between pairs of rodent species calculated from census data in different deserts

Desert	Number of possible pairs of interacting species	Number of pairs of competing species	Source
Chihuahua	36	3	Hallett (1982)
Northern Chihuahua	28	4	Our data
Negev (sandy habitats)	7	5	Rosenzweig et al. (1984)
Negev (rocky habitats)	30	6	Rosenzweig et al. (1985)
Negev Highlands	45	4	Our data
Thar	21	4	Our data
Southern Kyzylkum	28	10	Our data
Western Gobi	28	5	Our data
Eastern Gobi	55	18	Our data
Northern Gobi	55	17	Our data

have been collected over a comparatively short period. It is unreasonable to expect that populations of desert rodents would be at an equilibrium at the small time scale due to their high dynamics (see below). If so, the values presented appear to reflect inter-species negative spatial associations, i.e. interference rather than exploitative competition. In this context, it is interesting that in all cases, apart from the Negev and Thar communities, negative interactions are recorded for taxonomically distant species pairs. In all ecological theories it is common to assume that the closer species are taxonomically, the more similar they are in their manner of resource use. Consequently, competition between closely related species should be more intense than that between taxonomically distant species. However, this is not the case. The reason may be that co-evolved congenerics have developed mechanisms that prevent direct interactions. In this case, competition between congenerics can occur in the form of exploitative competition without interference and may not be detected by the regression method. Nevertheless, there is evidence that the importance of competition varies in different deserts. Kelt et al. (in press) analyzed the structure of rodent communities in different deserts of the world. This analysis demonstrated the overriding importance of competition in structuring the North American communities, whereas the role of competition in the Asian communities is much less. The results also showed that the latter communities are either non-interactive or are positively interactive, with species being found more frequently with similar species than expected by chance.

What are the structural consequences of competition in desert rodent communities? Obviously, these comprise changes in the pattern of niche overlap and/or the niche positions (niche shifts) of the competing species. There are three principal models of the effect of competition on these parameters, namely "distinct preferences", "shared preferences" and "centrifugal community organization" (Pimm and Rosenzweig 1981; Rosenzweig and Abramsky 1986; Rosenzweig 1987a, 1991, 1992). These models consider the situation with two competing species distributed in either two (distinct and shared preference models) or three (centrifugal model) habitats. All these models assume density-dependence and describe changes in habitat use with changes in the densities of competing species.

The "distinct preferences" model assumes that both species can exist in both habitats but one species prefers one of the habitats and the other species prefers another. The "shared preferences" model assumes that both competitors can exist in both habitats but that they prefer the same habitat. According to the centrifugal model, each of the species can exist in two of three habitats but the primary habitat preferences of the species are the same, whereas their secondary habitat preferences differ.

The consequences of competition in the "distinct preferences" model are (1) broadening of the niche of a species when its density increases or does not change and the density of a competitor does not change or decreases, (2) contraction of the niche of a species when its density does not change and the density of a competitor increases, and (3) stability of niche breadth when the densities of both species change simultaneously. We failed to find any empirical data on desert rodents supporting the "distinct preferences" model. Rosenzweig (1989) noted that the reason why distinct preferences are rarely found is that they are "haunted by the Ghost of Competition Past" (p. 17). This phrase is usually applied to all forms of competition that have not yet been shown to occur (Connell 1980, but see Rosenzweig 1979b, 1981; Pimm and Rosenzweig 1981). Our observations in the desert of the Negev Highlands show that Wagner's gerbil *Gerbillus dasyurus* and the house mouse *Mus musculus* can, at least partly, demonstrate distinct preferences. Both year-to-year and seasonal fluctuations in the densities of these species are parallel, but the breadth of their niches is stable.

The "shared preferences" model predicts that the dominant species either broadens its niche to use both habitats or contracts it to the optimal habitat only, depending on its own density as well as on density of the subordinate species. The subordinate species changes habitat use from the optimal habitat through both optimal and suboptimal habitats to the suboptimal habitat following changes both in its own density and in the density of the dominant species. In other words, the model predicts density dependent fluctuations in the niche breadth of both species and a niche shift in the subordinate species. The shared preference model of habitat selection fits the data for the pair *Meriones erythrourus* – *Allactaga severtzovi* in the southwestern Kyzylkum Desert with *M. erythrourus* as the dominant species. Both species prefer relatively soft sandy-clay soils with moderately dense vegetation. When the density of *M. erythrourus* increases, *A. severtzovi* shifts into takyrs which are presumably its suboptimal habitats.

The "centrifugal" model describes the narrowing of the niches with a decrease in competition. Each species will be restricted to the core habitat with a decline of the density of the competitor. A desert rodent example fitting this model, is explicitly described by Rosenzweig and Abramsky (1986; (*Gerbillus pyramidum* and *Gerbillus andersoni allenbyi*).

As mentioned above, there is a lot of evidence that competition leads to shifts in habitat/microhabitat distribution and, therefore, affects the spatial structure of desert rodent communities. However, it is unknown to what extent competitive interactions determine the structure of communities. The only rodent community well studied in this aspect is that of the northern Chihuahuan Desert (Munger and Brown 1981; Brown and Munger 1985; Heske et al. 1994), where studies demonstrated the profound role of competition as a structuring force. Nevertheless, it would not be valid to claim that the same processes govern rodent communities throughout the world's deserts.

8.8 Temporal Dynamics of Spatial Structure

The dynamic aspect of animal communities is well known. The main limiting resource in deserts is water, so the annual pattern of rainfall determines the timing of desert rodent breeding and density (Beatley 1969, 1976; Kenagy and Bartholomew 1985; Brown and Zeng 1989). As a result of species-specific responses to environmental variability, the composition of desert rodent communities demonstrates significant temporal fluctuations (Brown 1987; Brown and Kurzius 1989; Brown and Heske 1990).

Long-term field observations have demonstrated that the temporal dynamics of individual species in the same community are not always synchronous. For example, rodent species of the Negev Highlands can be divided into three groups according to the temporal pattern of their density dynamics (Fig. 8.20). Fluctuations in the densities of most species are positively correlated with the level of winter precipitation, whereas fluctuations in density of the golden spiny mouse *Acomys russatus* display the opposite. The densities of the lesser Egyptian jerboa *Jaculus jaculus* and the bush-tailed jird *Sekeetamys calurus* are temporally stable. Analogous results have been reported with the rodent communities of the northern Chihuahuan Desert (Brown and Heske 1990) and the southern border of the Atacama Desert (Meserve et al. 1995).

Competitive coexistence, based on temporal fluctuations in resource availability, population density, and foraging efficiencies has been suggested as an important mechanism of coexistence in desert rodent communities (Brown 1989). However, many ecologists have argued that competition can be considered as a structuring force only under equilibrium conditions, whereas environmental fluctuations over time disturb equilibria and, thus, alter the

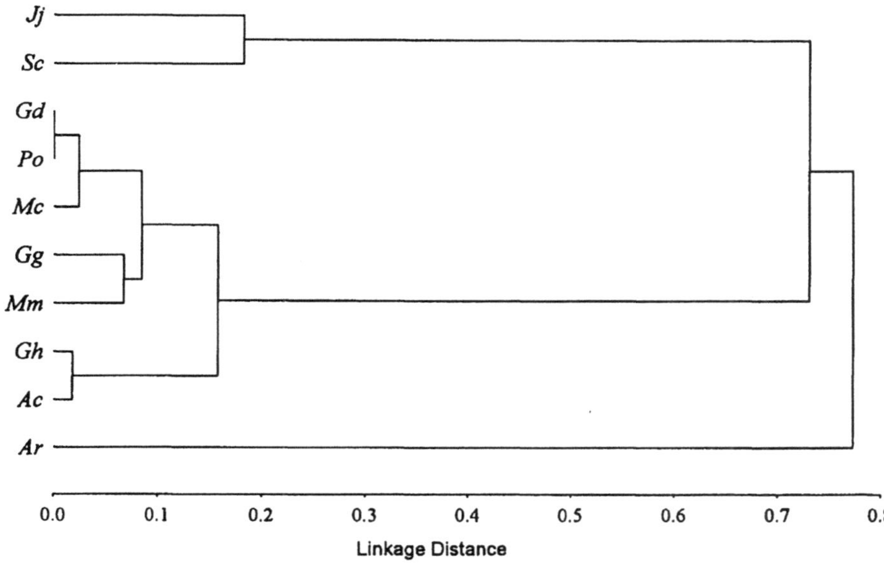

Fig. 8.20. Dendrogram showing rodent species of the Negev Highlands clustered according to similarity in their population dynamics. Species names are as follows: Jj *Jaculus juculus*, Mm *Mus musculus*, Ac *Acomys cahirinus*, Ar *A. russatus*, Gd *Gerbillus dasyurus*, Gg *G. gerbillus*, Gh *G. henleyi*, Mc *Meriones crassus*, Sc *Sekeetamys calurus*, Po *Psammomys obesus*

effects of competition (Hutchinson 1961; Huston 1979; Ricklefs 1987). Proponents of this concept suggest that environmental fluctuations lead to high niche overlap, coexistence of many similar species (Hubbell and Foster 1986) and, consequently, allow for the persistence of rich nonequilibrium communities (Hutchinson 1961). Chesson and Huntly (1997) have recently reexamined the role of environmental fluctuations. They argued that their effect must depend on particular circumstances because environmental fluctuations can provide opportunities for temporal niche partitioning. These opportunities allow species to be separated from each other in their relationships with the specific environmental conditions that occur over time. Species can coexist by having complementary relationships such as hibernation or asynchronous reproduction with the temporally fluctuating environment. Nonetheless, competition shapes these complementary patterns by competitive exclusion within groups of ecologically similar species. Environmental fluctuations do not therefore profoundly change the effect of competition. Considerations regarding the possible role of competition as a structuring force under a fluctuating environment remain theoretical. Nevertheless, our data on the rodent community of the Negev Highlands demonstrate that the interspecific competition which can be recorded from census data occurs only at high densities (Shenbrot and Krasnov 1997).

Table 8.5. Evaluations of habitat breadth and habitat shifts of rodent species at low and high density levels

Species	Habitat breadth			Habitat shift	
	Low density	High density	*t*-value[a]	Distance	*t*-value
Jaculus jaculus (summer)	1.49 <±> 0.15	1.51 <±> 0.01	0.08 ^{ns}	0.23 <±> 0.57	0.40 ^{ns}
J. jaculus (winter)	1.33 <±> 0.07	1.20 <±> 0.04	1.66 ^{ns}	0.31 <±> 0.23	1.39 ^{ns}
Gerbillus dasyurus	1.71 <±> 0.03	1.65 <±> 0.02	1.57 ^{ns}	0.07 <±> 0.13	0.55 ^{ns}
Gerbillus gerbillus	1.13 <±> 0.04	1.14 <±> 0.06	0.16 ^{ns}	0.24 <±> 0.22	1.07 ^{ns}
Gerbillus henleyi (summer)	1.40 <±> 0.07	1.80 <±> 0.09	**3.56****	1.11 <±> 0.35	**3.21****
G. henleyi (winter)	1.46 <±> 0.06	1.58 <±> 0.08	1.46 ^{ns}	0.51 <±> 0.37	1.36 ^{ns}
Meriones crassus	1.84 <±> 0.05	2.04 <±> 0.06	**2.51***	0.73 <±> 0.32	**2.27***
Psammomys obesus	1.26 <±> 0.04	1.43 <±> 0.03	**3.36****	0.16 <±> 0.26	0.62 ^{ns}
Mus musculus	1.18 <±> 0.33	1.17 <±> 0.03	0.03 ^{ns}	0.50 <±> 0.67	0.75 ^{ns}
Acomys cahirinus	0.92 <±> 0.06	1.13 <±> 0.09	1.93 ^{ns}	1.05 <±> 0.45	**2.33***
A. russatus	0.72 <±> 0.19	1.61 <±> 0.14	**3.82*****	0.13 <±> 0.65	0.20 ^{ns}

[a]Significance of *t*-values: * = $P < 0.05$, ** = $P < 0.01$, *** = $P < 0.001$, ^{ns} = not significant.

The processes of density dependent habitat selection may drastically change the niche breadth of co-occurring species on a temporal scale (see Chap. 6). Consequently, the pattern of niche overlap within a community also changes. This, in turn, will lead to changes in the guild structure of a community.

Temporal variation in niche breadth and position is illustrated by rodents in the Negev Highlands community. Values of habitat breadth measured as the standard deviation of a species observation points from its centroid in the ecological space are presented in Table 8.5. Density dependence of niche breadth is recorded in *Gerbillus henleyi* (in summer), *Meriones crassus, Psammomys obesus* and two *Acomys* species. The habitat breadth of other species appears to be density independent. Density dependent habitat shifts are indicated for three species only, namely *M. crassus, A. cahirinus* and *G. henleyi* (in summer). In addition, two species (*Jaculus jaculus* and *G. henleyi*) demonstrate pronounced seasonal differences in habitat distribution, shifting their preferences from open plains in summer to dry riverbeds in winter.

Shenbrot and Rogovin (1995) measured niche breadth and position of rodent species in the southwest Kyzylkum Desert for a period of two years, during which rodent abundance was either very low or very high. *Allactaga elater, Meriones meridianus* and *Meriones erythrourus* demonstrated density dependent changes in niche breadth. The density dependence of niche position was also recorded in *Alactaga severtzovi, Allactodipus bobrinskii, M. meridianus* and *M. erythrourus.*

Niche overlap also fluctuates with density. Figure 8.21 demonstrates the changes in average niche overlap between and within spatial guilds of

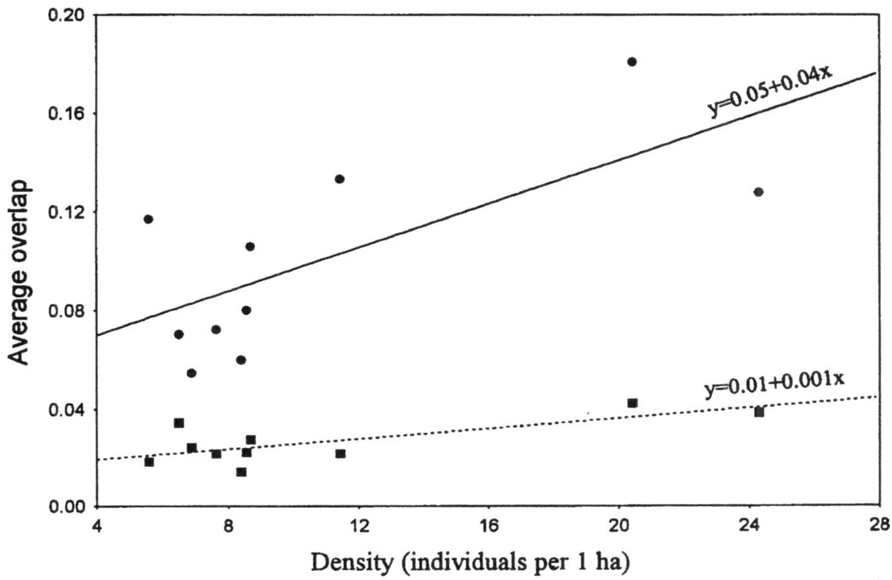

Fig. 8.21. Changes in within-guild (circles, solid line) and between-group (squares, dashed line) average overlap in the rodent community of the Negev Highlands relative to density

rodents in the Negev Highlands community, measured twice a year over 5 consecutive years. Values of both between- and within-guild niche overlap significantly increase with an increase in rodent density. The increase in within-guild overlap is steeper than in between-group overlap, indicating that within-guild relationships are relatively less stable than between-guild relationships. This suggests that guild structure is relatively stable in this community. In fact, it was distinguishable at all levels of rodent density, although expressed at different levels. This, however, appears not to be the rule for the rodent communities of other deserts. Indeed, the 330% decrease in rodent density in the community of the southwestern Kyzylkum Desert led to a 150% decrease in within-guild overlap and a 1400% decrease in between-guild overlap (calculations based on the data from Shenbrot and Rogovin 1995). As a result, guild structure that was clearly expressed at high density disappeared at low density.

These fragmentary data suggest that the spatial structure of a rodent community can be temporally either stable or unstable. Temporal instability may be manifested either in the periodic breakdown of structure (as in the case of the southwestern Kyzylkum Desert) or in the complete reorganization of guild composition, at least hypothetically. Lack of empirical data prevents us from analyzing the effects of temporal stability on spatial structure. Nevertheless, it can be assumed that the rodent community of the Negev Highlands

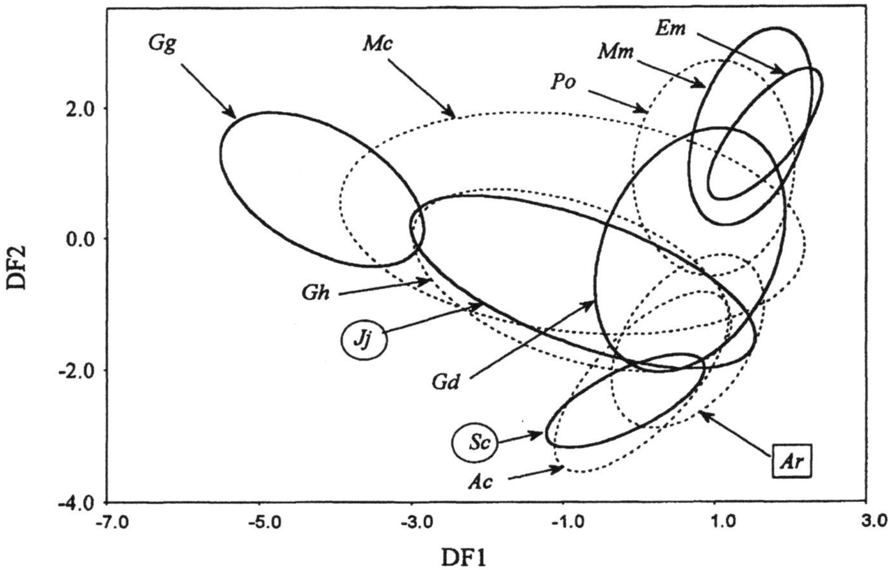

Fig. 8.22. Spatial relationships of niches of density dependent habitat selectors (solid lines) and density independent habitat selectors (dashed lines) in the rodent community of the Negev Highlands. Density changes of species with unmarked names follow the pattern of rainfall. Densities of species with names marked by circles are stable. Density changes of a species with name marked by a square is opposite to the pattern of rainfall. See Fig. 8.5 for abbreviations of species names

is temporally stable, because each guild comprises one or more species with density independent habitat selection (Fig. 8.22). In addition, most species have similar patterns of density dynamics.

The results described above demonstrate that long-term studies provide important perspectives for understanding processes and patterns in desert rodent communities. These studies provide knowledge about the responses not only of individual species, but also of communities as biological entities. Unfortunately, it is difficult to compare our results with those on rodent communities in other geographic regions because there is insufficient information. Data are available on only two other long-term projects. One is the research in the Northern Chihuahuan Desert by Brown and his colleagues (Brown 1987; Brown and Kurzius 1989; Brown and Zeng 1989; Brown and Heske 1990). The other is in semi-arid Chile, conducted by Jaksic and Meserve and colleagues (Jimenez et al. 1992; Jaksic et al. 1993; Meserve et al. 1995, 1996). Both studies concentrate on a single habitat type selected from the whole spectrum and the results are sometimes hard to interpret in terms of community structure.

CHAPTER 9

Conclusions and Perspectives for the Future

9.1 Historical and Ecological Components of Spatial Structure

The structure of a local community is determined by two main components: (1) the pool of species present, i.e. a subset of the regional fauna; (2) the environmental structure as it is perceived by the species. The composition of each regional fauna is determined by the history of its formation. This includes the biogeographic history of a region (invasions and/or extinctions of some taxa) as well as the processes of adaptive evolution of its constituent taxonomic elements. The environmental component of community structure is determined both by historical and current processes. Historical processes include the history of landscape evolution and the evolution of specific responses to this landscape. Current processes include relationships between landscape diversity and species distribution. In other words, community structure is determined by the occurrence of specific environmental diversity on one hand, and by the manner in which the local species perceive this diversity on the other. As a result, the structure of each local community is ecologically specific (the ratio between specialist and generalist species, types of ecological specialization, etc.). Historical processes play an important role in community structure and lead to the diversification of community structure. In contrast, there are common ecological rules governing desert rodent communities that lead to the convergence of community structure. Thus far we do not know the relative importance of these two processes either generally or in each particular case.

What mechanism supports the spatial structure of desert rodent communities? We have a theory that most spatial patterns within a community are determined by interspecific competition that is either currently active or occurred in the past. However, we do not know the degree of importance of competition in most real communities. The second mechanism supporting spatial structure is the relationship between each species and its environment. We do not know which mechanism prevails in each community. In the absence of knowledge concerning the relative importance of the different mechanisms, we have no basis for speculation on how their roles have changed during the evolution of a community.

9.2 Ecological Concepts and Comparative Approach

Observations on desert rodents have served as a basis for many theories and concepts of community ecology generally. At the same time, desert rodents are favorite objects for testing general ecological hypotheses. For example, Brown's (1973, 1975) assembly rule for size classes was based on studies of the seed-eating rodents of the Sonoran and Great Basin deserts. Early ideas about habitat selection have been exemplified by two species of *Dipodomys* in the Sonoran desert (Schroeder and Rosenzweig 1975) and further development of the theory (isocline and isoleg concepts) was achieved by studying gerbils in the Negev Desert (Rosenzweig and Abramsky 1985, 1986; Abramsky et al. 1990, 1991; Abramsky 1994). Concepts combining habitat selection and optimal foraging theories have been applied to studies of desert rodent communities in the Sonoran, Namib and Negev Deserts (Brown 1988, 1989; Hughes et al. 1994; Kotler and Brown 1990; Brown et al. 1992, 1994). Conclusions regarding the role of competition in structuring communities of desert rodents are based almost exclusively on the long-term experiments of Brown and his colleagues (Munger and Brown 1981; Brown and Munger 1985; Heske et al. 1994).

As can be seen from the above, these theoretical outputs are all based largely on data restricted to two regions, the southwestern USA and the Negev Desert of Israel. If these theories concern concrete mechanisms determining the relationships between a species and its environment or between two species, they can be tested in any region using suitable species. However, theoretical constructions considering community organization are valid only for the regions for which they were formulated and tested. This is a result of the geographic specificity mentioned above. Comparison of community organization of the rodents across the world's deserts demonstrates that there are striking regional differences. Some rules, which are true for the North American, Australian and Negev deserts are not valid for the Asian and South American deserts (Kelt et al. 1996). Thus, to date, the available theories can be related only to the North American and the Negev deserts. Comparative studies in other deserts are necessary for decisions as to whether these theories are globally universal or locally unique.

9.3 The Lack of Empirical Data

When writing this book, the first problems we faced involved compiling regional lists of the species of desert rodents and delineating their geographic ranges. The level of knowledge about the world's deserts differs drastically. Some regions have been investigated thoroughly. In these, the geographic ranges of all the rodent species have been studied in detail. It is difficult to expect new findings of either species or localities of species far from the known boundaries of geographic ranges in these deserts. We consider that the deserts of North America, the southern part of the Atacama Desert (Chile), the deserts of Turan and Kazakhstan, the Mongolian Desert, the northwest Arabian deserts (including the Negev, Sinai and western Jordanian deserts), the Thar Desert, and the South African deserts have all been well studied. There are also regions where well-studied spots alternate with large unstudied areas. The probability of finding new species here is low although it exists, and the geographic ranges of the known species may perhaps have to be refined. Understudied regions include the Saharan Desert, most of the Arabian Desert, the desert belt of southwest Asia from the Euphrates to the Indus valleys, most of the desert regions of China, and the South American deserts. The Australian desert is a special case. Here the distribution of the common species has been well studied, but that of numerous rare species is poorly known. Finally, the deserts of the Horn of Africa and of southwestern China (the Takla Makan and Beishan deserts), have not been studied satisfactorily. Species lists can be compiled, but they are undoubtedly full of gaps. Moreover, the details of species distributions in these regions are not clear.

These comments apply to biogeographical and faunal issues. The situation is even worse with regard to community ecology studies. It is easy to list regions for which at least some data about the composition of rodent communities can be found. These regions are the Sonoran, Chihuahuan and Great Basin deserts (the latter only partially) in the USA and Mexico, the southern Atacama in Chile, the northern Kalahari on the border of South Africa and Botswana, the Negev Desert, the Thar Desert, the Kyzylkum Desert in Uzbekistan, the Gobi Desert, and the Great Victoria Desert of Australia.

In many cases, the methodology of these studies is so different that their results are scarcely comparable. Furthermore, most investigations have been short-term. It is not possible to know whether the differences observed are regional or just result from studying different communities in different stages of density dynamics. This emphasizes the necessity for long-term observations on the community structure of desert rodents that are lacking to date.

Community ecologists who are interested in desert rodents still have a great deal of work to do. We hope that this book will serve as a stimulus to its readers to study patterns and processes in animal communities in order to understand better how all aspects of nature are interrelated. We would like to conclude this book with a quote from Professor Karl Rulie of Moscow University (1814-1858) “A phenomenon that exists per se, out of context with other phenomena, is nonsense, which cannot be ever imagined” (Rulie 1954, p. 24).

References

Abrams P (1975) Limiting similarity and the form of the competition coefficient. Theor Popul Biol 8:365-375

Abrams P (1980) Some comments on measuring niche overlap. Ecology 61:44-49

Abrams P (1990a) Mixed responses to resource densities and their implications for character displacement. Evol Ecol 4:93-102

Abrams P (1990b) Adaptive responses of generalist herbivores to competition: convergence or divergence. Evol Ecol 4:103-114

Abramsky Z (1980) Ecological similarity of *Gerbillus allenbyi* and *Meriones tristrami*. J Arid Environ 3:153-160

Abramsky Z (1984) Population biology of *Gerbillus allenbyi* in northern Israel. Mammalia 48:197-206

Abramsky Z (1988) The role of habitat and productivity in structuring desert rodent communities. Oikos 52: 107-114

Abramsky Z (1994) The shape of a *Gerbillus pyramidum* (Rodentia; Gerbillinae) isocline: an experimental field study. Oikos 69:318-326

Abramsky Z, Sellah C (1982) Competition and the role of habitat selection in *Gerbillus allenbyi* and *Meriones tristrami*: a removal experiment. Ecology 63:1242-1247

Abramsky Z, Rosenzweig ML (1984) Tilman's predicted productivity-diversity relationship shown by desert rodents. Nature 309:150-151

Abramsky Z, Brand S, Rosenzweig M (1985a) Geographical ecology of gerbilline rodents in sand dune habitats of Israel. J Biogeogr 12:363-372

Abramsky Z, Rosenzweig M, Brand S (1985b) Habitat selection of Israel desert rodents: comparison of a traditional and a new method of analysis. Oikos 45:79-88

Abramsky Z, Rosenzweig M, Pinshow B, Brown JS, Kotler BP, Mitchell WA (1990) Habitat selection: an experimental field test with two gerbil species. Ecology 71:2358-2369

Abramsky Z, Rosenzweig M, Pinshow B (1991) The shape of a gerbil isocline measured using principles of optimal habitat selection. Ecology 72:329-340

Aitken PF (1968) Observations on *Notomys fuscus* (Wood Jones) (Muridae: Pseudomyinae) with notes on new synonym. S Aust Nat 43:37-45

Alkon PU, Zaltz D (1985) Potatoes and the nutritional ecology of crested porcupines in a desert biome. J Appl Ecol 22:727-737

Alpers CN, Brimhall GH (1988) Middle Miocene climate change in the Atacama desert, northern Chile: evidence from supergene mineralization at La escondida. Geol Soc Am Bull 100:1640-1656

Alvarez T, Alvarez C (1991) Analysis of the rodent fauna from El Cedral, San Luis Potosi, Mexico. An Inst Biol Univ Nac Auton Mex Ser Zool 62:169-179

Alvarez T, Arroyo-Cabrales J (1990) Calculations of capture, density, and home range of three rodent species in an area of human influence in southeastern Durango, Mexico. An Esc Nac Cienc Biol Mex 33:185-210

Alvarez T, Lopez-Vidal JC, Polaco OJ (1988) Study of the nests of the wood rat *Neotoma mexicana* (Rodentia) in the La Michilia Biosphere Reserve, Durango, Mexico. An Esc Nac Cienc Biol Mex 32:131-154

Amarasekare P (1994) Spatial population structure in the banner-tailed kangaroo rat, *Dipodomys spectabilis*. Oecologia 100:166-176

Anstee SD, Roberts JD, Oshea JE (1997) Social structure and patterns of movement of the western pebble-mound mouse, *Pseudomys chapmani*, at Marandoo, western Australia. Wildl Res 24:295-304

Aragon PE, Baudoin C (1989) Some reproductive aspects of two squirrel species of the genus *Spermophilus* (Rodentia: Sciuridae) in a sympatric zone of the Chihuahua desert. Acta Zool Mex Nueva Ser 36:1-25

Armstrong AJ, Van Hensbergen HJ (1996) Small mammals in afforestable montane grasslands of the northern Eastern Cape Province, South Africa. S Afr J Wildl Res 26:11-18

Arrhenius O (1921) Species and area. J Ecol 9:95-99

Ascaray CM, McLachlan A (1990) Home range of *Gerbillurus paeba* in a southern African coastal dunefield. Z Saeugetierkd 55:399-406

Ascaray CM, Perrin M, McLachlan A, Els SF (1991) Population ecology of the hairy-footed gerbil, *Gerbillurus paeba*, in a coastal dunefield of South Africa. Z Saeugetierkd 56:296-305

Atallah SI (1970) A new subspecies of the Golden spiny mouse, *Acomys russatus* (Wagner), from Jordan. Occas Pap Univ Conn (Biol Sci Ser) 1:201-204

Atallah SI (1978) Mammals of the eastern Mediterranean region; their ecology, systematics and zoogeographical relationships. Part 2. Saeugetierkd Mitt 26:1-50

Atmar W, Patterson BD (1993) The measure of order and disorder in the distribution of species in a fragmented habitat. Oecologia 96:373-382

Augustine DJ, Manzon A, Klopp C, Elter J (1995) Habitat selection and group foraging of the springhare, *Pedetes capenis larvalis* Hollistier, in East Africa. Afr J Ecol 33:347-357

Austin ML (1985) Continuum concept, ordination methods and niche theory. Ann Rev Ecol Syst 16:39-61

Axelrod DI (1958) Evolution of the Madro-Tertiary geoflora. Bot Rev 24:433-509

Babaev AG, Drozdov NN, Zonn IS, Freikin ZG (1986) Deserts. Mysl Publ, Moscow (in Russian)

Babaev H, Ataev C (1962) On the occurrence of the Turkmenian jerboa (*Jaculus turkmenicus* Vinogr. et Bond.) in the southern Turkmenistan. Izv AN TSSR Ser Biol Sci N3:76-79 (in Russian)

Bahmanyar M, Lay DM (1975) First record of *Gerbillus henleyi* de Winton, 1903 and records of other rodents from northern Yemen. Mammalia 39:322-324

Baker RH, Shump KA (1978a) *Sigmodon fulviventer*. Mamm Species 94:1-4

Baker RH, Shump KA (1978b) *Sigmodon ochrognathus*. Mamm Species 97:1-2

Bannikov AG (1954) Mammals of the Mongolian People's Republic. USSR Acad Sci Press, Moscow (in Russian)

Barlow JC (1969) Observations on the biology of rodents in Uruguay. R Ont Mus Life Sci Contrib 75:1-59

Bashenina NV (1962) Ecology of the common vole and patterns of its geographic variation. Moscow Univ Press, Moscow (in Russian)

Bates PJJ (1994) The distribution of *Acomys* (Rodentia: Muridae) in Africa and Asia. Isr J Zool 40:199-214

Baudinette RV (1972) The impact of social aggregation on the respiratory physiology of Australian hopping mice. Comp Biochem Physiol 41A:35-38

Beatley JC (1969) Dependence of desert rodents on winter annuals and precipitation. Ecology 50:721-724

Beatley JC (1976) Environments of kangaroo rats (*Dipodomys*) and effects of environmental change on populations in southern Nevada. J Mammal 57:67-93

Begg RJ (1981) The small mammals of Little Nourlangie Rock, NT. IV. Ecology of *Zyzomys woodwardi*, the large rock-rat, and *Z. argurus*, the common rock-rat (Rodentia: Muridae). Aust Wildl Res 8:307-320

Bekenov AB, Eszhanov B (1984) Foraging of long-clawed ground squirrel in north Kyzylkum and south Balkhash. Bull Moscow Nat Soc Sec Biol 89:20-26 (in Russian)

Belabbas S, Butet A (1994) The diet of the merione, *Meriones shawi*, in the nature reserve of Mergueb, Algeria. Pol Ecol Stud 20:293-303

Belk MC, Smith HD (1990) *Ammospermophilus leucurus*. Mamm Species 368:1-8
Belovsky GE (1997) Optimal foraging and community structure: the allometry of herbivore food selection and competition. Evol Ecol 11:641-672
Ben Shahar R, Skinner JD (1988) Habitat preferences of African ungulates derived by univariate and multivariate analyses. Ecology 69:1479-1485
Berger PJ, Negus NC, Rowsemitt CN (1987) Effect of 6-methoxybenzoxazolinone on sex ratio and breeding performance in *Microtus montanus*. Biol Reprod 35:355-360
Berman SL (1985) Convergent evolution in the hind limb of bipedal rodents. Z Zool Syst Evol 23:1-80
Berry RJ (ed) (1981) Biology of the house mouse. Symp Zool Soc Lond 47:1-715
Best TL (1988a) *Dipodomys spectabilis*. Mamm Species 311:1-10
Best TL (1988b) *Dipodomys nelsoni*. Mamm Species 326:1-4
Best TL (1994) *Chaetodipus nelsoni*. Mamm Species 484:1-6
Best TL (1995) *Spermophilus mohavensis*. Mamm Species 509:1-7
Best TL, Skupski MP (1994) *Perognathus flavus*. Mamm Species 471:1-10
Best TL, Hildreth NJ, Jones C (1989) *Dipodomys deserti*. Mamm Species 339:1-8
Best TL, Lewis CL, Caesar K, Titus AS (1990a) *Ammospermophilus interpres*. Mamm Species 365: 1-6
Best TL, Titus AS, Caesar K, Lewis CL (1990b) *Ammospermophilus harrisii*. Mamm Species 366: 1-7
Best TL, Titus AS, Lewis CL, Caesar K (1990c) *Ammospermophilus nelsoni*. Mamm Species 367: 1-7
Best TL, Skupski MP, Smartt RA (1993) Food habits of sympatric rodents in the shinnery oak-mesquite grasslands of southeastern New Mexico. Southwest Nat 38:224-235
Birkenstock PJ, Nel JAJ (1977) Laboratory and field observations on *Zelotomys woosnami* (Rodentia: Muridae). Zool Afr 12:429-443
Black CC, Krishtalka L (1986) Rodents, bats and insectivores from the Plio-Pleistocene sediments to the east of Lake Turkana, Kenya. Contrib Sci Nat Hist Mus Los Ang County 372:1-15
Bobrinsky NA, Kuznetsov BA, Kuziakin AI (1965) A guide to mammals of the USSR. Prosvestchenye Press, Moscow (in Russian)
Bobrov VV, Neronov VM (1988) Zoogeographic analysis of rodent fauna of Turan. In: Kucheruk VV (ed) Problems of theriology. General and regional theriogeography. Nauka Press, Moscow, pp 238-246 (in Russian)
Bonaccorso FJ, Brown JH (1972) House construction of the desert woodrat, *Neotoma lepida lepida*. J Mammal 53:283-288
Borghi CE, Giannoni SM, Martinez Rica JP (1994) Habitat segregation of three sympatric fossorial rodents in the Spanish Pyrenees. Z Saeugetierkd 59:52-57
Bowers MA (1982) Foraging behavior of heteromyid rodents: field evidence of resource partitioning. J Mammal 63:361-367
Bowers MA, Brown JH (1982) Body size and coexistence in desert rodents: chance or community structure. Ecology 63:391-400
Boyer D (1985) Interactions of two *Gerbillurus* spp. in the Namib. S Afr J Sci 81:704
Bradley AJ, Kemper CM, Kitchener DJ, Humphreys WF, How RA, Schmitt LH (1988) Population ecology and physiology of the common rock rat, *Zyzomys argurus* (Rodentia: Muridae) in tropical northwestern Australia. J Mammal 69:749-764
Bradley WG (1968) Food habits of the antelope ground squirrel in southern Nevada. J Mammal 94:14-21
Bradley WG, Mauer RA (1971) Reproduction and food habits of Merriam's kangaroo rat, *Dipodomys merriami*. J Mammal 52:497-507
Branch LC (1993) Social organization and mating system of the plains viscacha (*Lagostomus maximus*) J Zool (Lond) 229:473-491
Branch LC, Villarreal D, Fowler GS (1994a) Factors influencing population dynamics of the plains viscacha (*Lagostomus maximus*, Mammalia, Chinchillidae) in scrub habitat of central Argentina. J Zool (Lond) 232:383-395
Branch LC, Villarreal D, Sbriller AP, Sosa RA (1994b) Diet selection of the plains viscacha (*Lagostomus maximus*, family Chinchillidae) in relation to resource abundance in semi-arid scrub. Can J Zool 72:2210-2216

Braun JK, Mares MA (1995) A new genus and species of phyllotine rodent (Rodentia: Muridae: Sigmodontinae: Phyllotini) from South America. J Mammal 76: 504-521

Breed WG (1992) Reproduction of the spinifex hopping mouse (*Notomys alexis*) in the natural environment. Aust J Zool 40:57-71

Briones MJI, Mascato R, Mato S (1992) Relationships of earthworms with environmental factors studied by means of detrended canonical correspondence analysis. Acta Oecol 13:617-626

Bronson FH (1989) Mammalian reproductive biology. Chicago Univ Press, Chicago

Brown JH (1973) Species diversity of seed-eating desert rodents in sand dune habitats. Ecology 54:775-787

Brown JH (1975) Geographical ecology of desert rodents. In: Cody ML, Diamond JM (eds) Ecology and evolution of communities. Harvard Univ Press, Harvard, pp 315-341

Brown JH (1984) On the relationship between abundance and distribution of species. Am Nat 124:255-279

Brown JH (1987) Variation in desert rodent guilds: patterns, processes, and scales. In: Gee JHR, Giller PS (eds) Organization of communities past and present: the 27th Symp British Ecological Society, Aberystwyth 1986. Blackwell, Oxford, pp 195-203

Brown JH (1995) Macroecology. Univ Chicago Press, Chicago

Brown JH, Harney BA (1993) Population and community ecology of heteromyid rodents in temperate habitats. In: Genoways HH, Brown JH (eds) Biology of Heteromyidae. Spec Publ No 10. Am Soc Mammal, Brigham Young University, Provo, Utah, pp 118-151

Brown JH, Heske EJ (1990) Temporal changes in a Chihuahuan desert rodent community. Oikos 59:290-302

Brown JH, Kurzius MA (1987) Composition of desert rodent faunas: combinations of coexisting species. Ann Zool Fenn 24:227-237

Brown JH, Kurzius MA (1989) Spatial and temporal variation in guilds of North American desert rodents. In: Morris DW, Abramsky Z, Fox BJ, Willig MR (eds) Patterns in the structure of mammalian communities. Spec Publ Mus Texas Tech Univ No 28. Texas Tech Univ Press, Lubbock, Texas, pp 71-90

Brown JH, Lieberman GA (1973) Resource utilization and coexistence of seed-eating desert rodents in sand dune habitats. Ecology 54:788-797

Brown JH, Munger JC (1985) Experimental manipulation of a desert rodent community: food addition and species removal. Ecology 66:1545-1563

Brown JH, Nicoletto PF (1991) Spatial scaling of species composition: body masses of North American land mammals. Am Nat 138:1478-1512

Brown JH, Zeng Z (1989) Comparative population ecology of eleven species of rodents in the Chihuahuan desert (Arizona, USA). Ecology 70:1507-1525

Brown JH, Davidson DW, Munger JK, Inouye RS (1986) Experimental community ecology: the desert granivore system. In: Diamond J, Case TJ (eds) Community ecology. Harper and Row, New York, pp 41-61

Brown JH, Marquet PA, Taper ML (1993) Evolution of body size: consequences of an energetic definition of fitness. Am Nat 142:573-584

Brown JS (1988) Patch use as an indicator of habitat preference, predation risk, and competition. Behav Ecol Sociobiol 22:37-47

Brown JS (1989) Desert rodent community structure: a test of four mechanisms of coexistence. Ecol Monogr 59:1-20

Brown JS, Arel Y, Abramsky Z, Kotler BP (1992) Patch use by gerbils (*Gerbillus allenbyi*) in sandy and rocky habitats. J Mammal 73:281-289

Brown JS, Kotler BP, Mitchell WA (1994) Foraging theory, patch use, and the structure of a Negev desert granivore community. Ecology 75:2286-2300

Buffenstein R (1984) The importance of microhabitat in thermoregulation and thermal conductance in two Namib rodents – a crevice dweller, *Aethomys namaquensis*, and a burrow dweller, *Gerbillurus paeba*. J Therm Biol 9:235-241

Butinsky TM (1973) Life history and economic value of the springhare (*Pedetes capensis* Forster) in Botswana. Botswana Not Rec 5:200-213

Butinsky TM (1979) Reproductive ecology of the springhare, *Pedetes capensis*, in Botswana. J Zool (Lond) 189:221-232

Calaby JH, Taylor JM (1983) Breeding in wild populations of the Australian rock rats, *Zyzomys argurus* and *Z. woodwardi*. J Mammal 64:610-616

Calder WA (1984) Size, function, and life history. Harvard Univ Press, Cambridge

Cameron GN, Rainey DC (1972) Habitat utilization by *Neotoma lepida* in the Mohave desert. J Mammal 53:251-266

Carleton MD, Musser GG (1984) Muroid rodents. In: Anderson S, Jones JK (eds) Orders and families of recent mammals of the world. John Wiley, New York, pp 289-379

Carnes BA, Slade NA (1982) Some comments on niche analysis in canonical space. Ecology 63:888-893

Carstairs JL (1976) Population dynamics and movement of *Rattus vilosissimus* (Waite) during the 1966-69 plague of Brunette Downs, Northern Territory. Aust Wildl Res 3:1-10

Cassini MH, Foeger B (1995) The effect of food distribution on habitat use of foraging hedgehogs and the ideal non-territorial despotic distribution. Acta Oecol 16:657-669

Chabovsky AV, Popov SV (1994) Interspecific relations of three species of gerbils in southeast Turkmenia. Bull Moscow Nat Soc Sec Biol 99:30-38 (in Russian)

Channing A (1984) Ecology of the namtap *Graphiurus ocularis* (Rodentia: Gliridae) in the Cedarberg, South Africa. S Afr J Zool 19:144-149

Chapman BR, Packard RL (1974) An ecological study of Merriam's pocket mouse in southern Texas. Southwest Nat 19:281-291

Charnov EL (1976) Optimal foraging: the marginal value theorem. Theor Popul Biol 9:129-136

Charnov EL (1993) Life history invariants: some explanations of symmetry in evolutionary ecology. Oxford Univ Press, New York

Chesson P, Huntly N (1997) The roles of harsh and fluctuating conditions in the dynamics of ecological communities. Am Nat 150:519-553

Christian DP (1978) Effects of humidity and body size on evaporative water loss in three desert rodents. Comp Biochem Physiol 60A:425-430

Christian DP (1979) Physiological correlates of demographic patterns in three sympatric Namib desert rodents. Physiol Zool 52:329-339

Cloudsley-Thompson JL (1991) Ecophysiology of desert arthropods and reptiles. Springer, Berlin Heidelberg New York

Cloudsley-Thompson JL (1993) The adaptational diversity of desert biota. Environ Conserv 20:227-231

Cloudsley-Thompson JL (1996) Biotic interactions in arid lands. Springer, Berlin Heidelberg New York

Cockrum EL (1960) The recent mammals of Arizona: their taxonomy and distribution. Univ Arizona Press, Tucson, Arizona

Cockrum EL (1977) Status of the hairy-footed gerbil, *Gerbillus latastei* Thomas and Trouessart. Mammalia 41:75-80

Cody ML (1974) Competition and structure in bird communities. Princeton Univ Press, Cambridge, Massachusetts

Cody ML, Walter H (1976) Habitat selection and interspecific interactions among Mediterranean sylviid warblers. Oikos 27:210-238

Coetzee CG (1969) The distribution of mammals in the Namib Desert and adjoining inland escarpment. Sci Pap Namib Desert Res Stn 40:23-36

Connell JH (1980) Diversity and the coevolution of competition, or the ghost of competition past. Oikos 35:131-138

Connell JH (1983) On the prevalence and relative importance of interspecific competition: evidence from field experiments. Am Nat 122:661-696

Connor EF, Simberloff D (1979) The assembly of species communities: Chance or competition? Ecology 60:1132-1140

Connor EF, Simberloff D (1984) Neutral models of species' co-occurrence patterns. In: Strong DR Jr, Simberloff D, Abele LG, Thistle AB (eds) Ecological communities: conceptual issues and the evidence. Princeton Univ Press, Princeton, New Jersey, pp 316-331

Contreras JR (1968) *Akodon molinae* una nueva especie de raton de campo del sur de la provincia de Buenos Aires. Zool Platense 1:9-12

Contreras JR, Torres-Mura JC, Yañez JL (1987) Biogeography of octodontid rodents: an eco-evolutionary hypothesis. Fieldiana Zool 39:401-411

Cooke R, Warren A, Goudie A (1993) Desert geomorphology. Univ College London Press, London

Corbet GB, Hill JE (1992) The mammals of the Indomalayan Region: a systematic review. Oxford Univ Press, New York

Cornely JE, Baker RJ (1986) *Neotoma mexicana*. Mamm Species 262:1-7

Cramer KL, Chapman JA (1990) Reproduction of three species of pocket mice (*Perognathus*) in the Bonneville Basin, Utah (USA). Great Basin Nat 50:361-366

Crowell KL, Pimm SL (1976) Competition and niche shifts of mice introduced onto small islands. Oikos 27:251-258

Currie DJ, Paquin V (1987) Large-scale biogeographical patterns of animal- and plant-species richness. Nature 329:326-327

Dalby PL (1975) Biology of pampa rodents. Michigan State Univ Mus Publ, Biol Ser 5:149-2722

Dalby PL, Mares MA (1974) Notes on the distribution of the coney rat, *Reithrodon auritus*, in northwestern Argentina. Am Midl Nat 92:205-206

Daly M (1979) On Libyan jirds and fat sand rats. Nat Hist 88:64-70

Daly M, Daly S (1975a) Socio-ecology of Saharan gerbils, especially *Meriones libycus*. Mammalia 39:289-312

Daly M, Daly S (1975b) Behaviour of *Psammomys obesus* (Rodentia: Gerbillinae) in the Algerian Sahara. Z Tierpsychol 37:289-321

Darlington PJ (1957) Zoogeography. John Wiley, New York

Darviche D, Orsini P (1982) Criteres de differenciation morphologique et biometrique deux especes de Souris sympatriques: *Mus spretus* et *Mus musculus domesticus*. Mammalia 46:205-217

Davies KC, Jarvis JUM (1986) The burrow systems and burrow dynamics of the mole-rats *Bathyergus suillis* and *Cryptomys hottentotus* in the fynbos of the south-western Cape, South Africa. J Zool (Lond) 209:125-147

Davis R, Dunford C (1987) An example of contemporary colonization of a montane island by small, nonflying mammals in the American southwest. Am Nat 129:398-406

Dawson TJ (1976) Energetic cost of locomotion in Australian hopping mice. Nature 259:305-307

Dawson TJ, Tailor CR (1973) Energy cost of locomotion in kangaroos. Nature 246: 313-314

Degen AA (1996) Ecophysiology of small desert mammals. Springer, Berlin Heidelberg New York

Degen AA, Kam M, Jurgrau D (1988) Energy requirements of fat sand rats (*Psammomys obesus*) and their efficiency of utilization of the saltbush *Atriplex halimus* for maintenance. J Zool (Lond) 215:443-452

De Graaff G (1981) The rodents of Southern Africa. Butterworths, Durban, Pretoria

Delany MJ, Farook SMS (1989) The small mammals of a coastal gravel plain in the Sultanate of Oman. J Zool (Lond) 218:319-321

De Meeûs T, Michalakis Y, Renaud F, Olivieri I (1993) Polymorphism in heterogeneous environments, evolution of habitat selection and sympatric speciation: soft and hard selection models. Evol Ecol 7:175-198

De Moor PP (1969) Seasonal variation in local distribution, age classes and population density of the gerbil, *Tatera brantsii*, on the South Africa highveld. J Zool (Lond) 157:399-411

Dempster ER (1955) Maintenance of genetic heterogeneity. Cold Spring Harbor Symp Quant Biol 20:25-32

Denys C (1994) Diet and dental morphology of two coexisting *Aethomys* species (Rodentia) in Mozambique. Implications for diet reconstruction in related extinct species from South Africa. Acta Theriol 39:357-364

Dewsbury DA, Lanier DL, Miglietta A (1980) A laboratory study of climbing behavior in 11 species of muroid rodents. Am Midl Nat 103:66-72

Dial KP (1988) Three sympatric species of *Neotoma*: Dietary specialization and coexistence. Oecologia 76:531-537

Diamond JM (1975) Assembly of species communities. In: Cody ML, Diamond JM (eds) Ecology and evolution of communities. Belknap Press, Harvard Univ Press, Cambridge, Massachusetts, pp 342-444

Diamond JM (1986) Overview: laboratory experiments, field experiments, and natural experiments. In: Diamond J, Case TJ (eds) Community ecology. Harper and Row, New York, pp 3-23

Diamond JM, Gilpin ME (1982) Examination of the "null" model of Connor and Simberloff for species co-occurrences on islands. Oecologia 52:64-74

Dice LR (1947) Effectiveness of selection by owls of deer mice (*Peromyscus maniculatus*) which contrast in color with their background. Contributions from the Laboratory of Vertebrate Biology, N 34, University of Michigan, Ann Arbor

Dickman CR, Predavec M, Downey FJ (1995) Long-range movements of small mammals in arid Australia: Implications for land management. J Arid Environ 31:441-452

Digby PGN, Kempton RA (1985) Multivariate analysis of ecological communities. Chapman and Hall, London

Din V, Anand R, Boursot P, Darviche D, Dod B, Jouvin M, Orth A, Talwar GP, Cazenave PA, Bonhomme F (1996) Origin and radiation of the house mouse: clues from nuclear genes. J Evol Biol 9:519-539

Dmitriev AI (1981) Craniometric variability and causes of extinction of *Eolagurus luteus* in the Volga-Ural interfluve. Zool Zh 60:1841-1848 (in Russian)

Dolnik VR (1986) Study of animals behavior by analysis of their time and energy budgets. In: Zykova LY, Panov EN (eds) Methods in ecology and ethology. Institute of Animal Evolutionary Morphology and Ecology, Pustchino, pp 94-106 (in Russian)

Downs CT, Perrin MR (1989) An investigation of the macro- and micro-environments of four *Gerbillurus* species. Cimbebasia 11:41-54

Downs CT, Perrin MR (1991) Urinary concentrating ability of four *Gerbillurus* spp. of southern African arid regions. J Arid Environ 20:71-82

Downs CT, Perrin MR (1996) The thermal biology of southern Africa's smallest rodent, *Mus minutoides*. S Afr J Sci 92:282-285

Drabek CM (1973) Home range and daily activity of the round-tailed ground squirrel, *Spermophilus tereticaudus neglectus*. Am Midl Nat 89:287-293

Drake JA (1990) Communities as assembled structures: do rules govern patterns? Trends Ecol Evol 5:159-164

Dregne HE (1976) Soils of arid regions. Elsevier, Amsterdam

Dubrovsky YA (1978) Jirds and natural nidi of the cutaneous leishmaniasis. Nauka Press, Moscow (in Russian)

Dueser RD, Hallett JG (1980) Competition and habitat selection in a forest-floor small mammal fauna. Oikos 35:293-297

Dueser RD, Shugart HH (1979) Niche pattern in a forest-floor small-mammal fauna. Ecology 55:73-83

Dueser RD, Porter JH, Dooley JL (1989) Direct tests for competition in North American rodent communities: synthesis and prognosis. In: Morris DW, Abramsky Z, Fox BJ, Willig MR (eds) Patterns in the structure of mammalian communities. Spec Publ Mus Texas Tech Univ, No 28. Texas Tech Univ Press, Lubbock, Texas, pp 105-125

Dunford C (1977) Social system of round-tailed ground squirrels. Anim Behav 25:885-906

Dunlop JN, Pound IR (1981) Observations on the pebble-mound mouse, *Pseudomys chapmani* Kitchener, 1980. Rec West Aust Mus 9:1-5

Du Plessis A, Kerley GI (1991) Refuge strategies and habitat segregation in two sympatric rodents *Otomys unisulcatus* and *Parotomys brantsii*. J Zool (Lond) 224:1-10

Du Plessis A, Kerley GI, Winter PE (1992) Refuge microclimates of rodents: A surface nesting *Otomys unisulcatus* and a burrowing *Parotomys brantsii*. Acta Theriol 37:351-358

Ebensperger LA, Simonetti JA (1996) Microclimate and microhabitat selection in nocturnal rodents of central Chile: a test between biophysical and experimental approaches. Mammalia 60:195-200

Eigelis YK (1980) Rodents of the eastern Caucasus and the problems of sanitation of the local plague nidi. Saratov State Univ Press, Saratov (in Russian)

Eisenberg JF (1963) The behavior of heteromyid rodents. Univ California Publ Zool 69:1-100
Eisenberg JF (1967) A comparative study in rodent ethology with emphasis on evolution of social behavior. Proc US Natl Mus 122:1-51
Eisenberg JF (1981) The mammalian radiations. Univ Chicago Press, Chicago
El Hilali M, Veillat JP (1975) *Jaculus orientalis*: a true hibernator. Mammalia 39:401-404
Elton CS (1927) Animal ecology. Sidgewick and Jackson, London
Elton CS, Miller RS (1954) The ecological survey of animal communities: with a practical system of classifying habitats by structural characteristics. J Ecol 42:460-496
Emlen JM (1966) The role of time and energy in food preference. Am Nat 100:611-617
Ernest KA, Mares MA (1987) *Spermophilus tereticaudus*. Mamm Species 274:1-9
Eshelkin II (1976) On the reproduction of *Alticola streltzovi* in the southeast Altai. Zool Zh 55:437-442
Evenari M (ed) (1985) Hot deserts and arid shrublands. Elsevier, Amsterdam
Feinsigner P, Spears EE, Poole RW (1981) A simple measure of niche breadth. Ecology 62:27-32
Felsenstein J (1979) Excursions along the interface between disruptive and stabilizing selection. Genetics 93:773-795
Field AC (1975) Seasonal changes in reproduction, diet and body composition of two equatorial rodents. East Afr Wildl J 13:221-235
Filippucci MG, Rodino E, Nevo E, Capanna E (1988) Evolutionary genetics and systematics of the garden dormouse, *Eliomys* Wagner, 1840. 2. Allozyme diversity and differentiation of chromosomal races. Boll Zool 55:47-54
Flint VE (1960) On the biology of the Roborovsky's hamster. Bull Moscow Nat Soc Sec Biol 65:98-101 (in Russian)
Flint VE (1966) On the biology of the long-tailed hamster. Zool Zh 45:471-474 (in Russian)
Flint VE, Golovkin AN (1961) Essay on the comparative ecology of the hamsters in Tuva. Bull Moscow Nat Soc Sec Biol 66:57-77 (in Russian)
Fokin IM (1963) Peculiarities of jerboas' bipedal run. Bull Moscow Nat Soc Sec Biol 68:22-28 (in Russian)
Fokin IM (1978) Locomotion and morphology of locomotor apparatus of jerboas. Nauka Press, Leningrad (in Russian)
Formozov AN (1929) Mammals of northern Mongolia according to collections of the expedition of 1926. In: Preliminary report of the zoological expedition to northern Mongolia in 1926. Materials of the commission for the study of the Mongolian and Tannu-Tuva People's Republics and Buriat-Mongolian ASSR. N 3. Acad Sci USSR Press, Leningrad, pp 1-144 (in Russian)
Formozov AN (1938) Notes on the extinction of some steppe rodent species in the late Quaternary and recent times. Zool Zh 17:260-270 (in Russian)
Formozov AN (1956) Biological forms of animals in arid and semiarid regions of Middle and Central Asia. In: Problems in geography. Selection of papers for XVIII Int Geographical Congr. Acad Sci USSR Press, Moscow Leningrad, pp 238-248 (in Russian)
Formozov AN (1964) Convergence in terrestrial vertebrates and geographic environment. In: Modern problems of geography. Scientific reports of the Soviet geographers for the program of 20th Int Geographic Congr. Nauka Publ, Moscow, pp 215-219 (in Russian)
Formozov AN (1976) The theory of life forms and its importance for ecological analysis of the faunas. In: Nasimovitch AA (ed) A.N. Formozov. Mammals and birds and their interrelations with the environment. Selected scientific papers. Nauka Publ, Moscow, pp 267-274 (in Russian)
Fox BJ (1987) Species assembly and the evolution of community structure. Evol Ecol 1:201-2133
Fox BJ (1989) Small-mammal community pattern in Australian heathlands: a taxonomically based rule for species assembly. In: Morris DW, Abramsky Z, Fox BJ, Willig MR (eds) Patterns in the structure of mammalian communities. Spec Publ Mus Texas Tech Univ, No 28. Texas Tech Univ Press, Lubbock, Texas, pp 91-104
Fox BJ, Brown JH (1993) Assembly rules for functional groups in North American desert rodent communities. Oikos 67:357-370

Fox BJ, Luo J (1996) Estimating competition coefficients from census data: a re-examination of the regression technique. Oikos 77:291-300

Frank DH, Heske EJ (1992) Seasonal changes in space use patterns in the southern grasshopper mouse, *Onychomys torridus torridus*. J Mammal 73:292-298

Freeland WJ (1991) Plant secondary metabolites: biochemical coevolution with herbivores. In: Palo RT, Robbins CT (eds) Plant defences against mammalian herbivory. CRC Press, Boca Raton

Fretwell SD (1972) Populations in a seasonal environment. Princeton Univ Press, Princeton

Fretwell SD, Lucas HL (1970) On territorial behavior and other factors influencing habitat distribution in birds. I. Theoretical development. Acta Biotheor 19:16-36

Fulk GW (1975) Population ecology of rodents in the semi-arid shrublands of Chile. Occas Pap Mus Texas Tech Univ 33:1-40

Funaioli U (1971) Guida breve dei mammiferi della Somalia. Inst Agronomico l'Oltremare, Firenze

Gambarian PP (1960) Adaptive features of locomotory apparatus of fossorial mammals. Armenian SSR Acad Sci Press, Erevan (in Russian)

Gambarian PP (1972) The running of mammals. Adaptational features of locomotory organs. Nauka Press, Leningrad (in Russian)

Gambarian PP, Martirosian BA (1960) On the ecology of the mouse-like hamster. Zool Zh 43:108-114 (in Russian)

Gambarian PP, Papanian SB, Martirosian BA (1960) Data on the ecology of the midday jird *Meriones meridianus dahli* Schidl. in Armenian SSR. Bull Moscow Nat Soc Sec Biol 65:17-22 (in Russian)

Gambarian PP, Zubtzova GE, Zykova LY, Krasts IV (1978) Special way of soil loosening by incisors in five-toed jerboas. Zool Zh 57:806-809 (in Russian)

Garrison TE, Best TL (1990) *Dipodomys ordii*. Mamm Species 353:1-10

Gauch HG (1982) Multivariate analysis in community ecology. Cambridge Univ Press, Cambridge

Gautun JC, Tranier M, Sicard B (1985) Liste preliminaire des rongeurs du Burkina Faso. Mammalia 49:537-542

Genoways HH, Brown JH (eds) (1993) Biology of Heteromyidae. Spec Publ 10. Am Soc Mammal, Brigham Young University, Provo, Utah

George W (1974) Notes on the ecology of gundis (F. Ctenodactylidae). Symp Zool Soc Lond 34:143-160

George W (1978) Reproduction of female gundis (Rodentia: Ctenodactylidae). J Zool (Lond) 185:57-71

George W (1981) The diet of *Petromus typicus* (Petromuridae, Rodentia) in the Augrabies Falls National Park. Koedoe 24:159-168

George W, Crowther G (1981) Space partitioning between two small mammals in a rocky desert. Biol J Linn Soc 15:195-200

Ghosh PK (1975) Thermo-regulation and water economy in Indian desert rodents. In: Prakash I, Ghosh PK (eds) Rodents in desert environments. Dr W Junk, The Hague, pp 397-412

Giller PS (1984) Community ecology and the niche. Chapman and Hall, London

Gilpin ME, Diamond JM (1984) Are species co-occurrences on islands non-random, and are null hypotheses useful in community ecology? In: Strong DR Jr, Simberloff D, Abele LG, Thistle AB (eds) Ecological communities: conceptual issues and the evidence. Princeton Univ Press, Princeton, New Jersey, pp 297-315

Gilpin ME, Hanski IA (eds) (1997) Metapopulation biology. Ecology, genetics and evolution. Academic Press, New York

Goltzman ME, Naumov NP, Nikolsky AA, Ovsianikov NG, Paskhina NM, Smirin VM (1977) Social behaviour of the great gerbil (*Rhombomys opimus* Licht.). In: Sokolov VE (ed) Mammalian behaviour. Nauka Press, Moscow, pp 5-69 (in Russian)

Gorbunov AK (1985) Characteristics of ecology of the porcupine in the eastern Caspian deserts. Ekologia No 4:66-70 (in Russian)

Goyal SP, Ghosh PK (1993) Burrow structure of two gerbil species of Thar desert, India. Acta Theriol 38:453-456

Grant PR (1972) Interspecific competition among rodents. Annu Rev Ecol Syst 3:79-106

Grant PR (1975) Population performance of *Microtus pennsylvanicus* confined to woodland habitat, and a model of habitat occupancy. Can J Zool 53:1447-1465

Grant PR (1978) Competition between species of small mammals. In: Snyder DP (ed) Populations of small mammals under natural conditions. Special Publ Series Pumatuning Lab of Ecol Univ Pittsburgh, Pittsburgh, pp 38-51

Green RH (1974) Multivariate niche analysis with temporally varying environmental factors. Ecology 52:73-83

Griffin M (1990) A review of taxonomy and ecology of gerbilline rodents of the central Namib desert, with keys to the species (Rodentia: Muridae). In: Seely MK (ed) Namib ecology: 25 years of Namib research. Transvaal Mus Monograph, Transvaal Mus, Pretoria, pp 83-98

Grinnell J (1917) Field test of theories concerning distribution control. Am Nat 51:115-128

Gromov IM, Erbajeva MA (1995) Mammals of the fauna of Russia and adjacent regions. Lagomorphs and rodents. Zool Inst Russian Acad Sci, Saint Petersburg (in Russian)

Gromov IM, Polyakov IY (1977) Fauna of USSR. Mammals. Vol 3. Rodents. Issue 8. Microtinae. Nauka Press, Moscow Leningrad (in Russian)

Gromov VS, Gromova LA (1996) The use of space, social interrelations and marking behavior of tamarisk gerbils (*Meriones tamariscinus*) under semi-free maintenance. Zool Zh 75:280-296 (in Russian)

Gromov IM, Bibikov DI, Kalabukhov NI, Meyer MN (1965) Fauna of USSR. Mammals. Vol 3. Rodents. Issue 2. Marmotinae. Nauka Press, Moscow Leningrad (in Russian)

Gromov VS, Chabovsky AV, Paramonov DV, Pavlov AP (1996) The seasonal dynamics of demographic and spatial structure in the populations of tamarisk gerbils (*Meriones tamariscinus*) in southern Kalmykia. Zool Zh 75:413-428 (in Russian)

Gurevitch J, Morrow LL, Wallace A, Walsh JS (1992) A meta-analysis of competition in field experiments. Am Nat 140:539-572

Guthmann N, Lozada M, Monjeau JA, Heinemann KM (1997) Population dynamics of five sigmodontine rodents of northwestern Patagonia. Acta Theriol 42:143-152

Hafner JC (1993) Macroevolutionary diversification in heteromyid rodents: heterochrony and adaptation in phylogeny. In: Genoways HH, Brown JH (eds) Biology of Heteromyidae. Spec Publ 10. Am Soc Mammal, Brigham Young University, Provo, Utah, pp 291-318

Hafner JC, Hafner MS (1983) Evolutionary relationships of heteromyid rodents. In: Reichman OJ, Brown JH (eds) Biology of desert rodents. Great Basin Nat Mem 7:3-29

Hagmeier EM (1966) A numerical analysis of the distributional patterns of North American mammals. II. Re-evaluation of the provinces. Syst Zool 15:279-299

Hagmeier EM, Stults CD (1964) A numerical analysis of the distributional patterns of North American mammals. Syst Zool 13:125-155

Haim A, Fluxman S (1996) Daily rhythms of metabolic rates: role of chemical signals in coexistence of spiny mice of the genus *Acomys*. J Chem Ecol 22:223-229

Haim A, Rozenfeld FM (1993) Temporal segregation in coexisting *Acomys* species: the role of odour. Physiol Behav 54:1159-1161

Haim A, Zisapel N (1995) Oxygen consumption and body temperature rhythms in the golden spiny mouse: responses to changes in day length. Physiol Behav 58:775-778

Halama KJ, Dueser RD (1994) Of mice and habitats: tests for density-dependent habitat selection. Oikos 69:107-114

Hall ER (1981) The mammals of North America, 2nd edn. John Wiley, New York

Hallett JG (1982) Habitat selection and the community matrix of a desert small-mammal fauna. Ecology 63:1400-1410

Hallett JG (1991) The structure and stability of small mammal faunas. Oecologia 88:383-393

Hallett JG, Pimm SL (1979) Direct estimation of competition. Am Nat 113:593-599

Hallett JG, O'Connell MA, Honeycutt RL (1983) Competition and habitat selection: test of a theory using small mammals. Oikos 40:175-181

Hanski I (1991) Single-species metapopulation dynamics: concepts, models and observations. Biol J Linn Soc 42:17-38
Happold DCD (1966) The mammals of Jebel Marra, Sudan. J Zool (Lond) 149:126-136
Happold DCD (1967a) *Gerbillus (Dipodillus) campestris* (Gerbillidae, Rodentia) from the Sudan. J Nat Hist 1:315-317
Happold DCD (1967b) Biology of jerboa *Jaculus jaculus* butleri (Rodentia: Dipodidae) in the Sudan. J Zool (Lond) 151:257-275
Happold DCD (1968) Observations on *Gerbillus pyramidum* (Gerbillinae: Rodentia) at Khartoum, Sudan. Mammalia 32:44-53
Happold M (1976a) Social behaviour of the conilurine rodents (Muridae) of Australia. Z Tierpsychol 40:113-182
Happold M (1976b) Reproductive biology and developments in the conilurine rodents (Muridae) of Australia. Aust J Zool 24:19-26
Hardin JW, Barbour RW, Davis WH (1970) Observations on the home range of a yellow-nosed cotton rat, *Sigmodon ochrognathus*. Southwest Nat 14:353-355
Harper DGC (1982) Competitive foraging in mallards: "ideal free" ducks. Anim Behav 30:575-584
Harris JH (1984) An experimental analysis of desert rodent foraging ecology. Ecology 65:1579-1584
Harrison DL, Bates P (1991) The mammals of Arabia, 2nd edn. Harrison Zool Mus Publ, Sevenoaks
Harvey PH, Colwell RK, Silvertown JW, May RM (1983) Null models in ecology. Annu Rev Ecol Syst 14:189-211
Hassinger JD (1973) A survey of the mammals of Afghanistan resulting from the 1965 Street Expedition (excluding bats). Fieldiana Zool 60:1-195
Hatough-Bouran A (1990) The burrowing habits of desertic rodents *Jaculus jaculus* and *Gerbillus dasyurus* in the Shaumari Reserve in Jordan. Mammalia 54:341-360
Hayssen V (1991) *Dipodomys microps*. Mamm Species 389:1-9
Hecnar SJ, M'Closkey RT (1997) Patterns of nestedness and species association in a pond-dwelling amphibian fauna. Oikos 80:371-381
Hedrick PW (1986) Genetic polymorphism in heterogenous environment: a decade later. Ann Rev Ecol Syst 17:535-566
Hegdal PL, Ward AL, Johnson AM, Tietjen HP (1965) Notes on the life history of the Mexican pocket gopher (*Cratogeopmys castanops*). J Mammal 46:334-335
Heim de Balsac H (1967) La distribution reelle de *Desmodilliscus* (Gerbillinae). Mammalia 31:160-164
Hengeveld R (1992) Dynamic biogeography. Cambridge University Press, Cambridge
Hershkovitz P (1962) Evolution of Neotropical cricetine rodents (Muridae) with special reference to the phyllotine group. Fieldiana Zool 46:1-524
Herzig-Straschil B (1978) On the biology of *Xerus inauris* (Zimmerman, 1980) (Rodentia, Sciurudae). Z Saeugetierkd 43:262-278
Herzig-Straschil B (1979) *Xerus inauris* (Rodentia, Sciurudae) – an inhabitant of the arid regions of Southern Africa. Folia Zool 28:119-124
Herzig-Straschil B, Herzig A (1989) Biology of *Xerus princeps* (Rodentia, Sciuridae). Madoqua 16:41-46
Heske EJ, Brown JH, Mistry S (1994) Long-term experimental study of a Chihuahuan desert rodent community: 13 years of competition. Ecology 75:438-445
Hickman GC (1977) Burrow system structure of *Pappogeomys castanops* (Geomyidae) in Lubbock County, Texas. Am Midl Nat 97:50-58
Hickman GC (1979) Burrow system structure of the bathyergid *Cryptomys hottentotus* in Natal, South Africa. Z Saeugetierkd 44:153-162
Hill M (1973) Diversity and evenness: an unifying notation and its consequences. Ecology 54:427-432
Hillyard JR, Phillips CJ, Birney EC, Monjeau JA, Sikes RS (1997) Mitochondrial DNA analysis

and zoogeography of two species of silky desert mice, *Eligmodontia*, in Patagonia. Z Saeugetierkd 62:281-292
Hoekstra RF, Bijlsma R, Dolman AJ (1985) Polymorphism from environmental heterogeneity: models are only robust if the heterozygote is close in fitness to the favoured homozygote in each environment. Genet Res 45:299-314
Hoffmeister DF (1986) Mammals of Arizona. Univ Arizona Press, Tucson, Arizona
Holling CS (1992) Cross-scale morphology, geometry, and dynamics of ecosystems. Ecol Monogr 62:447-502
Holt RD (1985) Population dynamics in two-patch environments: Some anomalous consequences of an optimal habitat distribution. Theor Popul Biol 28:181-208
Holt RD (1996) Adaptive evolution in source-sink environments: direct and indirect effects of density-dependence on niche evolution. Oikos 75:182-192
Holt RD, Gaines MS (1992) Analysis of adaptations in heterogeneous landscape: implications for the evolution of fundamental niches. Evol Ecol 6:433-447
Hoogstraal H (1963) A brief review of the contemporary land mammals of Egypt (including Sinai). 2. Lagomorpha and Rodentia. J Egypt Publ Health Assoc 38:1-36
Hooper ET, El Halili M (1972) Temperature regulation and habits in two species of jerboa, genus *Jaculus*. J Mammal 53:574-593
Hopf FA, Brown JH (1986) The bullseye method for testing for randomness in ecological communities. Ecology 67:1139-1155
Horner BE, Taylor JM (1968) Growth and reproductive behavior in the southern grasshopper mouse. J Mammal 49:644-660
Hoyt DF, Taylor CR (1981) Gait and the energetics of locomotion in horses. Nature 292:239-240
Hrachovy SK, Bradley RD, Jones C (1996) *Neotoma goldmani*. Mamm Species 545:1-3
Hubbell SP, Foster RB (1986) Biology, chance, history and the structure of tropical rainforest communities. In: Diamond J, Case TJ (eds) Community ecology. Harper and Row, New York, pp 314-329
Hubert B (1978) Modern rodent fauna of the lower Omo Valley, Ethiopia. Bull Carnegie Mus Nat Hist 14:48-52
Hughes JJ, Ward D, Perrin MR (1994) Predation risk and competition affect habitat selection and activity of Namib Desert gerbils. Ecology 75:1397-1405
Hughes JJ, Ward D, Perrin MR (1995) Effects of substrate on foraging decision by a Namib Desert gerbil. J Mammal 76:638-645
Huston M (1979) A general hypothesis of species diversity. Am Nat 113:81-101
Hutchinson GE (1957) Concluding remarks. Cold Spring Harbor Symp Quant Biol 22:415-427
Hutchinson GE (1959) Homage to Santa Rosalia or Why are there so many kinds of animals? Am Nat 93:145-159
Hutchinson GE (1961) The paradox of the plankton. Am Nat 95:137-145
Hutchinson GE, MacArthur RH (1959) A theoretical ecological model of size distributions among species of animals. Am Nat 93:117-125
Hutto RL (1978) A mechanism for resource allocation among sympatric heteromyid rodent species. Oecologia 33:115-126
Ilan M, Yom-Tov Y (1990) Dial activity pattern of a diurnal desert rodent, *Psammomys obesus*. J Mammal 71:66-69
Inger R, Colwell R (1977) Organization of contiguous communities of amphibians and reptiles in Thailand. Ecol Monogr 47:229-253
Iriarte JA, Contreras LC, Jaksic FM (1989) A long-term study of a small-mammal assemblage in the central Chilean matorral. J Mammal 70:79-87
Ismagilov MI (1960) Data on the ecology of the red-tailed jird (*Meriones erythrourus*) in the Bet-Pak-Dala desert. Zool Zh 39:1239-1249
Ivanter EV (1975) Population ecology of small mammals in the taiga of the northeastern USSR. Nauka Press, Leningrad (in Russian)
Ivanter EV, Makarov AM (1994) The spatial structure of shrew populations (*Sorex*, Insectivora) and its correlation with the feed biomass in communities. Zool Zh 73:124-138 (in Russian)

Jaksic FM, Feinsinger P, Jimenez JE (1993) A long-term study on the dynamics of guild structure among predatory vertebrates at a semi-arid Neotropical site. Oikos 67:87-96
James FC, McCulloch CE (1990) Multivariate analysis in ecology and systematics: panacea or Pandora's box? Annu Rev Ecol Syst 21:129-166
Jarvis JUM (1969) The breeding season and litter size of African mole-rats. J Reprod Fertil Suppl 6:237-248
Jimenez JE, Feinsinger P, Jaksic FM (1992) Spatiotemporal patterns of an irruption and decline of small mammals in northcentral Chile. J Mammal 73:356-364
Jingchuan M, Wang G, Wang C (1990) Cluster analysis on rodents community in the eastern part of desert-steppe. Acta Theriol Sin 10:145-150 (in Chinese)
Johnson DW, Armstrong DM (1987) *Peromyscus crinitus*. Mamm Species 287:1-8
Jones WT (1989) Dispersal distance and the range of nightly movements in Merriam's kangaroo rats. J Mammal 70:27-34
Jones WT (1993) The social systems of Heteromyid rodents. In: Genoways HH, Brown JH (eds) Biology of Heteromyidae. Spec Publ 10. Am Soc Mammal, Brigham Young University, Provo, Utah, pp 575-595
Jones WT, Waser PM, Elliott LF, Link NE, Bush BB (1988) Philopatry, dispersal, and habitat saturation in the banner-tailed kangaroo rat, *Dipodomys spectabilis*. Ecology 69:1466-1473
Jongman RHG, Ter Braak CJF, Van Tongeren OFR (1987) Data analysis in community and landscape ecology. Pudoc, Wageningen, Netherlands
Kalabukhov NI (1969) Periodic (seasonal and annual) changes in rodent organism, their causes and consequences. Nauka Press, Leningrad (in Russian)
Kam M, Degen AA (1989) Efficiency of use of saltbush (*Atriplex halimus*) for growth by fat sand rats (*Psammomys obesus*). J Mammal 70:485-493
Karasov WH (1989) Nutritional bottleneck in a herbivore, the desert wood rat (*Neotoma lepida*). Physiol Zool 62:1351-1382
Karim A, Nasher AM, Nader IA (1975) Natural history study of the Harrison's gerbil *Gerbillus dasyurus mesopotamiae* Harrison. Bull Nat Hist Res Cent Univ Baghdad 6:60-63
Kashkarov DN (1938) Fundamentals of animal ecology. Medgiz, Moscow Leningrad (in Russian)
Kaufman DW, Clark BK, Kaufamn GA (1990) Habitat breadth of nongame rodents in the mixed-grass prairie region of north central Kansas (USA). Prairie Nat 22:19-26
Kawecki TJ (1995) Demography of source-sink populations and the evolution of ecological niches. Evol Ecol 9:38-44
Kelt DA (1997) Assembly of local communities: consequences of an optimal body size for the organization of competitively structured communities. Biol J Linn Soc 62:15-37
Kelt DA, Meserve PL, Lang BK (1994) Quantitative habitat associations of small mammals in a temperate rainforests in southern Chile: empirical patterns and the importance of ecological scale. J Mammal 75:890-904
Kelt DA, Brown JH, Heske EJ, Marquet PA, Morton SR, Reid JR, Rogovin KA, Shenbrot G (1996) Community structure of desert small mammals: comparisons across four continents. Ecology 77:746-761
Kenagy GJ (1972) Saltbush leaves: excision of hypersaline tissue by a kangaroo rat. Science 178:1094-1096
Kenagy GJ (1973) Adaptations for leaf eating in the Great Basin kangaroo rat, *Dipodomys microps*. Oecologia 12:383-412
Kenagy GJ, Bartholomew GA (1985) Seasonal reproductive patterns in five coexisting California desert rodent species. Ecol Monogr 55:371-397
Kerley GI (1989) Diet of small mammals from the Karoo, South Africa. South Afr J Wildl Res 19:67-72
Kerley GI (1992) Trophic status of small mammals in the semi-arid Karoo, South Africa. J Zool (Lond) 226:563-572
Kerley GI, Knight KM, Erasmus T (1990) Small mammal microhabitat use and diet in the southern Kalahari, South Africa. S Afr J Wildl Res 20:123-126

Khidas K (1993) Distribution of rodents in Kabylie du Djurdjura (Algeria) Mammalia 57:207-212
Khodasheva KS (1953) Life forms of rodents of the Kazakhstan Lowlands and some patterns of their geographic distribution. Proc Inst Geogr Acad Sci USSR 54:33-194 (in Russian)
Khokhlova IS, Krasnov BR (1986) Mechanisms of density autoregulation in the house mouse populations with different types of spatial structure (*Mus musculus*). Zool Zh 65:407-415 (in Russian)
Khokhlova IS, Krasnov BR, Shenbrot GI, Degen A (1994) Seasonal body mass changes and habitat distribution in several rodent species from the Ramon erosion cirque, Negev Highlands, Israel. Zool Zh 73:115-121 (in Russian)
King JA (ed) (1968) Biology of *Peromyscus* (Rodentia). Spec Publ 2. Am Soc Mammal, Brigham Young University, Provo, Utah
Kingdon J (1974) East African mammals. An atlas of evolution in Africa. Vol II Part B (Hares and Rodents). Univ Chicago Press, Chicago
Kirmiz JP (1962) Adaptation to a desert environment: a study on the jerboa, rat and man. Butterworth, London
Kiseleva NK (1982) Evolution of the biogeocoenoses of the Caspian lowlands in the Holocene. Nauka Press, Moscow (in Russian)
Kleiman DG, Eisenberg JF, Maliniak E (1979) Reproductive parameters and productivity of caviomorph rodents. In: Eisenberg JF (ed) Vertebrate ecology in the northern Neotropics. Smithsonian Inst Press, Washington, pp 173-183
Klimova VN (1975) Morpho-functional and ecological analysis of subordinate hearing cavities of rodents. Zool Zh 59:273-282 (in Russian)
Knight MH, Skinner JD (1981) Thermoregulatory, reproductive and behavioral adaptations of the big-eared desert mouse, *Malacothrix typica* to its arid environment. J Arid Environ 4:137-145
Kock D (1977) Vergleichende Untersuchungen einiger Säugetiere im südlichen Niger. Senckenb Biol 58:113-136
Koffler BR (1972) *Meriones crassus*. Mamm Species 9:1-4
Koford CB (1968) Peruvian desert mice: water independence, competition, and breeding cycle near the equator. Science 160:552-553
Kofron CP (1987) Seasonal reproduction of the springhare (*Pedetes capensis*) in the southeastern Zimbabwe. Afr J Ecol 25:185-194
Komarova HV (1980) Geographic distribution and life history of the long-clawed ground squirrel. Materialy k poznaniyu fauny y flory SSSR [Materials to the knowledge of the fauna and flora of USSR] 51:201-231 (in Russian)
Korn GL, Korn U (1989) The effect of gerbils (*Tatera brantsii*) on primary production and plant species composition in a southern African savanna. Oekologia (Berl) 79:271-278
Kotler BP (1984a) Risk of predation and the structure of desert rodent communities. Ecology 65:689-701
Kotler BP (1984b) Effects of illumination on the rate of resource harvesting in a community of desert rodents. Am Midl Nat 11:383-389
Kotler BP (1985) Owl predation on desert rodents which differ in morphology and behavior. J Mammal 66:824-828
Kotler BP, Brown JS (1988) Environmental heterogeneity and the coexistence of desert rodents. Annu Rev Ecol Syst 19:281-307
Kotler BP, Brown JS (1990) Rates of seed harvest by two species of gerbilline rodents. J Mammal 71:591-596
Kotler BP, Brown JS, Smith RJ, Wirtz WO (1988) The effects of morphology and body size on rates of owl predation on desert rodents. Oikos 53:145-152
Kotler BP, Brown JS, Hasson O (1991) Factors affecting gerbil foraging behavior and rates of owl predation. Ecology 72:2249-2260
Kotler BP, Blaustein L, Brown JS (1992) Predator facilitation: the combined effect of snakes and owls on the foraging behavior of gerbils. Ann Zool Fenn 29:199-206
Kotler BP, Blaustein L, Dednam H (1993) The specter of predation: the effect of vipers on the foraging behavior of two species of gerbils. Isr J Zool 39:11-22

Kowalski K, Rzebik-Kowalska B (1991) Mammals of Algeria. Wydawnictwo Polsk Akad Nauk, Wroclaw

Krasnov BR (1983) Behavioral interactions as a factor determining spatial distribution of two rodent species in the semidesert of the northeastern Caspian lowlands. In: Zakharov AA (ed) Animal behavior in communities. Nauka Press, Moscow, pp 176-178 (in Russian)

Krasnov BR (1984) Co-habitation of the midday jird (*Meriones meridianus*) and the house mouse (*Mus musculus*) in the northwestern Caspian lowlands. Zool Zh 63:421-428 (in Russian)

Krasnov BR, Khokhlova IS (1989) Plasticity of spatial and social structure of house mouse populations as a strategy of adaptation to a changing environment. In: Sokolov VE, Kotenkova EV, Krasnov BR (eds) The house mouse, Nauka Press, Moscow, pp 223-253 (in Russian)

Krasnov BR, Khokhlova IS (1994) Spatial and behavioral structure of house mouse populations. In: Kotenkova EV, Bulatova NS (eds) The house mouse. Origin, distribution, systematics, behaviour. Nauka Press, Moscow, pp 190-215 (in Russian)

Krasnov BR, Khokhlova IS (1996) Discrimination of midday jird's odour by house mice. Anim Behav 52:659-665

Krasnov BR, Shenbrot GI (1996) Structure of darkling beetle communities and factors influencing their spatial distribution in the Negev highlands (Israel). Ecography 19:139-152

Krasnov BR, Shenbrot GI (1997) Seasonal variation in spatial organization of tenebrionid community in Negev Highlands, Israel. Environ Entomol 26:178-190

Krasnov BR, Shenbrot GI, Khokhlova IS, Ivanitskaya EY (1996a) Spatial structure of rodent community in the Ramon erosion cirque, Negev highlands (Israel). J Arid Environ 32:319-327

Krasnov BR, Shenbrot GI, Khokhlova IS, Degen AA, Rogovin KA (1996b) On the biology of Sundevall's jird (*Meriones crassus* Sundevall) in Negev Highlands, Israel. Mammalia 60:375-391

Krasnov BR, Shenbrot GI, Medvedev SG, Vatschenok VS, Khokhlova IS (1997) Host-habitat relation as an important determinant of spatial distribution of flea assemblages (Siphonaptera) on rodents in the Negev desert. Parasitology 114:159-173

Krasnov B, Shenbrot G, Khokhlova I, Medvedev S, Vatschenok V (1998) Habitat dependence of a parasite-host relationship: flea (Siphonaptera) assemblages in two gerbil species of the Negev desert. J Med Entomol 35:303-313

Krebs CJ (1994) Ecology. The experimental analysis of distribution and abundance, 4th edn. HarperCollins College Publ, New York

Krishnan MS (1952) Geological history of Rajasthan and its relation to present day conditions. Bull Natl Inst Sci 1:19-31

Kronfeld N, Dayan T, Zisapel N, Haim A (1994) Coexisting populations of *Acomys cahirinus* and *A. russatus*: a preliminary report. Isr J Zool 40:177-183

Kryzhanovsky OL (1965) Composition and origin of terrestrial fauna of Middle Asia. Nauka Press, Moscow (in Russian)

Kucheruk VV (1959) Steppe faunal complex of mammals and its place in Palaearctic fauna. In: Formosov AN (ed) Geography of terrestrial animal populations and methods of their study. USSR Acad Sci Press, Moscow, pp 45-87 (in Russian)

Kucheruk VV (1983) Mammal burrows – their structure, topology and use. Fauna and Ecology of Rodents 15:5-54 (in Russian)

Kucheruk VV, Kulik IL, Dubrovsky IA (1972) Great gerbil as a life form of a desert rodent. In: Kucheruk VV (ed) Fauna and ecology of rodents 11. Moscow Univ Press, Moscow, pp 5-70 (in Russian)

Kulik IL (1980) Distribution and types of geographic ranges in jerboas (Dipodidae, Rodentia). In: Voronov AG, Drozdov NN (eds) Modern problems of zoogeography. Nauka Press, Moscow, pp 152-166 (in Russian)

Lackey JA (1996) *Chaetodipus fallax*. Mamm Species 517:1-6

Lagos VO, Bozinovic F, Contreras LC (1995) Microhabitat use by a small diurnal rodent (*Octodon degus*) in a semiarid environment: Thermoregulatory constraints or predation risk? J Mammal 76:900-905

La Houérou HN (1997) Climate, flora and fauna changes in the Sahara over the past 500 million years. J Arid Environ 37:619-648
Lay DM (1967) A study of the mammals of Iran. Fieldiana Zool 54:1-183
Lay DM (1972) The anatomy, physiology, functional significance and evolution of specialized hearing organs of Gerbilline rodents. J Morphol 138:41-120
Lay DM, Nadler CF (1975) A study of *Gerbillus* (Rodentia, Muridae) east of the Euphrates river. Mammalia 39:423-445
Le Berre M (1990) Faune du Sahara. 2. Mammiferes. Lechevalier – R. Chabaud, Pariss
Lechowicz MJ (1982) The sampling characteristics of electivity indices. Oecologia 52:22-30
Lemen CA (1978) Seed size selection in heteromyids. Oecologia 35:13-19
Lemen CA, Freeman PW (1983) Quantification of competition among coexisting heteromyids in the Southwest. Southwest Nat 28:41-46
Levins R (1968) Evolution in changing environments. Princeton Univ Press, Princeton
Lidicker WZ Jr (1975) The role of dispersal in the demography of small mammals. In: Golley FB, Petrusewicz K, Ryszkowsky L (eds) Small mammals: their productivity and population dynamics. Cambridge Univ Press, New York, pp 103-128
Lima M, Marquet PA, Jaksic FM (1996) Extinction and colonization processes in subpopulations of five neotropical small mammal species. Oecologia 107:197-203
Lima M, Bozinovic F, Jaksic FM (1997) Body mass dynamics and growth patterns of leaf-eared mice *Phyllotis darwini* in a semi-arid region of the Neotropics. Acta Theriol 42:15-24
Linn IJ (1991) Influence of 6-methoxybenzoxazolinone and green vegetation on reproduction of the multimammate rat *Mastomys coucha*. S Afr J Wildl Res 21:33-37
Linn I, Key G (1996) Use of space by the African striped ground squirrel *Xerus erythropus*. Mammal Rev 26:9-26
Lively CM (1986) Canalization versus developmental conversion in a spatially variable environment. Am Nat 128:561-572
Livoreil B, Gouat P, Baudoin C (1993) A comparative study of social behaviour of two sympatric ground squirrels (*Spermophilus spilosoma*, *Spermophilus mexicanus*). Ethology 93:236-246
Lobachev VS, Shenbrot GI (1973) Comparative analysis of different methods of counting of jerboas. Bull Moscow Nat Soc Sec Biol 78:47-57 (in Russian)
Lobachev VS, Sabilaev AS, Shenbrot GI (1976) Habitat distribution and abundance of jerboas in the Aral Region. In: Kucheruk VV (ed) Fauna and ecology of rodents 13. Moscow Univ Press, Moscow, pp 13-40 (in Russian)
Loreau M, DeAngelis DL (1997) Source-sink dynamics and the coexistence of species on a single resource. Theor Popul Biol 51:79-93
Ma Y, Wang F, Jin S, Li S (1987) Glires (rodents and lagomorphs) of Northern Xinjiang and their zoogeographical distribution. Sci Press, Beijing (in Chinese)
MacArthur RH (1958) Population ecology of some warblers of northeastern coniferous forests. Ecology 39:599-619
MacArthur RH (1965) Patterns of species diversity. Biol Rev 40:510-533
MacArthur RH (1970) Species packing and competitive equilibrium among many species. Theor Popul Biol 1:1-11
MacArthur RH (1972) Geographical ecology: patterns in the distribution of species. Harper and Row, New York
MacArthur RH, Connell JH (1966) The biology of populations. John Wiley, New York
MacArthur RH, Levins R (1967) The limiting similarity of convergence and divergence of coexisting species. Am Nat 101:377-385
MacArthur RH, Pianka EL (1966) On optimal use of a patchy environment. Am Nat 100:603-609
Macedo RH, Mares MA (1988) *Neotoma albigula*. Mamm Species 310:1-7
Macholan M (1990) Systematics and evolutionary relationships between taxons of the house mouse (Mammalia, Rodentia). Biol Listy 55:216-229
MacMillen RE (1972) Water economy of nocturnal desert rodents. In: Maloiy GMO (ed) Comparative physiology of desert animals. Academic Press, New York, pp 147-174
MacMillen RE, Hinds DS (1983) Water regulatory efficiency in heteromyid rodents: a model and its application. Ecology 64:152-164

MacMillen RE, Lee AK (1969) Water metabolism of Australian hopping mice. Comp Biochem Physiol 28:493-514
Magurran AE (1988) Ecological diversity and its measurement. Chapman and Hall, London
Maley J (1977) Palaeoclimates of central Sahara during the Holocene. Nature 269:573-577
Manly BFJ, McDonald LL, Thomas DL (1993) Resource selection by animals. Statistical design and analysis for field studies. Chapman and Hall, London
Mares MA (1973) Desert rodent ecology. Review for origin and structure of ecosystems convergent evolution research program. Acta Zool Lilloana 30:207-225
Mares MA (1975a) Observations of Argentine desert rodent ecology, with emphasis on water relations of *Eligmodontia typus*. In: Prakash I, Ghosh PK (eds) Rodents in desert environments. Dr W Junk, The Hague, pp 155-175
Mares MA (1975b) South American mammal zoogeography: evidence for convergent evolution in desert rodents. Proc Natl Acad Sci 72:1702-1706
Mares MA (1977a) Water independence in a South American non-desert rodent. J Mammal 58:653-656
Mares MA (1977b) The comparative water relations and habitat requirements of three South American phyllotine rodents. J Mammal 58:514-520
Mares MA (1980) Convergent evolution among desert rodents: a global perspective. Bull Carnegie Mus Nat Hist 16:1-51
Mares MA (1993a) Desert rodents, seed consumption, and convergence. Evolutionary shuffling of adaptations. Bioscience 43:372-379
Mares MA (1993b) Heteromyids and their ecological counterparts: a pandesertic view of rodent ecology and evolution. In: Genoways HH, Brown JH (eds) Biology of Heteromyidae. Spec Publ 10. Am Soc Mammal, Brigham Young University, Provo, Utah, pp 652-715
Mares MA, Braun JK (1996) A new species of phyllotine rodent, genus *Andalgalomys* (Muridae: Sigmodontinae), from Argentina. J Mammal 77:928-941
Mares MA, Williams DF (1977) Experimental support for food particle size resource allocation in heteromyid rodents. Ecology 58:1186-1190
Mares MA, Ojeda RA, Kosko MP (1981) Observations on the distribution and ecology of the mammals of Salta province, Argentina. Ann Carnegie Mus 50:151-206
Mares MA, Ojeda RA, Barquez RM (1989) Guide to the mammals of Salta province. Univ Oklaoma Press, Norman
Mares MA, Ojeda RA, Borghi CE, Giannoni SM, Diaz GB, Braun JK (1997) How desert rodentsovercome halophytic plant defenses. Bioscience 47:699-704
Marshall JT, Sage RD (1981) Taxonomy of the house mouse. Symp Zool Soc Lond 47:15-25
Massoia E, Fornes A (1965) Nuevos datos sobre la morfologia, distribución geográfica y etoecología de *Calomys callosus callosus* (Rengger) (Rodentia: Cricetidae). Physis 25:325-331
Matiushkin EN (1972) Mixed character of the theriofauna of Ussury region: its common characters, historical roots and recent manifestations in the communities of the central Sikhote-Alin. Arch Zool Mus Moscow State Univ 13:86-144 (in Russian)
Matson J (1976) The distribution of rodents in Owens Lake region, Inyo County, California. Contrib Sci Nat Hist Mus Los Ang County 276:1-27
May RM, MacArthur RH (1972) Niche overlap as a function of environmental variability. Proc Natl Acad Sci USA 69:1109-1113
McCarty R (1975) *Onychomys torridus*. Mamm Species 59:1-5
McCarty R (1978) *Onychomys leucogaster*. Mamm Species 87:1-6
McClenaghan LR (1983) Notes on the population ecology of *Perognathus fallax* (Rodentia: Heteromyidae) in southern California. Southwest Nat 28:429-436
McClenaghan LR (1984) Comparative ecology of sympatric *Dipodomys agilis* and *Dipodomys merriami* populations in southern California. Acta Theriol 29:175-185
McGinnies WG, Goldman BJ, Paylore P (eds) (1968) Deserts of the world. An appraisal of research into their physical and biological environments. Univ Arizona Press, Tucson, Arizona
M'Closkey RT (1976) Community structure in sympatric rodents. Ecology 57:728-739
M'Closkey RT (1978) Niche separation and assembly in four species of Sonoran desert rodents. Am Nat 112:683-694

M'Closkey RT (1980) Spatial patterns in sizes of seeds collected by four species of heteromyid rodents. Ecology 61:486-489
M'Closkey RT (1983) Desert rodent activity: response to seed production by two perennial plant species. Oikos 41:233-238
M'Closkey RT (1985) Species pools and combinations of heteromyid rodents. J Mammal 66:132-134
McMahon T (1973) Size and shape in biology. Science 1979:1201-1204
McMahon T, Bonner JT (1983) On size and life. Scientific American Books, New York
McPeek MA (1996) Trade-offs, food web structure, and the coexistence of habitat specialists and generalists. Am Nat 148:S124-S138
Meffe GK, Sheldon AL (1988) The influence of habitat structure on fish assemblage composition in southeastern (USA) blackwater streams. Am Midl Nat 120:225-240
Meigs P (1953) World distribution of arid and semiarid homoclimates. In: Arid zone hydrology. UNESCO Arid Zone Res Ser 1:203-209
Mekaev YA (1987) Zoogeographic complexes of Eurasia. Nauka Press, Leningrad (in Russian)
Meliadou A, Troumbis AY (1997) Aspects of heterogeneity in the distribution of diversity of the European herpetofauna. Acta Oecol 18:393-412
Mello DA (1978) Biology of *Calomys callosus* (Rengger, 1830) under laboratory conditions (Rodentia, Cricetinae). Rev Bras Biol 38:293-295
Meserve PL (1976) Habitat and resource utilization by rodents of a California coastal sage scrub community. J Anim Ecol 45:467-466
Meserve PL (1981a) Resource partitioning in a Chilean semi-arid small mammal community. J Anim Ecol 50:745-757
Meserve PL (1981b) Trophic relationships among small mammals in a Chilean semiarid thorn scrub community. J Mammal 62:304-314
Meserve PL, Glanz WE (1978) Geographical ecology of small mammals in the northern Chilean arid zone. J Biogeogr 5:135-148
Meserve PL, Yunger JA, Gutierrez JR, Contreras LC, Mildstead WB, Lang BK, Cramer KL, Herrera S, Lagos VO, Silva SI, Tabilo EL, Torrealba MA, Jaksic FM (1995) Heterogeneous responses of small mammals to an El Nino southern oscillation event in northcentral semiarid Chile and the importance of ecological scale. J Mammal 76:580-595
Meserve PL, Gutierrez JR, Yunger JA, Contreras LC, Jaksic FM (1996) Role of biotic interactions in a small mammal assemblage in semiarid Chile. Ecology 77:133-148
Meyer MN, Skhol ED (1977) On some ecological parameters of the long-tailed hamster (*Cricetulus longicaudatus*) (Rodentia, Cricetidae). Zool Zh 56:484-485 (in Russian)
Milinski M (1984) Competitive resource sharing: an experimental test of a learning rule for ESSs. Anim Behav 32:233-242
Mishra AC, Dhanda V (1975) Review of the genus *Millardia* (Rodentia: Muridae), with description of a new species. J Mammal 56:76-80
Mitchell WA, Brown JS (1990) Density-dependent harvest rates by optimal foragers. Oikos 57:180-190
Mitchell WA, Abramsky Z, Kotler BP, Pinshow B, Brown JS (1990) The effect of competition on foraging activity in desert rodents: theory and experiments. Ecology 71:844-854
Mönkkönen M, Viro P (1997) Taxonomic diversity of the terrestrial bird and mammal fauna intemperate and boreal biomes of the northern hemisphere. J Biogeogr 24:603-612
Moran NA (1992) The evolutionary maintenance of alternative genotypes. Am Nat 139:971-989
Morris DW (1979) Microhabitat utilization and species distribution of sympatric small mammals in southwestern Ontario. Am Midl Nat 101:373-384
Morris DW (1987) Ecological scale and habitat use. Ecology 68:362-369
Morris DW (1988) Habitat-dependent population regulation and community structure. Evol Ecol 2:253-269
Morris DW (1990) Temporal variation, habitat selection and community structure. Oikos 59:303-312
Morris DW (1991) On the evolutionary stability of dispersal to sink habitats. Am Nat 137:907-911

Morris DW (1994) Habitat matching: alternatives and implications to populations and communities. Evol Ecol 8:387-406
Morris DW (1996) Temporal and spatial population dynamics among patches connected by habitat selection. Oikos 75:207-219
Morrison ML, Marcot BG, Mannan RW (1992) Wildlife-habitat relationships. Concepts and applications. Univ Wisconsin Press, Madison, Wisconsinn
Morse DR, Stock NE, Lawton JH (1988) Species numbers, species abundance, and body length relationships of arboreal beetles in Bornean lowland rain forest trees. Ecol Entomol 13:25-37
Morton SR (1993) Determinants of diversity in animal communities of arid Australia. In: Ricklefs RE, Schluter D (eds) Species diversity in ecological communities. Univ Chicago Press, Chicago, pp 159-169
Morton SR, Baynes A (1985) Small mammal assemblages in arid Australia: a reappraisal. Aust Mammal 8: 159-169
Morton SR, Brown JH, Kelt DA, Reid JRW (1994) Comparisons of community structure among small mammals of North American and Australian deserts. Aust J Zool 42:501-525
Mugo DN, Lombard AT, Bronner GN, Gelderblom CM, Benn GA (1995) Distribution and protection of endemic or threatened rodents, lagomorphs and macrosceledids in South Africa. S Afr J Zool 30:115-126
Munger JC, Brown JH (1981) Competition in desert rodents: an experiment with semipermeable enclosures. Science 211:510-512
Munger JC, Bowers MA, Jones WT (1983) Desert rodent populations: factors affecting abundance, distribution and genetic structure. Great Basin Nat Mem 7:91-116
Murray BR, Dickman CR (1994) Granivory and microhabitat use in Australian desert rodents: are seeds important? Oecologia (Berl) 99:216-225
Myers AA, Giller PS (1988) Process, pattern and scale in biogeography. In: Myers A, Giller P (eds) Analytical biogeography. Chapman and Hall, London, pp 3-12
Nader IA, Kock D, Al Khalili AK (1983) *Eliomys melanurus* (Wagner 1839) and *Praomys funatus* (Peters 1878) from the Kingdom of Saudi Arabia (Mammalia: Rodentia). Senckenb Biol 63:313-324
Naumov NP (1937) On the distribution of small rodents by stations (the common vole – *Microtus arvalis* Pall., the social vole – *Microtus socialis* Pall., the steppe lemming – *Lagurus lagurus* Pall., the mound-building mouse – *Mus musculus hortulanus* Nordm.). Arch Moscow Univ 13: 3-38 (in Russian)
Naumov NP (1948) Essays on comparative ecology of small rodents. USSR Acad Sci Press, Moscow (in Russian)
Naumov NP (1965) Animal ecology, 2nd edn. Vysshaya Shkola Publ, Moscow (in Russian)
Naumov NP (1975) The role of rodents in ecosystems of the northern deserts of Eurasia. In: Golley FB, Petrusewicz K, Ryszkowsky L (eds) Small mammals: their productivity and population dynamics. Cambridge Univ Press, New York, pp 299-309
Naumov NP, Lobachev VS (1975) Ecology of desert rodents of the USSR. In: Prakash I, Ghosh PK (eds) Rodents in desert environments. Dr W Junk, The Hague, pp 465-598
Naumov NP, Lobachev VS, Dmitriev PP, Smirin VM (1972) Natural plague nidus of the north Aral desert. Moscow Univ Press, Moscow (in Russian)
Naumova EI (1981) Functional morphology of the alimentary system of rodents and lagomorphs. Nauka Press, Moscow (in Russian)
Negus NC, Berger PJ (1977) Experimental triggering of reproduction in a natural population of *Microtus montanus*. Science 196:1230-1231
Negus NC, Printer AJ (1966) Reproductive responses of *Microtus montanus* to plants and plant extracts in the diet. J Mammal 47:596-601
Nel JAJ (1967) Burrow systems of *Desmodillus auricularis* in the Kalahari Gemsbok National Park. Koedoe 10:188-121
Nel JAJ (1975) Aspects of the social ethology of some Kalahari rodents. Z Tierpsychol 37:322-331
Nel JAJ (1978) Habitat heterogeneity and changes in small mammal community structure and resource utilization in the southern Kalahari. Bull Carnegie Mus Nat Hist 6:118-131

Nel JAJ, Rautenbach LL (1974) Notes on the activity patterns, food and feeding behaviour of *Parotomys brantsii* (Smith, 1840). Mammalia 38:7-15
Nelson G, Platnick N (1981) Systematics and biogeography: cladistics and vicariance. Columbia Univ Press, New York
Neronov VM (1976) Zoogeographic analysis of rodent fauna of Iran. Bull Moscow Nat Soc Ser Biol 81:32-47 (in Russian)
Neronov VM, Arsenieva LP (1980) Zoogeographic analysis of rodent fauna of Afgahanistan. In: Voronov AG, Drozdov NN (eds) Modern problems of zoogeography. Nauka Press, Moscow, pp 254-272 (in Russian)
Neronov VM, Bobrov VV (1990) Zoogeographic analysis of rodent fauna of Pakistan. Zool Zh 69: 98-109 (in Russian)
Neronov VM, Lushchekina AA (1980) Regional biogeography of the Mongolian People's Republic. Adv Sci Technol Ser Biogeogr 2: 41-64 (in Russian)
Neronov VM, Malkhazova SM (1985) Regional geography of leishmaniases. Part 2. Adv Sci Technol Ser Med Biogeogr 13: 1-200 (in Russian)
Nevo E (1979) Adaptive convergence and divergence of subterranean mammals. Annu Rev Ecol Syst 10:269-308
Nowak RM (1991) Walker's mammals of the World. 5th edn. Johns Hopkins Univ Press, Baltimore
Oaks EC, Young PJ, Kirkland GR, Schmidt DF (1987) *Spermophilus variegatus*. Mamm Species 272:1-8
Odum EP (1983) Basic ecology. Saunders, Philadelphia
O'Farrell TP (1978) Home range dynamics of rodents in a sagebush community. J Mammal 59:657-668
O'Farrell TP, Blaustein AR (1974a) *Microdipodops megacephalus*. Mamm Species 46:1-3
O'Farrell TP, Blaustein AR (1974b) *Microdipodops pallidus*. Mamm Species 47:1-2
O'Farrell TP, Olson RJ, Gilbert RO, Hedlund JD (1975) A population of Great Basin pocket mice, *Perognathus parvus*, in the shrub-steppe of south-central Washington. Ecol Monogr 45:1-28
Ognev SI (1948) Mammals of the USSR and adjacent countries vol7. Rodents Acad Sci USSR Press, Moscow (in Russian)
Ojeda RA, Gonnet JM, Borghi CE, Giannoni SM, Campos CM, Diaz GB (1996) Ecological observations of the red viscacha rat, *Tympanoctomys barrerae* in desert habitats of Argentina. Mastozool Neotrop 3:183-191
Ortega JC (1990a) Reproductive biology of the rock squirrel (*Spermophilus variegatus*) in southeastern Arizona (USA). J Mammal 71:448-457
Ortega JC (1990b) Home-range size of adult rock squirrels (*Spermophilus variegatus*) in southeastern Arizona (USA). J Mammal 71:171-176
Ortega JC (1991) The annual cycles of activity and weight of rock squirrels (*Spermophilus variegatus*) in southeastern Arizona (USA). Am Midl Nat 126:159-171
Osborn DJ, Helmy I (1980) The contemporary land mammals of Egypt (including Sinai). Fieldiana Zool 5:1-579
Ovadia O, Abramsky Z (1995) Density-dependent habitat selection: evaluation of the isodar method. Oikos 73:86-94
Owen JG (1988) On productivity as a predictor of rodent and carnivore diversity. Ecology 69:1161-1165
Owen JG (1990) Patterns of mammalian species richness in relation to temperature, productivity, and variance in elevation. J Mammal 71:1-13
Packard RL (1968) An ecological study of the fulvous harvest mouse in eastern Texas. Am Midl Nat 79:68-88
Palomo LJ, Vargas JM (1988) On the irregular molt in the Algerian mouse, *Mus spretus* Lataste, 1883. Acta Theriol 33:67-78
Panteleev PA (1968) Population ecology of the water vole and control activities. Nauka Press, Moscow (in Russian)
Patterson BD (1987) The principle of nested subsets and its implications for biological conservation. Conserv Biol 1:323-334

Patterson BD (1990) On the temporal development of nested subset patterns of species composition. Oikos 59:330-342
Patterson BD, Atmar W (1986) Nested subsets and the structure of insular mammalian faunas and archipelagos. Biol J Linn Soc 28:65-82
Patterson BD, Brown JH (1991) Regionally nested patterns of species composition in granivorous rodent assemblages. J Biogeogr 18:395-402
Paulson DD (1988a) *Chaetodipus baileyi*. Mamm Species 297:1-5
Paulson DD (1988b) *Chaetodipus hispidus*. Mamm Species 320:1-4
Pavlinov IY (1988) Evolution of the mastoid part of the *bulla tympani* in specialized arid rodents. Zool Zh 67:739-750 (in Russian)
Pavlinov IY, Dubrovsky YA, Rossolimo OL, Potapova EG (1990) Gerbils of the world's fauna. Nauka Press, Moscow (in Russian)
Pavlinov IY, Jakhontov EA, Agadjanian AK (1995) Mammals of Eurasia. I. Rodentia. Arch Zool Mus Moscow State Univ 32-239 (in Russian)
Pearson OP (1988) Biology and feeding dynamics of a South American herbivorous rodent, *Reithrodon*. Stud Neotrop Fauna Environ 23:25-39
Pearson OP, Martin S, Bellati J (1987) Demography and reproduction of the silky desert mouse (*Eligmodontia*) in Argentina. Fieldiana Zool 5:433-446
Peet RK, Loucks OL (1977) A gradient analysis of southern Wiskonsin forests. Ecology 58:485-499
Perrin MR, Slotow RH, Mendelsohn JM (1992) The population dynamics of rodents at Settlers, Transvaal, South Africa. Z Saeugetierkd 57:238-247
Peters RH (1983) The ecological implications of body size. Cambridge Univ Press, Cambridge
Petraitis PS (1979) Likelihood measures of niche breadth and overlap. Ecology 60:703-710
Petrov MP (1976) Deserts of the world. John Wiley, New York
Petter F (1961) Répartition géographique et écologie des rongeurs désertiques (du Sahara occidental à l'Iran oriental). Mammalia 21:1-222
Pianka ER (1978) Evolutionary ecology, 2nd edn. Harper and Row, New York
Pickett STA (1980) Non-equilibrium coexistence of plants. Bull Torrey Bot Club 107:238-248
Pielou EC (1984) The interpretation of ecological data. John Wiley, New York
Pierotti R (1982) Habitat selection and its effect on reproductive output in the herring gull in Newfoundland. Ecology 63:854-868
Pimm SL (1985) Estimating competition coefficients from census data. Oecologia 67:588-590
Pimm SL, Rosenzweig ML (1981) Competitors and habitat use. Oikos 37:1-6
Pine RH, Miller SD, Schamberger ML (1979) Contribution to the mammalogy of Chile. Mammalia 43:339-376
Podolsky RH, Price MV (1990) Patch use by *Dipodomys deserti* (Rodentia: Heteromyidae): profitability, preference, and depletion dynamics. Oecologia (Berl) 83:83-90
Polop JJ (1989) Distribution and ecological observations of wild rodents in Pampa de Achala, Cordoba, Argentina. Stud Neotrop Fauna Environ 24:53-60
Popov SV, Chabovsky AV (1996) Factors affecting body mass and ventral gland size in great gerbil (*Rhombomys opimus*) in the south-eastern Karakum desert. Zool Zh 75:1404-1411 (in Russian)
Poulet AR (1984) Quelques observations sur la biologie de *Desmodilliscus braueri* Wettstein (Rodentia: Gerbillidae) dans le sahel du Senegal. Mammalia 48:59-64
Prakash I (1962) Ecology of the gerbils of the Rajasthan desert, India. Mammalia 26:311-331
Prakash I (1971) Breeding season and litter size of Indian desert rodents. Z Angew Zool 58:441-454
Prakash I (1973) A study of field populations of rodents in the Indian deserts. III. Sand dunes in 100 mm rainfall zone. Z Angew Zool 60:31-41
Prakash I (1975) The population ecology of the rodents of the Rajasthan Desert, India. In: Prakash I, Ghosh PK (eds) Rodents in desert environments. Dr W Junk, The Hague, pp 75-116
Prakash I (1981) Ecology of the indian desert gerbil *Meriones hurrianae*. Central Arid Zone Research Institute Press, Jodhpur, India
Prakash I, Purohit KG (1966) Some observations on the hairy-footed gerbil *Gerbillus gleadowi* Murray in the Rajasthan desert. J Bombay Nat Hist Soc 63:431-434

Prakash I, Rana BD (1972) A study of filed populations of rodents in the Indian desert. II. Rocky and piedmont zone. Z Angewandte Zool 62:339-348

Prakash I, Saravanan A, Singh P (1995) Ecology and taxonomy of the field mice in the Aravalli ranges. J Bombay Nat Hist Soc 92:372-377

Price MW (1978) The role of microhabitat in structuring desert rodent communities. Ecology 59:910-921

Price MW (1981) Body size and patch choice in desert rodents: a model for searching predators. Bull Ecol Soc Am 62:109

Price MW (1983) Ecological consequences of body size: a model of patch choice in desert rodents. Oecologia 59:384-392

Price MW (1986) Structure of desert rodent communities: a critical review of questions and approaches. Am Zool 26:39-49

Price MW, Brown JH (1983) Patterns of morphology and resource use in North American desert rodent communities. In: Reichman OJ, Brown JH (eds) Biology of desert rodents. Great Basin Nat Mem 7:117-134

Price MW, Longland WS (1989) Use of artificial seed patches by heteromyid rodents. J Mammal 70:316-322

Price MW, Waser NM (1985) Microhabitat use by heteromyid rodents: effects of artificial seed patches. Ecology 66:211-219

Price MV, Waser NM, Bass TA (1984) Effects of moonlight on microhabitat use by desert rodents. J Mammal 65:353-356

Pulliam HR (1988) Sources, sinks, and population regulation. Am Nat 132:652-661

Pulliam HR, Danielson BJ (1991) Sources, sinks and habitat selection: a landscape perspective on population dynamics. Am Nat 137:S50-S66

Qumsiyeh M (1996) Mammals of the Holy Land. Texas Tech Univ Press, Lubbock, Texas

Rall YM (1938) Introduction to ecology of the midday jirds *Pallasiomys meridianus* Pall. I. General notes, density dynamics, burrowing activity. Vestn Microbiol Epidemiol Parasitol 17:331-363 (in Russian)

Rall YM (1940a) Introduction to ecology of the midday jirds *Pallasiomys meridianus* Pall. II. Reproduction. Vestn Microbiol Epidemiol Parasitol 18:139-167 (in Russian)

Rall YM (1940b) Introduction to ecology of the midday jirds *Pallasiomys meridianus* Pall. III. Feeding, age patterns, longevity and mortality. Vestn Microbiol Epidemiol Parasitol 18:320-358 (in Russian)

Rall YM (1940c) Introduction to ecology of the midday jirds *Pallasiomys meridianus* Pall. IV.Importance and control of jirds. Jirds as a life form (conclusion). Vestn Microbiol Epidemiol Parasitol 19:135-143 (in Russian)

Rall YM (1941) Essay on the ecology of the tamarisk jird *Meriones tamariscinus* Pall. Rodents and their control No 1:179-207 (in Russian)

Ranck GL (1968) The rodents of Libya. Taxonomy, ecology and zoogeographical relationships. United States Nat Mus Bull, 275. Smithsonian Inst Publ, Washington

Randall JA (1981) Olfactory communication at sandbathing loci by sympatric species of kangaroo rat. J Mammal 62:12-19

Randall JA (1993) Behavioral adaptations of desert rodents (Heteromyidae). Anim Behav 45:263-287

Randall JA (1994) Convergences and divergences in communication and social organization of desert rodents. Aust J Zool 42:405-433

Rautenbach IL (1978) A numerical reappraisal of the southern African biotic zones. Bull Carnegie Mus Nat Hist 6:175-187

Rebar CE (1995) Ability of *Dipodomys merriami* and *Chaetodipus intermedius* to locate resource distributions. J Mammal 76:437-447

Rebar CE, Conley W (1983) Interactions in microhabitat use between *Dipodomys ordii* and *Onychomys leucogaster*. Ecology 64:984-988

Redford KH, Eisenberg JF (1992) Mammals of the Neotropics. The southern cone, vol 2. Chile, Argentine, Uruguay, Paraguay. Univ Chicago Press, Chicago

Reichman OJ (1975) Relation of desert rodent diets to available resources. J Mammal 56:731-751

Reichman OP (1981) Factors influencing foraging in desert rodents. In: Kamil A, Sargent T (eds) Foraging behavior: ecological, ethological, and psychological approaches. Garland-STMP Press, New York, pp 192-213
Reichman OP (1991) Desert mammal communities. In: Polis GA (ed) The ecology of desert communities. Univ Arizona Press, Tucson, Arizona, pp 311-347
Reichman OP, Oberstein D (1977) Selection of seed distribution types by *Dipodomys merriami* and *Perognathus amplus*. Ecology 58:636-643
Reichman OJ, Price MW (1993) Ecological aspects of heteromyid foraging. In: Genoways HH, Brown JH (eds) Biology of Heteromyidae. Spec Publ 10. Am Soc Mammal, Brigham Young University, Provo, Utah, pp 539-574
Reiss MJ (1989) The allometry of growth and reproduction. Cambridge Univ Press, Cambridge
Reynolds HG, Turkowski F (1972) Reproductive variations in the round-tailed ground squirrel as related to winter rainfall. J Mammal 53:893-898
Rickart EA (1987) *Spermophilus townsendii*. Mamm Species 268:1-6
Ricklefs RE (1987) Community diversity: relative roles of local and regional processes. Science 235:167-171
Ricklefs RE, Schluter D (eds) (1994a) Species diversity in ecological communities: historical and geographical perspectives. Univ Chicago Press, Chicago.
Ricklefs RE, Schluter D (1994b) Species diversity: regional and historical influences. In: Ricklefs RE, Schluter D (eds) Species diversity in ecological communities: historical and geographical perspectives. Univ Chicago Press, Chicago, pp 350-363
Riddle BR (1995) Molecular biogeography in the pocket mice (*Perognathus* and *Chaetodipus*), and grasshopper mice (*Onychomys*): the Late Cenozoic development of a North American aridlands rodent community. J Mammal 76:283-301
Roberts TJ (1977) The mammals of Pakistan. Ernest Benn, London
Roche J, Petter F (1968) Faits nouveaux concernant trois gerbillides mal connus de Somalia: *Ammodillus imbellis* (De Winton), *Microdillus peeli* (De Winton), *Monodia juliani* (Saint Leger). Monit Zool Ital N S 2 Suppl:181-1988
Rogovin KA (1981) Spatial use and group structure in *Alactagulus acontion* according to mark-recapture data. Zool Zh 60:568-578 (in Russian)
Rogovin KA (1985) Comparative analysis of behaviour and supergeneric classification of jerboas (Rodentia, Dipodidae). Zool Zh 64:1702-1711 (in Russian)
Rogovin KA (1988) Space interrelation between symbiotopic species of three-toed pygmy jerboas (*Salpingotus kozlovi* and *S. crassicauda*) and the Roborovsky's dwarf hamster (*Phodopus roborovskii*) in the south of Mongolia. Zool Zh 67:751-759 (in Russian)
Rogovin KA (1995) Size differences between sympatric species of jerboas (Dipodidae, Rodentia): once more about the limits of morpho-ecological similarity in natural communities. Zool Zh 74:101-108 (in Russian)
Rogovin KA (1996) Social behavior of *Psammomys obesus* (Rodentia, Gerbillidae) under seminatural conditions: a settlement of a high density. Zool Zh 75:399-412 (in Russian)
Rogovin KA, Shenbrot GI (1995) Geographical ecology of Mongolian desert rodent communities. J Biogeogr 22:111-128
Rogovin KA, Surov AV (1990) Morpho-ecological structure of desert rodent communities in Central Asia and southwestern North America: a multivariate approach. Acta Theriol 35:225-239
Rogovin KA, Surov AV, Serrano V (1985) Niche convergence in the desert rodents of two geographically isolated communities. Acta Zool Mex N S 10:1-36
Rogovin KA, Kulikov VF, Surov AV, Vasilieva NY (1986) Small mammals. In: Sokolov VE, Gunin PD (eds) Deserts of Trans-Altai Gobi. Nauka Press, Moscow, pp 122-131 (in Russian)
Rogovin KA, Shenbrot GI, Surov AV (1991) Analysis of spatial organization of a desert rodent community in Bolson de Mapimi, Mexico. J Mammal 72:347-359
Rogovin KA, Shenbrot GI, Surov AV, Idris M (1994) Spatial organization of a rodent community in the western Rajasthan desert (India). Mammalia 58:234-260
Rohde K (1992) Latitudinal gradients in species diversity: the search for the primary cause. Oikos 65:514-527

Roldan ERS, Acuna AM, Vercellini O, Horgan C, Merani MS (1984) Parametros reproductivos de *Akodon molinae*, Contreras, 1968, *Akodon dolores*, Thomas, 1916 y sus hibridos (Rodentia, Cricetidae). Rev Mus Argent Cienc Nat Bernardino Rivadavia Ser Zool 13:485-490

Root RB (1967) The niche exploitation pattern of the blue-gray gnatcatcher. Ecol Monogr 37:317-350

Rosen BR (1988) Biogeographic patterns: a perceptual overview. In: Myers A, Giller P (eds) Analytical biogeography. Chapman and Hall, London, pp 23-55

Rosenzweig ML (1973) Habitat selection experiments with a pair of coexisting heteromyid species. Ecology 54:111-117

Rosenzweig ML (1974) On the evolution of habitat selection. In: Proc of the 1st Int Congr of Ecology, Aberdeen 1973, pp 401-404

Rosenzweig ML (1979a) Three probable evolutionary causes for habitat selection. In: Patil GP, Rosenzweig ML (eds) Contemporary quantitative ecology and related ecometrics. Statistical Ecology series, vol 12. Int Co-operative Publ House, Burtonsville, Maryland, pp 49-60

Rosenzweig ML (1979b) Optimal habitat selection in two-species competitive systems. Fortschr Zool 25:283-293

Rosenzweig ML (1981) A theory of habitat selection. Ecology 62:327-335

Rosenzweig ML (1987a) Community organization from the point of view of habitat selectors. In: Gee JHR, Giller PS (eds) Organization of communities: past and present. 27th Symp British Ecological Society. Aberystwyth 1986. Blackwell, Oxford, pp 437-458

Rosenzweig ML (1987b) Habitat selection as a source of biological diversity. Evol Ecol 1:315-330

Rosenzweig ML (1989) Habitat selection, community organization and small mammal studies. In: Morris DW, Abramsky Z, Fox BJ, Willig MR (eds) Patterns in the structure of mammalian communities. Spec Publ Mus Texas Tech Univ No 28. Texas Tech Univ Press, Lubbock, Texas, pp 5-21

Rosenzweig ML (1991) Habitat selection and population interactions: the search of mechanism. Am Nat 137 Suppl:5-28

Rosenzweig ML (1992) Species diversity gradients: we know more and less than we thought. J Mammal 73:715-730

Rosenzweig ML (1995) Species diversity in space and time. Cambridge Univ Press, Cambridge

Rosenzweig, ML, Abramsky Z (1985) Detecting density-dependent habitat selection. Am Nat 126:405-417

Rosenzweig ML, Abramsky Z (1986) Centrifugal community organization. Oikos 46:339-348

Rosenzweig ML, Abramsky Z (1993) How are diversity and productivity related? In: Ricklefs RE, Schluter D (eds) Species diversity in ecological communities: historical and geographical perspectives. Univ Chicago Press, Chicago, pp 52-65

Rosenzweig ML, Abramsky Z, Brand S (1984) Estimating species interactions in heterogeneous environments. Oikos 43:329-340

Rosenzweig ML, Abramsky Z, Kotler B, Mitchell W (1985) Can interaction coefficients be determined from census data? Oecologia 66:194-198

Rosi MI (1983) Notas sobre la ecología, distribución y systematica de *Graomys griseoflavus griseoflavus* (Waterhouse, 1837) (Rodentia, Cricetidae) en la provincia de Mendoza. Hist Nat 3:161-167

Ross PD (1994) *Phodopus roborovskii*. Mamm Species 469:1-4

Ross PD (1995) *Phodopus campbelli*. Mamm Species 503:1-7

Rotenberry JT, Wiens JA (1980) Habitat structure, patchiness and avian communities in North American steppe vegetation: a multivariate analysis. Ecology 61:1228-1250

Roughgarden J (1976) Resource partitioning among competing species – a coevolutionary approach. Theor Popul Biol 9:388-424

Rowe-Rowe DT, Meester JAJ (1982) Habitat preferences and abundance relations of some mammals in the Natal Drakensberg. S Afr J Zool 17:202-209

Rowlands IW (1974) Mountain viscacha. Symp Zool Soc Lond 34:131-141

Rulie KF (1954) The selected biological publications. USSR Acad Sci Press, Moscow (in Russian)

Rundel PW, Dillon MO, Palma B, Mooney HA, Gulmon SL. Ehleringer JR (1991) The phytogeography and ecology of the coastal Atacama and Peruvian deserts. Aliso 13:1-49
Rusch G, Van der Maarel E (1992) Species turnover and seedling recruitments in limestone grasslands. Oikos 63:131-146
Rushton SP, Hill D, Carter SP (1994) The abundance of river corridor birds in relation to their habitats: a modelling approach. J Appl Ecol 31:313-328.
Rust RW (1989) Opportunistic foraging by the kangaroo rat *Dipodomys deserti* Stephens (Rodentia: Heteromyidae). Great Basin Nat 49:396-397
Ruzik A (1967) Contribution à l'écologie de *Spermophilopsis leptodactylus* Licht dans le desert au sud de Buchara. Mammalia 31:173-185
Schall JJ, Pianka ER (1978) Geographic ranges of Australian mammals. Science 201:679-686
Schlitter DA (1973) A new species of gerbil from South West Africa with remarks of *Gerbillurus tytonis* Bauer & Niethammer 1959 (Rodentia: Gerbillinae). Bull South Calif Acad Sci 72:13-18
Schmidly DJ (1974) *Peromyscus pectoralis*. Mamm Species 49:1-3
Schmidly DJ, Wilkins KT, Derr JN (1993) Biogeography. In: Genoways HH, Brown JH (eds) Biology of Heteromyidae. Spec Publ 10. Am Soc Mammal, Brigham Young University, Provo, Utah, pp 319-356
Schmidt-Nielsen K (1964) Desert animals: physiological problems of heat and water. Clarendon Press, Oxford
Schmidt-Nielsen K (1975a) Animal physiology: adaptations and environment. Cambridge Univ Press, London
Schmidt-Nielsen K (1975b) Desert rodents: physiological problems of desert life. In: Prakash I, Ghosh PK (eds) Rodents in desert environments. Dr W Junk, The Hague, pp 397-388
Schmidt-Nielsen K (1984) Scaling, why is animal size so important? Cambridge Univ Press, Cambridge
Schmidt-Nielsen K, Schmidt-Nielsen B, Schneiderman H (1948) Salt excretion in desert mammals. Am J Physiol 154:163-166
Schoener TW (1968) Nonsynchronous spatial overlap of lizards in patchy habitats. Ecology 51:408-418
Schoener TW (1971) Theory of feeding strategies. Annu Rev Ecol Syst 2: 369-404
Schoener TW (1974) Resource partitioning in ecological communities. Science 185:27-39
Schoener TW (1983) Field experiments on interspecific competition. Am Nat 122:240-285
Schoener TW (1985) On the degree of consistency expected when different methods are used to estimate competition coefficients. Oecologia 67: 591-592
Schoener TW (1986) Mechanistic approaches to community ecology: a new reductionism? Am Zool 26:81-106
Schroeder GD, Rosenzweig ML (1975) Perturbation analysis of competition and overlap in habitat utilization between *Dipodomys ordii* and *Dipodomys merriami*. Oecologia 19:9-28
Schroepfer R, Klenner-Fringes B, Naumer E (1985) Locomotion pattern and habitat utilization of the two jerboas *Jaculus jaculus* and *Jaculus orientalis* (Rodentia, Dipodidae). Mammalia 49:445-454
Seely MK (1977) Sand sodified by gemsbok urine as selected burrow sites of gerbils. Zool Afr 12:247-249
Seguignes M, Vernet R (1996) Record of *Ctenodactylus gundi* (Rodentia, Ctenodactylidae) in two isolated sites of southern Tunisia. Mammalia 60:151-153
Setzer HW (1969) A review of the African mice of the genus *Desmodilliscus* Wettstein, 1916. Misc Publ Mus Nat Hist Univ Kansas 51:283-288
Severtzov NA (1855) Periodic phenomena in the life of mammals, birds, reptiles and amphibians of Voronezh Region. Saint Petersburg (in Russian)
Shalmon B (1993) A field guide to the land mammals of Israel, their tracks and signs. Keter Publ House, Jerusalem (in Hebrew)
Shantz HL (1956) History and problems of arid land development. In: White GF (ed) The future of arid lands. American Association for the Advancement of Science, Washington, pp 3-25

Sheikher C, Chauhan MS, Guraya SS (1989) A comparative study of seasonal changes in the body and gonad weights in mice (*Mus musculus, Mus booduga, Mus platythrix* and *Mus cervicolor*) in relation to some environmental factors of Garhwal Himalayas (India). Z Angew Zool 76:287-302

Shenbrot GI (1989) Review of the book: Y.A. Mekaev *Zoogeographic complexes of Eurasia.* Zool Zh 68:154-157 (in Russian)

Shenbrot GI (1992) Spatial structure and niche patterns of a rodent community in the south Bukhara desert (Middle Asia). Ecography 15:347-357

Shenbrot GI, Krasnov BR (1996) Habitat relationships of the lizard fauna in the Ramon erosion cirque, Negev Highlands (Israel). J Zool 234:429-440

Shenbrot GI, Krasnov BR (1997) Estimation of competition from census data in a Central Negev desert rodent community. Isr J Zool 43:116-117

Shenbrot GI, Rogovin KA (1988) Comparative ecology of pygmy jerboas of southern Mongolia. In: Sokolov VE (ed) Jerboas of the USSR fauna. Fan Press, Tashkent, pp 125-130 (in Russian)

Shenbrot GI, Rogovin KA (1995) Temporal variation in spatial organization of a rodent community in the southwestern Kyzylkum desert (Middle Asia). Ecography 18:370-383

Shenbrot GI, Rogovin KA, Surov AV (1991) Comparative analysis of spatial organization of desert lizard communities in Middle Asia and Mexico. Oikos 61:157-168

Shenbrot GI, Krasnov BR, Khokhlova IS (1994a) On the biology of *Gerbillus henleyi* (Rodentia: Gerbillidae) in the Negev Highlands, Israel. Mammalia 58:581-589

Shenbrot GI, Rogovin KA, Heske EJ (1994b) Comparison of niche-packing and community organization in desert rodents in Asia and North America. Aust J Zool 42:479-499

Shenbrot GI, Sokolov VE, Heptner VG, Kovalskaya YM (1995) Mammals of Russia and adjacent regions. Dipodoid rodents. Nauka Press, Moscow (in Russian)

Shenbrot GI, Krasnov BR, Khokhlova IS (1997) On the biology of Wagner's gerbil *Gerbillus dasyurus* (Wagner, 1842) (Rodentia: Gerbillidae) in the Negev Highlands, Israel. Mammalia 61:467-486

Shevyreva NS (1989) New rodents (Ctenodactyloidea, Rodentia, Mammalia) from the lower Eocene in Mongolia. Paleontol Zh 3:60-72 (in Russian)

Shevyreva NS (1994) A new ctenodactyloid (Rodentia, Mammalia) from the Neogen of the Zaisan depression (eastern Kazakhstan). Paleontol Zh 2:126-131 (in Russian)

Shevyreva NS (1996) The age of rodent fauna (Rodentia, Mammalia) from the Buran formation of the Zaisan depression, eastern Kazakhstan. Stratigr Geol Correl 3:167-175

Shilov IA (1977) Ecological and physiological basis of relationships within animal populations. Moscow Univ Press, Moscow (in Russian)

Shkolnik A (1971) Diurnal activity in a small desert rodent. Intern J Biometeorol 15: 115-120

Shkolnik A, Borut A (1969) Temperature and water relations in two species of spiny mice (*Acomys*). J Mammal 50:245

Shriner WM, Stacey PB (1991) Spatial relationships and dispersal patterns in the rock squirrel, *Spermophilus variegatus.* J Mammal 72:601-606

Shubin IG (1974) Ecology of *Lagurus luteus* in the Zaisan hollow. Zool Zh 53:272-277 (in Russian)

Shvartz EA, Chernyshev NV, Popov IY (1997) Do shrews have an impact on soil invertebrates in Eurasian forests? Ecoscience 4:158-162

Sidiyhene EA, Tranier M (1990) Recent data on the mammals of the Adrar des Iforas (Mali). Mammalia 54:471-478

Simberloff D (1983) Competition theory, hypothesis testing, and other community-ecological buzzwords. Am Nat 122:626-635

Simberloff D, Boecklen W (1981) Santa Rosalia reconsidered: Size ratios and competition. Evolution 35:1206-1228

Simberloff D, Dayan T (1991) The guild concept and the structure of ecological communities. Annu Rev Ecol Syst 22:115-143

Simonetta A, Roche I, Manicci P (1978) Primo tentativo di descrizione e di identificazione dei Roditori della Republica Democratica Somala. Stud Richerche 2:157-185

Simonetti JA (1989) Microhabitat use by small mammals in central Chile. Oikos 56:309-318

Simpson GG (1980) Splendid isolation. Yale Univ Press, New Haven
Singh G (1988) History of aridland vegetation and climate: a global perspective. Biol Rev 63:159-195
Singh P, Prakash I (1997) Food preference and calorific requirements of the cutch rock-rat, *Cremnomys cutchicus medius*. Ann Arid Zone 36:65-72
Sinitzin VM (1962) Palaeogeography of Asia. USSR Acad Sci Press, Moscow (in Russian)
Skinner JD, Smithers HN (1990) The mammals of the Southern African subregion. 2nd edn. Univ Pretoria, Pretoria
Skupski MP (1995) Population ecology of the western harvest mouse, *Reithrodontomys megalotis*: a long-term perspective. J Mammal 76:358-367
Sludsky AA, Varshavsky SN, Ismagilov MI, Kapitonov BI, Shubin IG (1969) Mammals of Kazakhstan vol 1. Rodents. Part 1. Nauka Kazakh SSR Press, Alma-Ata (in Russian)
Sludsky AA, Bekenov AB, Borisenko BA, Grachev YA, Ismagilov MI (1977) Mammals of Kazakhstan vol 1. Rodents. Part 2. Nauka Kazakh SSR Press, Alma-Ata (in Russian)
Sludsky AA, Borisenko BA, Kapitonov BI, Makhmutov S, Mokrousov NY (1978) Mammals of Kazakhstan vol 1. Rodents. Part 3. Nauka Kazakh SSR Press, Alma-Ata (in Russian)
Smigel BW, Rosenzweig ML (1974) Seed selection in *Dipodomys merriami* and *Perognathus penicillatus*. Ecology 55:329-339
Smirin VM, Orlov OY (1973) Some peculiarities of external morphology and behavior of rodents relating to vision. In: Sokolov VE (ed) Proc of 2nd USSR Mammal Conf. Moscow Univ Press, Moscow, pp 204-209 (in Russian)
Smith EP, Zaret TM (1982) Bias in estimating niche overlap. Ecology 63:1248-1253
Smith FA (1995) Den characteristics and survivorship of woodrats (*Neotoma lepida*) in the eastern Mojave desert. Southwest Nat 40:366-372
Smith HD, Jorgensen CD (1975) Reproductive biology of North American desert rodents. In: Prakash I, Ghosh PK (eds) Rodents in desert environments. Dr W Junk, The Hague, pp 305-330
Smithers RHN (1971) The mammals of Botswana. Mus Mem Natl Mus Monum Rhodesia 4:1-340
Smolen MJ, Genoways HH, Baker RJ (1980) Demographic and reproductive parameters of the yellow-cheeked pocket gopher (*Pappogeomys castanops*). J Mammal 61:224-236
Sokal RR, Rohlf FJ (1995) Biometry, 3rd edn. WH Freeman, New York
Sokolov VE (1973) The mammalian skin. Nauka Press, Moscow (in Russian)
Sokolov VE, Gunin PD (eds) (1986) Deserts of Trans-Altai Gobi. Nauka Press, Moscow (in Russian)
Sokolov VE, Kuznetzov GV (1978) Daily rhythms of activity in mammals. Cytological and ecological aspects. Nauka Press, Moscow (in Russian)
Sokolov VE, Orlov VN (1980) A guide to mammals of the Mongolian People's Republic. Nauka Press, Moscow (in Russian)
Sokolov VE, Shenbrot GI (1987) A new species of the thick-tailed jerboa, *Stypodipus sungorus* sp. n. (Rodentia, Dipodidae) from western Mongolia. Zool Zh 66:579-587 (in Russian)
Sokolov VE, Shenbrot GI (1988) Data on the geographic variability and taxonomy of pygmy jerboas (Rodentia, Cardiocraniinae) in Mongolia. Zool Zh 67:1561-1569 (in Russian)
Sokolov VE, Kotenkova EV, Krasnov BR (eds) (1989) The house mouse. Nauka Press, Moscow (in Russian)
Sokolov VE, Lobachev VS, Orlov VN (1996) The mammals of Mongolia. Family Dipodidae. Moscow, Nauka Press (in Russian)
Southgate R, Masters P (1996) Fluctuations of rodent populations in response to rainfall and fire in a Central Australian hummock grassland dominated by *Plectrachne schinzii*. Wildl Res 23:289-303
Southgate R, Masters P, Seely MK (1996) Precipitation and biomass changes in the Namib desert dune ecosystem. J Arid Environ 33:267-280
Southwood TRE (1978) Ecological methods, 2nd edn. Chapman and Hall, London
Southwood TRE (1987) The concept and nature of the community. In: Gee JHR, Giller PS (eds) Organization of communities: Past and present. Blackwell, Oxford, pp 3-27
Spencer SR, Cameron GN (1982) *Reithrodontomys fulvescens*. Mamm Species 174:1-7

Stamp N, Ohmart R (1978) Resource utilization by desert rodents in the lower Sonoran desert. Ecology 59:700-707

Stangl FB Jr, Dalquest WW, Kuhn S (1993) Mammals from the Beach mountains of Culberson County, Trans- Pecos Texas. Texas J Sci 45:87-96

Stephens DW, Krebs JR (1986) Foraging theory. Princeton Univ Press, Princeton

Stevens GC (1989) The latitudinal gradient in geographic range: how so many species coexist in the tropics. Am Nat 133:240-256

Stones RC, Hayward CL (1968) Natural history of the desert woodrat, *Neotoma lepida*. Am Midl Nat 80:458-476

Strahan R (ed) (1983) The complete book of Australian mammals. Angus and Robertson, Australia

Streubel DP, Fitzgerald JP (1978) *Spermophilus spilosoma*. Mamm Species 101: 1-4

Strong DR (1986) Null hypotheses in ecology. Synthese 43:271-285

Tabachnik BG, Fidell LS (1996) Using multivariate statistics, 3rd edn. HarperCollins College Publ, New York

Taber AB, Macdonald DW (1992) Spatial organization and monogamy in the mara *Dolichotis patagonum*. J Zool (Lond) 227:417-438

Tchernov E (1988) The biogeographical history of the southern Levant. In: Yom-Tov Y, Tchernov E (eds) The zoogeography of Israel: distribution and abundance at zoogeographical crossroad. Dr W Junk, Dordrecht, pp 159-250

Tchernov E (1992) The Afro-Arabian component in the Levantine mammalian fauna – a short biogeographical review. Isr J Zool 38:155-192

Ter Braak CJF, Prentice IC (1988) A theory of gradient analysis. Adv Ecol Res 18:271-317

Thomas D (ed) (1989) Arid zone geomorphology. Belhaven Press, London

Thompson RS, Mead JI (1982) Late Quaternary environments and biogeography in the Great Basin. Quat Res 17:39-55

Thompson SD (1982a) Microhabitat utilization and foraging behavior of bipedal and quadrupedal heteromyid rodents. Ecology 63:1303-1312

Thompson SD (1982b) Structure and species composition of desert heteromyid species assemblages: effects of a simple habitat manipulation. Ecology 63:1313-1321

Thompson SD (1985) Bipedal hopping and seed dispersion selection by heteromyid rodents: the role of locomotion energetics. Ecology 66:220-229

Thompson SD, Macmillen RE, Burke E, Taylor CR (1980) The energetic cost of bipedal hopping in small mammals. Nature 287:223-229

Thornthwaite CW (1948) An approach towards a rational classification of climate. Geogr Rev 38:55-94

Tilman D (1987) The importance of the mechanisms of interspecific competition. Am Nat 129:769-764

Tilman D (1988) Plant strategies and the dynamics and structure of plant communities. Princeton Univ Press, Princeton, New Jersey

Tognelli MF, Campos CM, Ojeda RA, Roig VG (1995) Is *Microcavia australis* (Rodentia: Caviidae) associated with a particular plant structure in the Monte desert of Argentina? Mammalia 59:327-333

Torres-Contreras H, Silva-Aranguiz E, Marquet PA, Camus PA, Jaksic FM (1997) Spatiotemporal variability of rodent subpopulations at a semiarid neotropical locality. J Mammal 78:505-513

Torres-Mura JC, Lemus ML, Contreras LC (1989) Herbivorous specialization of the South American desert rodent *Tympanoctomys barrerae*. J Mammal 70:646-648

Tregenza T, Parker GA, Thompson DJ (1996) Interference and the ideal free distribution: models and tests. Behav Ecol 7:379-386

Turaev AK (1984) On the ecology of the gray hamster (*Cricetulus migratorius* Pallas, 1770) in the Gissar Range of Tajikistan. Ann Acad Sci Tajikistan SSR Sec Biol 1:74-75 (in Russian)

Turelli M (1978) A reexamination of stability in random versus deterministic environments with comments on the stochastic theory of limiting similarity. Theor Popul Biol 13:244-267

Valdez M, Ceballos G (1991) Natural history, diet, and reproduction of the ground squirrel (*Spermophilus mexicanus*) in an intermontane meadow. Acta Zool Mex Nueva Ser 43:1-31
Valone TJ, Brown JS (1989) Measuring patch assessment abilities of desert granivores. Ecology 70:1800-1810
Van Aarde RJ (1987) Demography of a Cape porcupine, *Hystrix africaeaustralus*, population. J Zool (Lond) 213:205-212
Vandermeer JH (1972) Niche theory. Annu Rev Ecol Syst 3:107-132
Van Devender TR, Spaulding WG (1979) Development of vegetation and climate in the southwestern united States. Science 204:701-710
Van Hersbergen HJ, Channing A (1989) Habitat preference and use of space by the namtap *Graphyurus ocularis* (Rodentia: Gliridae). Mammalia 53:25-34
Van Horne B (1981) Demography of *Peromyscus maniculatus* populations in serial stages of coastal coniferous forest in southeastern Alaska. Can J Zool 59:1045-1061
Van Horne B (1983) Density as a misleading indicator of habitat quality. J Wildl Manage 47:893-901
Van Horne B, Ford RG (1982) Niche breadth calculation based on discriminant analysis. Ecology 63:1172-1174
Van Tienderen PH (1991) Evolution of generalists and specialists in spatially heterogeneous environment. Evolution 45:1317-1331
Van Tienderen PH (1997) Generalists, specialists, and the evolution of phenotypic plasticity in sympatric populations of distinct species. Evolution 51:1372-1380
Van Valen L (1968) Morphological variation and width of ecological niche. Am Nat 99:377-390
Vargas JM, Palomo LJ, Palmquist P (1991) Reproduction of the Algerian mouse (*Mus spretus* Lataste, 1883) in the south of the Iberian Peninsula (Spain). Bonn Zool Beitr 42:1-10
Varshavsky AA, Karimova TY, Neronov VM (1995) Zoogeographic analysis of rodent fauna of North Africa. Zool. Zh 74:136-151 (in Russian)
Varshavsky AA, Neronov VM, Zhang YZ (1997) Zoogeographic regionalization of an arid area of China (based on data on rodents' distribution). Zool Zh 76:473-486 (in Russian)
Vasquez RA (1996) Patch utilization by three species of Chilean rodents differing in body size and mode of locomotion. Ecology 77:2343-2351
Veal R, Caire W (1979) *Peromyscus eremicus*. Mamm Species 118:1-6
Verevkin MV (1985) Reproductive biology of the midday jird (*Meriones meridianus*). Zool Zh 64:276-281 (in Russian)
Verts BJ, Kirkland GL (1988) *Perognathus parvus*. Mamm Species 318:1-8
Vesey-Fitzgerald D (1953) Notes on some rodents from Saudi Arabia and Kuwait. J Bombay Nat Hist Soc 51:424-428
Villafañe G (1981) Reproducci6n y crecimiento de *Calomys musculinus* (Thomas, 1916). Hist Nat 1:193-204
Villafañe G, Bonaventura SM (1987) Ecological studies in crop fields of the endemic area of Argentine hemorrhagic fever: *Calomys musculinus* movements in relation to habitat and abundance. Mammalia 51:233-248
Vinogradov BS (1937) Fauna of USSR. Mammals vol 3, part 4. Dipodid rodents. USSR Acad Sci Press, Leningrad (in Russian)
Von Bertalanffy L (1968) General system theory; foundations, developments, applications. G Braziller, New York
Vorontsov NN (1967) Evolution of the alimentary system of myomorph rodents. Nauka Press, Novosibirsk (in Russian)
Wadia DN (1960) The post glacial desiccation of Central Asia. Monogr 10, Natl Inst Sci India:1-25
Wahlert JH (1993) The fossil records. In: Genoways HH, Brown JH (eds) Biology of Heteromyidae. Spec Publ 10. Am Soc Mammal, Brigham Young University, Provo, Utah, pp 1-37
Walter H (1979) Vegetation of the earth and ecological systems of the geobioshere. 2nd edn. Springer, New York
Wang BY (1997) The mid-Tertiary Ctenodactylidae (Rodentia, Mammalia) of eastern and central Asia. Bull Am Mus Nat Hist 234:1-88

Wang SQ, Yang HF, Hao SS, Zhang ZB, Xu TQ (1992) Food preference and consumption of striped hamsters (*Cricetulus barabensis*). Acta Theriol Sin 38:156-164 (in Chinese)
Ward D, Olsvig-Whittaker L (1993) Plant species diversity at the junction of two desert biogeographic zones. Biodivers Lett 1:172-185
Waser PM, Jones WT (1991) Survival and reproductive effort in banner-tailed kangaroo rats. Ecology 72:771-777
Wassif K, Hoogstraal H (1954) The mammals of south Sinai, Egypt. Proc Egypt Acad Sci 22:77-97
Waterman JW (1995) The social organization of the cape ground squirrel (*Xerus inauris*; Rodentia: Sciuridae). Ethology 101:130-147
Waterman JM (1996) Reproductive biology of a tropical, non-hibernating ground squirrel. J Mammal 77:134-146
Watkinson AR, Sutherland WJ (1995) Sources, sinks and pseudo-sinks. J Anim Ecol 64:126-1300
Webb SD (1977) A history of savanna vertebrates in the New World. Part I: North America. Annu Rev Ecol Syst 8:355-380
Webster DB (1962) A function of the enlarged middle ear cavities of the kangaroo rat, *Dipodomys*. Physiol Zool 35:248-255
Webster DB, Plassman W (1992) Parallel evolution of low-frequency sensitivity in Old World and New World desert rodents. In: Fay RR, Popper AN (eds) The evolutionary biology of hearing. Springer, New York, pp 633-636
Webster DB, Strother DB (1972) Middle-ear morphology and auditory sensitivity of heteromyid rodents. Am Zool 12: 727
Webster DB, Webster M (1971) Adaptive value of hearing and vision in kangaroo rat predator avoidance. Brain Behav Evol 4:310-322
Webster WD, Jones JK (1982) *Reithrodontomys megalotis*. Mamm Species 167:1-5
Weisserberg S, Shkolnik A (1994) Metabolic rate and water economy in the desert and Mediterranean populations of the common spiny mouse (*Acomys cahirinus*) in Israel. Isr J Zool 40:135-143
Wells PV (1979) An aquable glaciopluvial in the West: pleniglacial evidence of increased precipitation on a gradient from Great Basin to the Sonoran and Chihuahuan Deserts. Quat Res 12:311-325
Werner EE, McPeek MA (1994) The roles of direct and indirect effects on the distribution of two frog species along an environmental gradient. Ecology 75:1368-1382
White RM, Kerley GIH, Bernard RTF (1997) Pattern and controls of reproduction of the southern African rodent *Gerbillurus paeba* in the semi-arid Karoo, South Africa. J Arid Environ 37:529-549
Whittaker RH (1972) Evolution and measurement of species diversity. Taxon 21:213-251
Whittaker RH (1975) Communities and ecosystems, 2nd edn. Macmillan, New York
Wiens JA (1989) The ecology of bird communities, vols. 1 and 2. Cambridge Univ Press, Cambridge
Williams DF, Mares MA (1978) A new genus and species of phyllotine rodent (Mammalia: Muridae) from northwestern Argentina. Ann Carnegie Mus 47:193-221
Wilson DE, Reeder DM (eds) (1992) Mammal species of the world: a taxonomic and geographic reference, 2nd edn. Smithsonian Inst Press, Washington
Wilson RC, Vacek T, Lanier DL, Dewsbury DA (1976) Open-field behavior in muroid rodents. Behav Biol 17:495-506
Winemiller KO, Pianka ER (1990) Organization in natural assemblages of desert lizards and tropical fishes. Ecol Monogr 60:27-55
Winkelman JR, Getz LL (1962) Water balance in the Mongolian gerbil. J Mammal 43:150-1544
Withers PC (1979) Ecology of a small mammal community on a rocky outcrop in the Namib desert. Madoqua II:229-246
Withers PC (1983) Seasonal reproduction by small mammals of the Namib desert. Mammalia 47:195-204
Withers PC, Louw GN, Henschel JR (1980) Energetics and water relationships of Namib desert rodents. S Afr J Zool 15:131-137

Wondolleck JT (1978) Forage-area separation and overlap in heteromyid rodents. J Mammal 59:510-518
Wright DH, Patterson BD, Mikkelson GM, Cutler A, Atmar W (1998) A comparative analysis of nested subset patterns of species composition. Oecologia 113:1-20
Wynne-Edwards K, Lisk RD (1987) Behavioral interactions differentiate Djungarian (*Phodopus campbelli*) and Siberian (*Phodopus sungorus*) hamsters. Can J Zool 65:2229-2235
Yakimenko LV (1984) Distributional map with gazetteer of geographical ranges of the common (*Ellobius talpinus* Pall.) and the Zaisan (*E. tancrei* Blasius) mole-voles. In: Krasilov VA (ed) Problems of variability and zoogeography of mammals. Inst Biol Pedol, Far East Branch Acad Sci USSR, Vladivostok, pp 76-102 (in Russian)
Yanushevich AI, Aisin BM, Kydyraliev AK (1972) Mammals of Kyrgyzia. Ilym Press, Frunze (in Russian)
Yensen E, Quinney CA (1992) Can Townsend's ground squirrels survive on diet of exotic annuals? Great Basin Nat 52:269-277
Yom-Tov Y (1991) Character displacement in the psammophile Gerbillidae of Israel. Oikos 60:173-179
Young CJ, Jones JK (1982) *Spermophilus mexicanus*. Mamm Species 164:1-4
Yu SX, Orloci L (1990) On niche overlap and its measurement. Coenoses 5:159-165
Yu SX, Orloci L (1992) Niche breadth: An index of species environmental fitness. Coenoses 7:121-125
Yu SX, Orloci L (1993) Index of species environmental fitness: the niche breadth. Coenoses 8:159-164
Zahavi A, Wahrman J (1957) The cytotaxonomy, ecology and evolution of the gerbils and jirds of Israel (Rodentia, Gerbillinae). Mammalia 21:341-380
Zeng Z (1991) Population ecology of a desert rodent *Dipodomys spectabilis* in the Chihuahuan desert of North America. Acta Theriol Sin 11:87-98 (in Chinese)
Zhu S, Qin Z (1991) Studies on the population dynamics of rat-like hamster and striped hamster in rural districts of Northern Huaihe River, Anhui. Acta Theriol Sinica 11:99-108 (in Chinese)
Ziv Y, Abramsky Z, Kotler BP, Subach A (1993) Interference competition and temporal and habitat partitioning in two gerbil species. Oikos 66:237-246
Zubtzova GE (1978) Some morphological features of incisor structure of some fossorial rodents in connection with their different manners of burrowing. Report I. *Spalax*, *Ellobius*, Allactaginae. Proc Zool Inst Acad Sci USSR 79:105-108 (in Russian)
Zubtzova GE (1980) Peculiarities of the form of incisors of fossorial rodents in connection with different manners of burrowing. Report II. Jerboas (Dipodidae). Proc Zool Inst Acad Sci USSR 91:63-68 (in Russian)

Appendix:
List of Rodent Species Considered to be Desert-Dwellers

Species	Main sources of information on the geographic distribution and biology
Sciuridae	
Marmotinae	
Ammospermophilus harrisii	Cockrum (1960); Hoffmeister (1986); Best et al. (1990b)
Ammospermophilus interpres	Best et al. (1990a); Stangl et al. (1993)
Ammospermophilus leucurus	Bladley (1968); Smith and Jorgensen (1975); Hoffmeister (1986); Belk and Smith (1990)
Ammospermophilus nelsoni	Best et al. (1990c)
Spermophilus mexicanus	Young and Jones (1982); Valdez and Ceballos (1991); Livoreil et al. (1993)
Spermophilus mohavensis	Best (1995)
Spermophilus spilosoma	Streubel and Fitzgerald (1978); Aragon and Baudoin (1989); Best et al. (1993); Livoreil et al. (1993)
Spermophilus tereticaudus	Reynolds and Turkowski (1972); Drabek (1973); Dunford (1977); Ernest and Mares (1987)
Spermophilus townsendi	Rickart 1987; Yensen and Quinney 1992
Spermophilus variegatus	Oaks et al. (1987); Ortega (1990a,b, 1991); Shriner and Stacey (1991)
Spermophilus alashanicus	Bannikov (1954)
Spermophilus erythrogenis	Gromov et al. (1965); Sludsky et al. (1969)
Spermophilus fulvus	Gromov et al. (1965); Sludsky et al. (1969)
Spermophilus pygmaeus	Gromov et al. (1965); Sludsky et al. (1969)
Xerinae	
Atlantoxerus getulus	Le Berre (1990); Kowalski and Rzebik-Kowalska (1991); Nowak (1991)
Spermatophilopsis leptodactylus	Ruzic (1967); Komarova (1980); Bekenov and Eszhanov (1984)
Xerus erythropus	Kingdon (1974); Linn and Key (1996)
Xerus inauris	Herzig-Straschil (1978, 1979); Waterman (1995, 1996)
Xerus princeps	Smithers (1971); Herzig-Straschil and Herzig (1989)
Geomyidae	
Thomomys umbrinus	Patton in Wilson and Reeder (1992)
Pappogeomys castanops	Hegdal et al. (1965); Hickman (1977); Smolen et al. (1980)

Species	Main sources of information on the geographic distribution and biology
Heteromyidae	
Perognathus parvus	O'Farrell et al. (1975); Verts and Kirkland (1988); Cramer and Chapman (1990); Genoways and Brown (1993)
Perognathus flavus	Chapman and Packard (1974); Best and Skupski (1994); Genoways and Brown (1993)
Perognathus longimembris	O'Farrel (1978); Cramer and Chapman (1990); Genoways and Brown (1993)
Perognathus amplus	Wondolleck (1978); Genoways and Brown (1993)
Perognathus formosus	Cramer and Chapman (1990); Genoways and Brown (1993)
Chaetodipus baileyi	Paulson (1988a); Genoways and Brown (1993)
Chaetodipus hispidus	Paulson (1988b); Kaufman et al. (1990); Genoways and Brown (1993)
Chaetodipus penicillatus	Hoffmeister (1986); Genoways and Brown (1993)
Chaetodipus intermedius	Hoffmeister (1986); Genoways and Brown (1993); Rebar (1995)
Chaetodipus nelsoni	Genoways and Brown (1993); Best (1994)
Chaetodipus fallax	McClenaghan (1983); Genoways and Brown (1993); Lackey (1996)
Microdipodops megacephalus	O'Farrell and Blaustein (1974a); Genoways and Brown (1993)
Microdipodops pallidus	O'Farrell and Blaustein (1974b); Genoways and Brown (1993)
Dipodomys ordii	O'Farrell (1978); Garrison and Best (1990); Genoways and Brown (1993); Best et al. (1993)
Dipodomys microps	Hayssen (1991); Genoways and Brown (1993)
Dipodomys spectabilis	Jones et al. (1988); Best (1988a); Waser and Jones (1991); Zeng (1991); Amarasekare (1994)
Dipodomys nelsoni	Best (1988b); Alvarez and Alvarez (1991); Genoways and Brown (1993)
Dipodomys merriami	Bradley and Mauer (1971); Reichman (1975); McClenaghan (1984); Jones (1989); Genoways and Brown (1993)
Dipodomys deserti	Best et al. (1989); Rust (1989); Podolsky and Price (1990); Genoways and Brown (1993)
Pedetidae	
Pedetes capensis	Butinsky (1973, 1979); Kofron (1987); Skinner and Smithers (1990); Augustine et al. (1995)
Muridae	
Sigmodontinae	
Neotoma albigula	Dial (1988); Macedo and Mares (1988)
Neotoma lepida	Stones and Hayward (1968); Cameron and Rainey (1972); Smith (1995)
Neotoma goldmani	Hrachovy et al. (1996)
Neotoma mexicana	Cornely and Baker (1986); Alvarez et al. (1988)
Onychomys leucogaster	McCarty (1978); Kaufman et al. (1990)
Onychomys torridus	Horner and Taylor (1968); McCarty (1975); Frank and Heske (1992)
Reithrodontomys megalotis	O'Farrell (1978); Webster and Jones (1982)
Reithrodontomys fulvescens	Packard (1968); Spencer and Cameron (1982)

Species	Main sources of information on the geographic distribution and biology
Peromyscus eremicus	King (1968); Veal and Caire (1979)
Peromyscus merriami	King (1968)
Peromyscus crinitus	King (1968); Johnson and Armstrong (1987)
Peromyscus pectoralis	King (1968); Schmidly (1974)
Peromyscus boylii	King (1968); Alvarez and Arroyo-Cabrales (1990)
Akodon iniscatus	Redford and Eisenberg (1992)
Akodon molinae	Contreras (1968); Roldan et al. (1984)
Calomys callosus	Massoia and Fornes (1965); Mello (1978); Mares et al. (1981)
Calomys musculinus	Mares (1977a); Villafane (1981); Villafane and Bonaventura (1987)
Andalgalomys olrogi	Mares (1977a);Williams and Mares (1978)
Andalgalomys roigi	Mares and Braun (1996)
Eligmodontia moreni	Mares et al. (1981, 1989)
Eligmodontia morgani	Guthmann et al. (1997)
Eligmodontia typus	Mares (1975a, 1977a); Pearson et al. (1987)
Salinomys delicatus	Braun and Mares (1995)
Phyllotis amicus	Musser and Carleton in Wilson and Reeder (1992)
Phyllotis darwini	Hershkovitz (1962); Fulk (1975); Meserve (1981a,b); Simonetti (1989); Ebensperger and Simonetti (1996); Vaskuez (1996); Lima et al. (1997); Torres-Contreras et al. (1997)
Phyllotis gerbillus	Musser and Carleton in Wilson and Reeder (1992)
Phyllotis xanthopigus	Hershkovitz (1962); Musser and Carleton in Wilson and Reeder (1992)
Graomys edithae	Redford and Eisenberg (1992)
Graomys griseoflavus	Hershkovitz (1962); Mares et al. (1981); Rosi (1983)
Reithrodon physodes	Barlow (1969); Dalby and Mares (1974); Dalby (1975); Pine et al. (1979); Pearson (1988); Polop (1989)
Sigmodon fulviventer	Baker and Shump (1978a)
Sigmodon ochrognathus	Hardin et al. (1970); Davis and Dunford (1987); Baker and Shump (1978b)
Cricetinae	
Calomyscus bailwardi	Gambarian and Martirosian (1960); Lay (1967); Hassinger (1973); Roberts (1977);
Phodopus campbelli	Flint and Golovkin (1961); Wynne-Edwards and Lisk (1987); Ross (1995)
Phodopus roborovskii	Flint (1960); Wynne-Edwards and Lisk (1987); Ross (1994)
Cricetulus barabensis	Flint and Golovkin (1961); Zhu and Qin (1991); Wang et al. (1992)
Cricetulus longicaudatus	Flint (1966); Meyer and Skhol (1977)
Cricetulus migratorius	Hassinger (1973); Harrison and Bates (1991); Turaev (1984)
Allocricetulus curtatus	Bannikov (1954); Jingchuan et al. (1990)
Allocricetulus eversmanni	Gromov and Erbajeva (1995)

Species	Main sources of information on the geographic distribution and biology
Otomyinae	
Parotomys brandsii	Nel and Rautenbach (1974); Kerley (1989); Du Plessis and Kerley (1991); Du Plessis et al. (1992)
Parotomys littledalei	Coetzee (1969); De Graaf (1981); Skinner and Smithers (1990)
Gerbillinae	
Gerbillus agag	Kock (1977); Gautun et al. (1985); Pavlinov et al. (1990)
Gerbillus amoenus	Ranck (1968); Osborn and Helmy (1980); Pavlinov et al. (1990)
Gerbillus andersoni	Abramsky (1980, 1984); Harrison and Bates (1991); Brown et al. (1992)
Gerbillus aquillus	Lay and Nadler (1975); Roberts (1977); Hassinger (1993); Pavlinov et al. (1990)
Gerbillus campestris	Petter (1961); Gappold (1967a); Osborn and Helmy (1980); Harrison and Bates (1991); Khidas (1993)
Gerbillus cheesmani	Lay and Nadler (1975); Harrison and Bates (1991)
Gerbillus dasyurus	Zahavi and Wahrman (1957); Shenbrot et al. (1997)
Gerbillus famulus	Bahmanyar and Lay (1975); Harrison and Bates (1991)
Gerbillus gerbillus	Ranck (1968); Daly and Daly (1975a); Osborn and Helmy (1980); Harrison and Bates (1991)
Gerbillus gleadowi	Prakash and Purohit (1966); Prakash (1973); Lay and Nadler (1975); Roberts (1977)
Gerbillus henleyi	Bahmanyar and Lay (1975); Abramsky et al. (1985a); Shenbrot et al. (1994a)
Gerbillus latastei	Ranck (1968); Cockrum (1977)
Gerbillus mackilligini	Osborn and Helmy (1980)
Gerbillus mesopotamiae	Karim et al. (1975); Lay and Nadler (1975); Harrison and Bates (1991)
Gerbillus nanus	Vesey-Fitzgerald (1953); Sahavi and Wahrman (1957); Lay (1967); Kowalski and Rzebik-Kowalska (1991); Harrison and Bates (1991)
Gerbillus poecilops	Bahmanyar and Lay (1975); Harrison and Bates (1991)
Gerbillus pusillus	Hubert (1978); Pavlinov et al. (1990)
Gerbillus pyramidum	Petter (1961); Happold (1968); Abramsky (1994)
Gerbillus rosalinda	Pavlinov et al. (1990)
Gerbillus simoni	Petter (1961); Hoogstraal (1963); Ranck (1968); Osborn and Helmy (1980); Kowalski and Rzebik-Kowalska (1991)
Gerbillus watersi	Pavlinov et al. (1990)
Gerbillurus paeba	Seely (1977); Downs and Perrin (1989); Ascaray and McLachlan (1990); Ascaray et al. (1991); White et al. (1997)
Gerbillurus setzeri	Schlitter (1973); Downs and Perrin (1989); Griffin (1990)
Gerbillurus tytonis	Boyer (1985); Downs and Perrin (1989); Skinner and Smithers (1990); Griffin (1990); Hughes et al. (1995)
Gerbillurus vallinus	Downs and Perrin (1989); Skinner and Smithers (1990); Griffin (1990)
Tatera brantsii	De Moor (1969); Korn and Korn (1989)
Tatera indica	Petter (1961); Prakash (1962, 1975); Roberts (1977); Harrison and Bates (1991)

Species	Main sources of information on the geographic distribution and biology
Desmodillus auricularis	Nel (1967, 1975); Rautenbach (1978); Christian (1979)
Monodia mauritaniae	Pavlinov et al. (1990)
Microdillus peeli	Roche and Petter (1968); Funaioli (1971)
Desmodilliscus braueri	Heim de Balsac (1967); Setzer (1969); Poulet (1984)
Pachyuromys duprasi	Petter (1961); Ranck (1968); Daly (1979); Osborn and Helmy (1980); Kowalski and Rzebik-Kowalska (1991)
Ammodillus imbellis	Roche and Petter (1968); Pavlinov et al. (1990)
Sekeetamys calurus	Zahavi and Wahrman (1957); Osborn and Helmy (1980); Harrison and Bates (1991); Krasnov et al. (1996a)
Meriones crassus	Koffler (1972); Krasnov et al. (1996b)
Meriones erythrourus	Ismagilov (1960); Naumov and Lobachev (1975)
Meriones grandis	Petter (1961); Pavlinov et al. (1990)
Meriones hurrianae	Prakash (1981); Goyal and Ghosh (1993)
Meriones libycus	Petter (1961); Daly and Daly (1975a); Daly (1979)
Meriones meridianus	Rall (1938, 1940a,b,c); Naumov and Lobachev (1975); Sludsky et al. (1978); Verevkin (1985)
Meriones persicus	Naumov and Lobachev (1975); Roberts (1977); Eigelis (1980)
Meriones rex	Vesey-Fitzgerald (1953); Bahmanyar and Lay (1975); Harrison and Bates (1991)
Meriones sacramenti	Osborn and Helmy (1980); Abramsky (1989); Harrison and Bates (1991)
Meriones shawi	Petter (1961); Osborn and Helmy (1980); Neronov and Malkhazova (1985), Belabbas and Butet (1994)
Meriones tamariscinus	Rall (1941); Naumov and Lobachev (1975); Gromov and Gromova (1996); Gromov et al. (1996)
Psammomys obesus	Petter (1961); Daly and Daly (1975b); Daly (1979); Osborn and Helmy (1980); Ilan and Yom-Tov (1990)
Rhombomys opimus	Naumov and Lobachev (1975); Dubrovsky (1978); Popov and Chabovsky (1996)
Arvicolinae	
Alticola barakschin	Bannikov (1954)
Alticola strelzowi	Eshelkin (1976); Sludsky et al. (1978)
Eolagurus luteus	Shubin (1974); Sludsky et al. (1978)
Eolagurus przewalskii	Gromov and Polyakov (1977)
Ellobius talpinus	Yakimenko (1984); Gromov and Erbajeva (1995)
Ellobius tancrei	Yakimenko (1984); Gromov and Erbajeva (1995)
Dendromurinae	
Malacothrix typica	Smithers (1971); Knight and Skinner (1981); Kerley (1992)
Petromyscinae	
Petromyscus collinus	Withers (1979, 1983); Withers et al. (1980)
Murinae	
Rattus villosissimus	Carstairs (1976); Redhead in Strahan (1983)
Zyzomys argurus	Begg (1981); Calaby and Taylor (1983); Bradley et al. (1988)
Zyzomys pedunculatus	Begg in Strahan (1983)

Species	Main sources of information on the geographic distribution and biology
Pseudomys australis	Watts in Strahan (1983)
Pseudomys chapmani	Dunlop and Pound (1981); Anstee et al. (1997)
Pseudomys desertor	Happold (1976a,b)
Pseudomys hermannsburgensis	Dunlop and Pound (1981); Dickman et al. (1995); Southgate and Masters (1996)
Leggadina forresti	Morton in Strahan (1983)
Notomys alexis	Breed (1992); Murray and Dickman (1994)
Notomys fuscus	Aitken (1968); Watts in Strahan (1983)
Notomys cervinus	Watts in Strahan (1983)
Millardia gleadowi	Prakash (1971); Mishra and Dhanda (1975)
Cremnomys cutchicus	Prakash (1971); Prakash and Rana (1972); Singh and Prakash (1997)
Zelotomys woosnami	Smithers (1971); Birkenstock and Nel (1977); Mugo et al. (1995)
Acomys cahirinus	Vesey-Fitzgerald (1953); Osborn and Helmy (1980); Delany and Farook (1989); Harrison and Bates (1991); Haim and Rozenfeld (1993); Kronfeld et al. (1994); Khokhlova et al. (1994)
Acomys cinereacus	Happold (1966); Bates (1994)
Acomys louisae	Simonetta et al. (1978); Bates (1994)
Acomys russatus	Shkolnik (1971); Osborn and Helmy (1980); Qumsiyeh (1996) Haim and Rozenfeld (1993); Kronfeld et al. (1994); Krasnov et al. (1996b)
Rhabdomys pumilio	Rowe-Rowe and Meester (1982); Kerley (1989); Kerley et al. (1990); Perrin et al. (1992)
Aethomys chrysophilus	Skinner and Smithers (1990); Denys (1994)
Aethomys namaquensis	Coetzee (1969); Buffenstein (1984); Kerley et al. (1990)
Mus booduga	Prakash (1975); Sheikher et al. (1989)
Mus minutoides	Kerley (1989); Armstrong and Van Hersbergen (1996); Downs and Perrin (1996)
Mus musculus	Berry (1981); Khokhlova and Krasnov (1986); Sokolov et al. (1989); Krasnov and Khokhlova (1994)
Mus saxicola	Prakash et al. (1995)
Mus spretus	Palomo and Vargas (1988); Kowalsi and Rzebik-Kowalska (1991); Vargas et al. (1991)
Myoxidae	
Eliomys melanurus	Vesey-Fitzgerald (1953); Wassif and Hoogstraal (1954); Atallah (1978); Nader et al. (1983)
Eliomys quercinus	Khidas (1993)
Graphiurus ocularis	Channing (1984); Van Hersbergen and Channing (1989)
Selevinia betpakdalensis	Sludsky et al. (1977); Gromov and Erbajeva (1995)
Dipodidae	
Euchoreutinae	
Euchoreutes naso	Ma et al. (1987); Sokolov et al. (1996)
Cradiocraniinae	

Species	Main sources of information on the geographic distribution and biology
Cardiocranius paradoxus	Shenbrot et al. (1995)
Salpingotus michaelis	Roberts (1977)
Salpingotus pallidus	Shenbrot et al. (1995)
Salpingotus heptneri	Shenbrot et al. (1995)
Salpingotus crassicauda	Shenbrot et al. (1995)
Salpingotus kozlovi	Shenbrot and Rogovin (1988); Sokolov and Shenbrot (1988); Sokolov et al. (1996)
Dipodinae	
Dipus sagitta	Shenbrot et al. (1995)
Stylodipus telum	Shenbrot et al. (1995)
Stylodipus sungorus	Sokolov and Shenbrot (1987)
Stylodipus andrewsii	Sokolov and Shenbrot (1987)
Eremodipus lichtensteinii	Shenbrot et al. (1995)
Jaculus jaculus	Happold (1967b); Hooper and El Halili (1972); Schroepfer et al. (1985)
Jaculus orientalis	Hooper and El Halili (1972); El Hilali and Veillat (1975); Schroepfer et al. (1985)
Jaculus blanfordi	Shenbrot et al. (1995)
Paradipodinae	
Paradipus ctenodactylus	Shenbrot et al. (1995)
Allactaginae	
Allactaga balikunica	Ma et al. (1987)
Allactaga bullata	Ma et al. (1987)
Allactaga elater	Shenbrot et al. (1995)
Allactaga euphratica	Shenbrot et al. (1995)
Allactaga hotsoni	Hassinger (1973); Roberts (1977)
Allactaga major	Shenbrot et al. (1995)
Allactaga severtzovi	Shenbrot et al. (1995)
Allactaga sibirica	Shenbrot et al. (1995)
Allactaga tetradactyla	Osborn and Helmy (1980)
Allactaga vinogradovi	Shenbrot et al. (1995)
Allactodipus bobrinskii	Shenbrot et al. (1995)
Pygeretmus pumilio	Shenbrot et al. (1995)
Pygeretmus zhitkovi	Shenbrot et al. (1995)
Pygeretmus platyurus	Shenbrot et al. (1995)
Hystricidae	
Hystrix africaeaustralius	Van Aarde (1987); Skinner and Smithers (1990)
Hystrix cristata	Kowalski and Rzebik-Kowalska (1991)
Hystrix indica	Alkon and Zaltz (1985); Gorbunov (1985)
Caviidae	
Microcavia australus	Mares (1973); Kleiman et al. (1979); Tognelli et al. (1995)

Species	Main sources of information on the geographic distribution and biology
Dolichotis patagonum	Mares (1973); Taber and Macdonald (1992)
Dolichotis salinicola	Mares et al. (1981); Kleiman et al. (1979)
Chinchillidae	
Lagostomus maximus	Mares et al. (1981); Branch (1993); Branch et al. (1994a,b)
Lagidium viscacia	Rowlands (1974)
Octodontidae	
Octodon degus	Meserve and Glanz (1978); Meserve (1981a,b); Contreras et al. (1987); Lagos et al. (1995)
Octodon lunatus	Contreras et al. (1987); Redford and Eisenberg (1992)
Octomys mimax	Mares (1973)
Tympanoctomys barrerae	Mares (1973); Torres-Mura et al. (1989); Ojeda et al. (1996)
Ctenomyidae	
Ctenomys emilianus	Redford and Eisenberg (1992)
Ctenomys latro	Redford and Eisenberg (1992)
Ctenomys mendocinus	Mares et al. (1981); Redford and Eisenberg (1992)
Ctenomys occultus	Redford and Eisenberg (1992)
Ctenomys saltarius	Mares et al. (1981); Redford and Eisenberg (1992)
Abrocomidae	
Abrocoma benetti	Meserve and Glanz (1978); Meserve et al. (1986); Iriarte et al. (1989)
Petromuridae	
Petromus typicus	George (1981); George and Crowther (1981); Withers (1983)
Bathyergidae	
Cryptomys hottentotus	Hickman (1979); Davies and Jarvis (1986)
Bathyergus janetta	Jarvis (1969); Skinner and Smithers (1990); Mugo et al. (1995)
Ctenodactylidae	
Pectinator spekei	George (1974, 1978)
Felovia vae	George (1974, 1978)
Massoutiera mzabi	George (1978); Sidiyhene and Tranier (1990)
Ctenodactylus gundi	George (1978); Seguignes and Vernet (1996)
Ctenodactylus vali	Petter (1961)

Subject Index

Springer and the environment

At Springer we firmly believe that an international science publisher has a special obligation to the environment, and our corporate policies consistently reflect this conviction.

We also expect our business partners – paper mills, printers, packaging manufacturers, etc. – to commit themselves to using materials and production processes that do not harm the environment. The paper in this book is made from low- or no-chlorine pulp and is acid free, in conformance with international standards for paper permanency.

MIX
Papier aus verantwortungsvollen Quellen
Paper from responsible sources
FSC® C105338

If you have any concerns about our products,
you can contact us on
ProductSafety@springernature.com

In case Publisher is established outside the EU,
the EU authorized representative is:
Springer Nature Customer Service Center GmbH
Europaplatz 3, 69115 Heidelberg, Germany

Printed by Libri Plureos GmbH
in Hamburg, Germany